Graduate Texts in Contemporary Physics

Springer
New York
Berlin
Heidelberg
Barcelona
Budapest
Hong Kong
London
Milan
Paris
Tokyo

Graduate Texts in Contemporary Physics

R.N. Mohapatra: **Unification and Supersymmetry: The Frontiers of Quark-Lepton Physics, 2nd Edition**

R.E. Prange and S.M. Girvin (eds.): **The Quantum Hall Effect**

M. Kaku: **Introduction to Superstrings**

J.W. Lynn (ed.): **High-Temperature Superconductivity**

H.V. Klapdor (ed.): **Neutrinos**

J.H. Hinken: **Superconductor Electronics: Fundamentals and Microwave Applications**

M. Kaku: **Strings, Conformal Fields, and Topology: An Introduction**

A. Auerbach: **Interacting Electrons and Quantum Magnetism**

Yu.M. Ivanchenko and A.A. Lisyansky: **Physics of Critical Fluctuations**

Yuli M. Ivanchenko
Alexander A. Lisyansky

Physics of Critical Fluctuations

With 121 Illustrations

Yuli M. Ivanchenko
Department of Physics
Polytechnic University
Brooklyn, NY 11201, USA

Alexander A. Lisyansky
Department of Physics
Queens College of CUNY
Flushing, NY 11367-0904, USA

Library of Congress Cataloging-in-Publication Data
Ivanchenko, Yuli M.
Physics of critical fluctuations / Yuli M. Ivanchenko and Alexander A. Lisyansky.
p. cm. — (Graduate texts in contemporary physics)
Includes bibliographical references and index.
ISBN 0-387-94414-1
1. Phase transformations (Statistical physics) 2. Fluctuations (Physics) I. Lisyansky, Alexander A. II. Title. III. Series.
QC175.16.P5I96 1995
530.4′14—dc20 94-39124

Printed on acid-free paper.

Photocomposed copy prepared from the authors' LaTeX file.
Printed and bound by Maple-Vail, York, PA.
Printed in the United States of America.

9 8 7 6 5 4 3 2 1

ISBN 0-387-94414-1 Springer-Verlag New York Berlin Heidelberg

Preface

In the last 25 years, the problem of phase transitions has turned from a narrow, special topic into one of the central problems of condensed matter physics. This period can be characterized by enormous progress in the understanding of processes related to the problem. Achievements were primarily due to qualitative comprehension of the physical nature of a phase transition. This resulted in the creation of a new, powerful mathematical approach. This approach is not only important to the theory of phase transitions but also happens to be extremely useful in a number of areas only quite remotely related to this theory. Application of the approach to diverse fields has accelerated progress in phase transition physics, which is continuing successfully today.

The start of the contemporary physics of phase transitions was due to the pioneering works by Patashinskii and Pokrovskii (1964, 1966), Widom (1965), and Kadanoff (1966). These works formulated the main ideas of the fluctuation theory of phase transitions, which predefined the appearance of the outstanding works by Wilson (1971 a,b). In these works, Wilson developed a special form of the renormalization group (RG) approach applicable to the theory of phase transitions. Now this method is considered the most appropriate for the study of critical behavior of fluctuating systems.

At present, there are a number of excellent books related to this subject. These books, as a rule, rely on a particular type of RG formulation or are dedicated to a specific method of analyzing critical behavior. Among these books is the now-classic monograph by Ma (1976) and texts on the field theoretical RG by Patashinskii and Pokrovskii (1982), Amit (1984), Zinn-Justin (1989) and numerical studies by Baker (1990). In this text we consider in a concise manner quite a wide spectrum of phase transition theory problems, revealing their common features and the distinctions between them. Formal mathematical aspects of the theory are treated as a necessary instrument of application to particular physical systems. Wherever possible, the results of the theory are compared with experimental results.

Naturally, within the scope of a single book one cannot treat all aspects of phase transition physics. In this text we have not considered such fundamental topics as the real space RG, phase transitions in finite systems, numerical methods in the theory of critical phenomena and some others. Special reviews and monographs have been written on these topics (see Swendsen (1982), Burchardt and Leeuwen (1988), Cardy (1988), Privman

(1990)) to which we refer the reader.

The content of this monograph is closely related to the book "Fluctuation Effects in Systems with Competing Interactions" published in the former USSR in 1989 (see Ivanchenko, Lisyansky and Filippov (1989 a)). The above book was co-authored by A. Filippov. This text is substantially different from its predecessor in that Chapters 1–5 were written completely anew. Chapters 6–8, based on the material in the above monograph, were essentially remade, corrected, and updated. Unfortunately, Dr. Filippov was unable to participate in this rewriting due to a number of obstacles imposed by the revolutionary process taking place in the now-independent states of the former USSR.

The structure of this book is as follows. A concise review of the classical approach is given in Chapter 1. The main aspects of a selfconsistent treatment of systems with broken symmetry can be found in this chapter. The introduction of the Gaussian approximation naturally leads to Ginzburg's criterion bounding the fluctuation region. The next step, ending this chapter, is the scaling hypothesis. This hypothesis emerges as an endeavor to overcome inconsistencies in the classical approximation, therefore, it naturally demarcates two qualitatively different periods in the development of phase transition physics. The following chapter is intended to fill the gap between the phenomenological formulation of the classical approach and microscopic essence of a system undergoing a phase transition. In Chapter 2 we have shown on a variety of microscopic models how a complicated microscopic motion can be reduced, in the vicinity of a critical point, to the evolution of a slowly varying phenomenological entity: the Ginzburg-Landau functional. The first two chapters do not require any knowledge of the renormalization group process considered in Chapters 3 and 4. At present there are a number of different formulations of RG theory. Nonetheless, they all lead to practically equivalent results. Though the field theoretical RG was developed long before Wilson's RG formulation, which is specific to the theory of critical behavior, we consider the latter prior to the former. In this we not only pay tribute to pioneering works by Wilson but also to the historical logic of the scientific development caused by Kadanoff's outstanding hypothesis (1966). The following chapters are based on the scaling equation method, which is deeply rooted in both approaches and is introduced in Chapter 5. This method neither uses Kadanoff's scale invariance, the main hypothesis of the Wilson's approach, nor does it use the requirement of the renormalizability of the Ginzburg-Landau functional, which is fundamental to the field theoretical approach. Nonetheless, this method is closely related to RG, namely: the exact RG equation, in a form similar to the initial formulation by Wilson, follows from the scale equations. These equations help to solve mathematical problems inherent in the traditional approach. In particular, they solve the problem of elimination of redundant operators and help to formulate a new perturbational approach using the small critical exponent η as an expansion parameter. The procedure of derivation of the RG equation is very simple in the framework of this approach. The latter is demonstrated in Chapter 6 on a number of models popular in the theory of critical behavior. Finally, with the help of the scale-equation method, results which could not be obtained in traditional approaches are found. Thus, in Sect. 5.5.1 the structure of the two-point correlation function is revealed at the critical point for the whole range of the momentum variable.

Shortly after its first appearance, the fluctuation theory was mainly applied to simple systems. The results, then obtained, were only quantitatively different from those obtained using the phenomenological theory. Later, application of RG theory to real physical systems led to a number of surprising results. Macroscopic behavior of a real system emerges as a result of a number of interactions, often competing with each other. Competition of critical fluctuations in such systems may crucially affect the picture of phase transitions. The latter is demonstrated in Chapter 7 for different physical situations. All the qualitatively new effects considered in Chapters 6 and 7 are anticipated on the basis of an RG analysis. As mentioned above, the RG approach has proved to be very fruitful. This approach has qualitatively changed phase transition theory. However, the RG analysis is quite complicated, and cannot be applied without a number of approximations. The latter often raises doubts, especially when one encounters a new result which is qualitatively contradicting to the much simpler selfconsistent approach. Thus, the interpretation of the RG analysis results is often obscure. The main drawback of the RG method emerges from the absence of a small parameter in the theory. Practically all the results of the theory are obtained using perturbation theory with an expansion parameter of the order of unity. This means that "qualitatively new" results may be consequences of the approximations made. In other words, the fluctuation theory so far needs an independent confirmation either from experiment or theory. Such confirmation can be obtained on the basis of the simple exactly solvable models considered in Chapter 8. The results of this chapter can be easily interpreted and have a clear physical sense. On the other hand, these results in general coincide with the results of the RG analysis. The similarity of the results, obtained using unrelated approaches, is an essential argument justifying the predictions of the fluctuation theory not as "imaginary" but as really existent physical effects. Finally, in Chapter 9 RG analysis as well as methods based on the exactly-solvable-model approach are applied to the consideration of phase transitions in high-T_c superconductors.

This book is intended for a broad audience including graduate and undergraduate students in physics. Although we have tried to make the book self-contained, we assumed the reader has a good undergraduate background in physics as well as some knowledge of statistical mechanics at a graduate level, such as can be found in "Statistical and Thermal Physics" by F. Reif (1965) and in "Statistical Physics" by Landau and Lifshitz (1976).

A few words should also be said in guiding the reader through the book. The chapters are very different in their level of difficulty. Depending on the background of the reader, we would like to suggest a few ways of dealing with the book. Experienced workers in the field may start reading this book from Chapter 5 using Chapters 3 and 4 as reference material. Chapter 5 contains results which lead directly to the forefront of fluctuational theory. For such an experienced reader the results of the following chapters may be considered as illustrations of the method, though in Chapter 8 he or she may find instructive material that can be used in the work on original problems. Readers working in condensed matter theory, but less familiar with phase transition theory, will certainly acquire useful information by reading the book from the very beginning. The material delivered in the book can be successfully applied in other areas of research. An experimentalist may find the material in the first two chapters enlightening. Then, he may skip Chapters 3–5 and resume reading from Chapter 6. Experimentalists, working in this area, may substantially benefit from the material of Chapter 8. The models considered in this chapter help to build appropriate analogues that qualitatively describe fluctuation effects observed in experiment, without referring to the RG theory. We would suggest that a graduate student might first read Chapters 1,2 and the first three sections from Chapter 3 up to page 68. The student should then read the beginning of Chapter 4 including the section "Scaling Laws" and browse over the rest of this chapter. The beginning of Chapter 5 up to page 154 should also be read carefully. The rest of this chapter may require wrestling with difficult material without appropriate background and should be relegated to second reading. Only a few applications, from the following two chapters, may be considered appropriate on first reading. Their selection should be influenced by the scientific interests of the student. Chapter 8, as well as the last chapter, should also be read at this stage. The second reading should fill in the gaps and provide an opportunity for review. An undergraduate student majoring in physics may read Chapters 1, 2 and 8. The most advanced students may also follow the book according to the first stage recommended for graduate students.

This book is essentially linked to its predecessor (Ivanchenko *et al* 1989a) in the topics considered and in general motifs. A number of people contributed explicitly or implicitly to the creation of this book. We would like to thank A.E. Filippov with whom we had a fertile collaboration in critical phenomena theory over the years. Our special thanks to P. S. Riseborough who read the manuscript carefully and made a number of comments which were gratefully accepted. We indebted to A.Z. Genack who also made useful comments about the manuscript. One of us (Yu.I.) is grateful to E.L. Wolf for his hospitality and encouragement during the work on the manuscript. The final preparation of the manuscript was supported in part by the US Department of Energy, Office of Basic Energy Sciences, through Grant # DE-EG02-87ER45301.

Contents

Chapter 1

Classical Approach

1.1 Introduction

At present it is difficult to trace the beginning of research in the field of phase transition physics. It might have been started by the experimental work by Andrews (1869) in which he studied the liquid–vapor critical point. That study motivated van der Waals theory (1873). At about the same time, a systematic study of magnetic ordering was undertaken. A cornerstone of that research was the molecular field theory by Weiss (1907). Later a number of new phase transitions were discovered. They were phase transitions in binary alloys, structural phase transitions, superfluid and superconducting transitions, and some others. For some of these transitions, latent heat was released or absorbed along with the occurrence of a discontinuous change of density. Such phase transitions are called transitions of the first order. Below, we will be considering, as a rule, continuous phase transitions. For these phase conversions, the first derivatives from thermodynamic functions like energy, free energy, and so on, are continuous with respect to thermodynamic variables in the vicinity of the phase transition. The second derivatives (specific heat, susceptibility) are singular. The discussions of discontinuous transitions, in this book, are limited to the case when systems in the vicinity of a second-order transition are influenced by fluctuations to change the order of the transition.

A phenomenological theory of the second-order phase transitions was developed by Landau (1937). That was an important step towards the solution of the problem of phase transitions. This theory managed to describe from the general point of view any kind of second-order phase transition independent of its nature. In this chapter the fundamentals of the Landau theory (mean field theory) will be concisely presented and some of the constructs needed in the following chapters will be introduced. This chapter also deals with all important aspects of phase transitions which can be understood either on the basis of Landau theory or with the help of natural generalizations of this theory like Gaussian approach, extension for different space dimensionality, scaling hypothesis, and scaling laws. All these aspects of phase transition physics were understood on the grounds of classical methods of statistical physics without recourse to the application of the renormalization group approach.

1.2 Landau Theory

Landau was the first to notice (1937) that all second-order phase transitions have an important common constituent. At the transition point a new element of symmetry first appears. As the symmetry is a qualitative feature, the new element can appear only abruptly at some fixed values of thermodynamic variables. The latter means that the second-order phase transition breaks symmetry spontaneously and the symmetry of the system in the ordered state is lower than a symmetry of the Hamiltonian. To describe such a phenomenon Landau introduced a new entity — the order parameter. This parameter is equal to zero in the high-temperature (non ordered) phase (i.e. at temperatures above some critical value). Below the critical temperature in an ordered phase it acquires a finite value. For instance, in the case of a magnetic transition into a ferromagnetic state, the magnetization of this state can serve as an order parameter, the liquid–vapor transition order parameter can be characterized by the difference in the densities of the liquid and gas phases. From these examples it is seen that the order parameter is not always a scalar entity but can be a vector or even a tensor quantity. In the most general case, one may consider it as some macroscopic quantity having several components n. Sometimes, it is convenient to treat the order parameter as a vector in some isotopic space, not related to real configurational space.

The concept of the order parameter ϕ enabled Landau to write the free energy of any system in the form of an expansion in powers of ϕ. A particular representation of this expansion does not depend on the nature of forces acting in the system nor on the character of the phase transition. It is completely defined by the symmetry of the system and the number of components of the order parameter.[1]

[1] Due to this reason, for the sake of clarity, we will, as a rule, consider magnetic phase transitions. With the same effect, we could speak about structural or any other continuous phase transitions.

1.2.1 The Scalar Order Parameter

In the case of an isotropic system, at temperature T and in an external field h, the Landau expansion for the free energy can be written[2]

$$\mathcal{F}(T,h,\phi) = \mathcal{F}_0(T,h) + V\left\{-h\phi + \frac{1}{2}a(T,h)\phi^2 + \frac{1}{4}b(T,h)\phi^4 + \cdots\right\}, \quad (1.1)$$

where V is the volume of the system. An equilibrium value of the order parameter is defined from the free energy minimum conditions

$$\frac{\partial\mathcal{F}(T,h,\phi)}{\partial\phi} = 0, \quad \frac{\partial^2\mathcal{F}(T,h,\phi)}{\partial\phi^2} > 0. \quad (1.2)$$

Evidently, if $h = 0$ and $a, b > 0$ the only solution of Eq. (1.2) is $\phi = 0$. This means that the system is in the non-ordered state. A nontrivial solution for the order parameter can be obtained if one considers that there is some critical temperature T_c below which $a(T) < 0$, $b(T) > 0$. Then in a zero field the minimum value of free energy is achieved for

$$\phi^2 = -\frac{a(T)}{b(T)}. \quad (1.3)$$

Until now the description of a phase transition has been quite general and rigorous. In what follows, such generality and rigor are no longer present. Landau suggested that $a(T)$ is an analytic function in the vicinity of T_c and it can be expanded in the power series

$$a(T) = \alpha\tau + \cdots, \quad \alpha > 0, \quad \tau = (T - T_c)/T_c. \quad (1.4)$$

The function $b(T)$ in the T_c vicinity can be considered as a constant. In such a case Eq. (1.3) has the form

$$\phi_0 = (-\alpha\tau/b)^{1/2}, \quad \tau < 0. \quad (1.5)$$

This equation shows that in the vicinity of T_c the order parameter depends nonlinearly on the value τ. As we will see later, all other physically measurable quantities have power law dependencies in the critical region. The powers in these laws are called critical exponents or critical indices. The exponent in the ϕ dependence on temperature is denoted by β and in Landau theory, as is seen from Eq. (1.5), it equals 1/2. If $h \neq 0$, Eq. (1.2) with the help of the expansion, Eq. (1.4) can be reduced to

$$\alpha\tau\phi + b\phi^3 = h. \quad (1.6)$$

[2] In this chapter we deal with expansions having only even powers (excepting the term proportional to the external field). Such an expansion corresponds to the second-order phase transition.

From this expression some other thermodynamic functions can be obtained. When $\tau = 0$ $\phi = (h/b)^{1/3}$, or $\phi \propto h^{1/\delta}$, here a new critical exponent δ is introduced. Landau theory gives $\delta = 3$. The susceptibility in zero field is defined by the derivative $\chi = \partial\phi/\partial h \mid_{h=0}$. From Eq. (1.6) one can find

$$\alpha\tau\chi + 3b\phi^2\chi = 1. \tag{1.7}$$

When $\tau > 0$ $\phi = 0$, if $\tau < 0$ $\phi = (-\alpha\tau/b)^{1/2}$. In either case we have $\chi \propto \mid \tau \mid^{-\gamma}$, where $\gamma = 1$. Eq. (1.5) enables one to calculate the entropy $S = -\partial\mathcal{F}/\partial T$ which is continuous at $\tau = 0$, but the second derivative of $\mathcal{F}$, the specific heat $C = T\partial S/\partial T$ is a singular function. This singularity introduces in the general case a new critical exponent α as $C \propto \mid \tau \mid^{-\alpha}$. In Landau theory C has a discontinuous jump of $\delta C = -\alpha^2/2bT_c$, which corresponds to the value $\alpha = 0$.

1.2.2 The Vector Order Parameter

Now we will generalize considerations given in the previous section in three aspects. The first one is: the scalar order parameter ϕ will be replaced by a vector one $\boldsymbol{\phi}$. If one only makes this substitution in Eq. (1.1) and considers the case having spherical symmetry in the isotopic and real spaces,[3] then not much essentially new information about our system will emerge. In reality, however, the situation is different and involves more deep physical considerations. The expansion, Eq. (1.1) is made in the powers of the order parameter characterizing ordering in our system as a whole. For instance, if we deal with ferromagnetic ordering, ϕ will represent the full magnetization of a sample. It is credible, however, to take into account possible local fluctuations of the order parameter. Thus, the second generalization accounts for the local space variations. This can be achieved if we replace Eq. (1.1) with the functional

$$\begin{aligned}\mathcal{F}(T,\{\boldsymbol{h}\},\{\boldsymbol{\phi}\}) &= \mathcal{F}_o(T,\{\boldsymbol{h}\}) \\ &+ \int d\boldsymbol{r}\left\{\tfrac{1}{2}\alpha\tau\boldsymbol{\phi}^2(\boldsymbol{r}) + \tfrac{1}{2}c\left(\nabla\boldsymbol{\phi}(\boldsymbol{r})\right)^2 + \tfrac{1}{4}b\left(\boldsymbol{\phi}^2(\boldsymbol{r})\right)^2 - \boldsymbol{h}(\boldsymbol{r})\cdot\boldsymbol{\phi}(\boldsymbol{r})\right\},\end{aligned} \tag{1.8}$$

which was introduced by Ginzburg and Landau (1950) and is traditionally called the Ginzburg-Landau functional. The second term in the integral of Eq. (1.8) is introduced to suppress rapid configurational fluctuations in the function $\boldsymbol{\phi}(\boldsymbol{r})$, which occurs in real physical situations. Applying

[3] A system with this kind of symmetry is usually referred to as an $\mathcal{O}(n)$ symmetrical one.

the Landau approximation to a system described by the thermodynamic function, Eq. (1.8), one should define an equilibrium value of the order parameter by

$$\frac{\delta \mathcal{F}}{\delta \boldsymbol{\phi}(\boldsymbol{r})} = \alpha\tau\boldsymbol{\phi}(\boldsymbol{r}) - c\Delta\boldsymbol{\phi}(\boldsymbol{r}) + b\boldsymbol{\phi}^2(\boldsymbol{r})\boldsymbol{\phi}(\boldsymbol{r}) - \boldsymbol{h}(\boldsymbol{r}) = 0. \tag{1.9}$$

The last aspect of the generalization is the introduction of statistical mechanics. From the point of view of statistical mechanics the probability density to find a given configuration of the field $\{\boldsymbol{\phi}(\boldsymbol{r})\}$ is defined by

$$\rho\{\boldsymbol{\phi}\} = Z^{-1}\exp(-\mathcal{F}\{\boldsymbol{\phi}\}/T). \tag{1.10}$$

The quantity $\mathcal{F}\{\boldsymbol{\phi}\}$ in this equation should be treated as the free energy of a nonequilibrium state, with the given configuration $\boldsymbol{\phi}(\boldsymbol{r})$. This statement follows from the fact that the equilibrium state of a system is usually considered to be realized at fixed temperature T and volume V. The way to obtain the probability density Eq. (1.10) will be seen in Chapter 2. There on examples of some particular systems, Gibbs canonical distributions are integrated over all degrees of freedom excepting the field $\boldsymbol{\phi}(\boldsymbol{r})$ itself. This also gives a microscopic definition of $\mathcal{F}\{\boldsymbol{\phi}\}$.

As one can see from Eq. (1.10), different realizations of the field $\boldsymbol{\phi}(\boldsymbol{r})$ have essentially different probabilities. Evidently, the most probable configuration corresponds to the minimum value of the functional $\mathcal{F}\{\boldsymbol{\phi}\}$. Under the conditions we will consider in Sect. 1.6, this configuration gives the main contribution to all physically measurable quantities. In this case they can be obtained by retaining only terms containing $\boldsymbol{\phi}(\boldsymbol{r})$ which corresponds to the minimum configuration, defined by $\delta\mathcal{F}\{\boldsymbol{\phi}\}/\delta\boldsymbol{\phi}(\boldsymbol{r}) = 0$. This approximation is called "mean field" approximation. As is seen from Eq. (1.9) it coincides with Landau approximation. Historically, however, it was first introduced by Weiss in 1907 to explain magnetic ordering. Weiss's idea consisted of suggesting that individual magnetic momenta feel some resulting magnetic field constituted by the environment and an external field. This approximation is called, after Weiss, "molecular field" approximation. In fact, all three different terms — Landau, mean field, and molecular field approximations — represent the same physical approach in the theory of phase transitions.

1.3 Broken Symmetry and Condensation

Let us now consider the homogeneous solution of Eq. (1.9), which, for the space independent external field $\boldsymbol{h} = const$, is the solution ensuring the

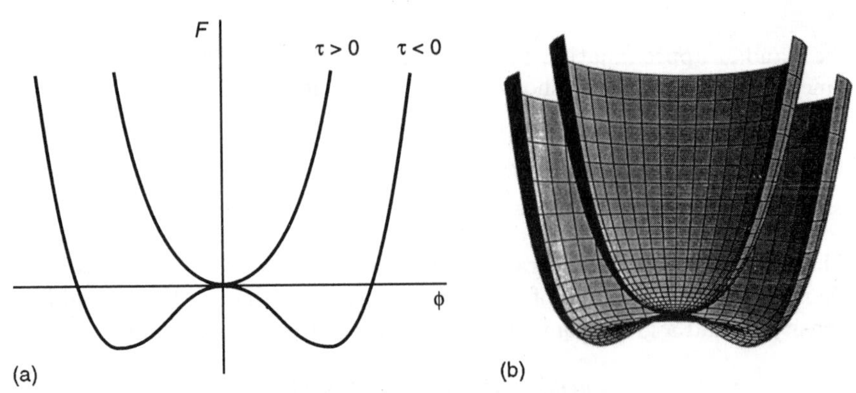

Figure 1.1: The Landau free energy dependence on the value of the order parameter a) $n = 1$, b) $n = 2$. Note the difference between $\tau > 0$ and $\tau < 0$. For the case with $n = 1$, when $\tau > 0$ the free energy is twice degenerate for any value of ϕ excepting $\phi = 0$. The same holds if $\tau < 0$, but the ground state corresponds to the degenerate state with $\phi \neq 0$. The situation is similar for the case with $n = 2$, but here the degeneracy is continuous.

free energy minimum. Evidently, if $h \neq 0$, the order parameter will be directed along the applied field $\boldsymbol{\phi} = \boldsymbol{n}\phi$, where $\boldsymbol{n}$ is a unit vector along the external field direction. This statement is independent of temperature. The situation is not so clear when $h = 0$. The question which should be answered is again, what is the direction of the order parameter vector? When $\tau > 0$, there is no such question, since $\phi = 0$, but it stands when $\tau < 0$. An endeavor to find an answer from the condition of the free energy minimum fails because the system with $n > 1$ is continuously degenerate. In Fig. 1.1 this continuous degeneracy is demonstrated for the case $n = 2$ in comparison with the case $n = 1$. As is seen from this figure, even for $n = 1$ we have two equivalent free energy minima leading to the twofold degeneracy. Nonetheless, the break of the inherent symmetry (replacement ϕ on $-\phi$) and the choice of the sign of the function ϕ can be understood as a result of an action of arbitrary fluctuations. If our system, at the moment of time $t = 0$, was in the symmetrical state ($\phi = 0$ and $\delta\mathcal{F} = \mathcal{F} - \mathcal{F}_l = 0$), then with time it will eventually arrive at one of the minima due to fluctuations. Applying this consideration to the system having a two-component order parameter, one can see that after arriving at the state with minimum free energy the vector $\boldsymbol{\phi}$ will continue to move chaotically along the circle $\phi_1^2 + \phi_2^2 = \phi^2$.[4] This movement has equal probability for occurring in either direction and therefore the expectation value

[4]Here and everywhere below, ϕ_i $(i = 1, \ldots, n)$ are components of the vector ϕ.

of the order parameter should be zero. The latter, however, does not agree under experiment. To understand the reason for this disagreement, let us return to the statistical definition of magnetization. Expectation quantities of different thermodynamic functions can be represented with the help of the normalization constant Z in the definition of the probability density, Eq. (1.10). Traditionally, this value is called "partition function" and this function is very convenient, because all thermodynamic quantities can be obtained as derivatives with respect to the thermodynamic variables. The most general definition for Z is

$$Z = \prod \int \frac{d\phi_q}{\sqrt{2\pi V}} \exp[-\mathcal{F}\{\phi\}/T], \tag{1.11}$$

where the product is taken over all modes ϕ_q defined by the Fourier transform

$$\phi(\boldsymbol{r}) = \frac{1}{\sqrt{V}} \sum_q \phi_q e^{i\boldsymbol{qr}}, \tag{1.12}$$

Keeping in mind the problem of the expectation value of the magnetization for degenerate systems, let us calculate partition function under the assumption that only the homogeneous contributions are essential. For this case in Eq. (1.11) only the modes with $\boldsymbol{q} = 0$ should be retained. Then, after angular integration in the isotopic space, we arrive at the expression

$$Z \propto \zeta^{1-\frac{n}{2}} \int_0^\infty dx x^{n/2} I_{\frac{n}{2}-1}(\zeta x) \exp[-\varepsilon(2\mathrm{sign}\tau x^2 + x^4)], \tag{1.13}$$

where I_ν is Bessel's function of an imaginary variable, ζ and ε are dimensionless parameters defined by

$$\zeta = \frac{Vh(\alpha \mid \tau \mid)^{1/2}}{T}, \tag{1.14}$$

$$\varepsilon = \frac{V\alpha^2\tau^2}{bT}. \tag{1.15}$$

If $\tau > 0$, the procedure of integration due to the very large value of the parameter ε, is quite evident.[5] As the above case does not contain any information about the ordered phase let us return to the case with $\tau < 0$. Further integration in Eq. (1.13) can be carried out simply in the two limiting cases, when ζ is either very small or very large. With the help of the partition function, the magnetization can be reduced to

$$\langle\phi\rangle = \frac{T}{V}\frac{\partial \ln Z}{\partial h} = n\phi_0 \frac{\partial \ln Z}{\partial \zeta}, \tag{1.16}$$

[5] This parameter is the condensation energy related to temperature.

where ϕ_0 is a value defined by Eq. (1.5). When $\zeta \ll 1$ in Eq. (1.13), Bessel's function can be expanded to the first nonvanishing order in ζ. Then, from Eq. (1.16), and making use of the saddle point method in the integral of Eq. (1.13), we obtain

$$\langle \boldsymbol{\phi} \rangle = \boldsymbol{n}\phi_0\zeta. \tag{1.17}$$

In the opposite case $\zeta \gg 1$, using asymptotic representation for Bessel's functions $I_\nu(z) \propto z^{-1/2} \exp z$ and the saddle point method, we find

$$\langle \boldsymbol{\phi} \rangle = \boldsymbol{n}\overline{\phi}, \tag{1.18}$$

where $\overline{\phi}$ should be taken as a solution of Eq. (1.2). Let us consider the limit $h \to 0$. From Eq. (1.17) one can see that in this limit $\boldsymbol{\phi} = 0$, which corresponds to the intuitive idea of the fluctuation order parameter suppression. On the other hand, Eq. (1.18) in the limit $h \to 0$ gives $\langle \boldsymbol{\phi} \rangle = \boldsymbol{n}\phi_0$, which corresponds to the real experimental situation. Of course, in this limit, care should be taken not to change the condition $\zeta \gg 1$. The latter can be kept in order, if along with the assumption $h \to 0$ we, also, assume $V \to \infty$ fixing ζ in the Eq. (1.15). The phenomenon that the results obtained from Eq. (1.17) and Eq. (1.18) do not coincide in the same limit means that the limits $h \to 0$ and $V \to \infty$ cannot be interchanged.

$$\begin{aligned} \lim\nolimits_{V\to\infty} \lim\nolimits_{h\to 0} \langle \boldsymbol{\phi} \rangle &= 0, \\ \lim\nolimits_{h\to 0} \lim\nolimits_{V\to\infty} \langle \boldsymbol{\phi} \rangle &= \boldsymbol{n}\phi_0. \end{aligned} \tag{1.19}$$

From this equation one may conclude that the partition function defined by Eq. (1.11) and its derivatives are no longer thermodynamic functions. By the conventional definition,[6] the change of thermodynamic functions is only determined by the initial and final state of thermodynamic variables, irrespective of the way along which one had to pass. Nonetheless, it is possible to restore the usual thermodynamic sense of all the functions used in the theory of phase transitions. One of the feasible ways was proposed by Bogoliubov in 1946. He introduced a nonconventional definition for expectation values. Every averaged quantity should be treated as a limiting value obtained for a system with infinite volume having an infinitesimally small external field. Such a definition preserves those quantities which arise due to the breaking of the inherent symmetry of the system, leaving all other mean values unchanged. This approach is not always convenient, because it requires apparent introduction of an external field which sometimes does

[6]See, for instance, Bloch's textbook "Fundamentals of Statistical Mechanics".

not have a clear physical meaning. In the case of superconductivity or superfluidity we deal with a two-component complex order parameter, which could be combined with a complex field. This field cannot be considered a physical quantity in a homogeneous situation. Another way to deal with broken symmetry is the one most clearly revealed by the case of the Bose-Einstein condensation of noninteracting particles.

1.3.1 Bose-Einstein Condensation

Since a particle of an ideal Bose gas has only kinetic energy $\epsilon_q = q^2/2m$, the energy of a container, having volume V and density of particles $\rho = N/V$, will be zero at $T = 0$. This means that the macroscopically large number of particles N are condensed in the state with the lowest energy. A small increase of temperature does not destroy the macroscopic population of the lowest energy state; though the number of particles in this state, N_0, will decrease. There is a critical temperature T_c at which N_0 becomes a number of the order of unity, not proportional to the volume of the system. Therefore, at $T > T_c$ the energy level with $\epsilon_q = 0$ is occupied on an equal footing with other levels — microscopically.

Let us suppose that the density ρ of the gas is fixed and $T > T_c$, then chemical potential μ should be defined from the following equation

$$\rho = \frac{1}{V} \sum_q \frac{1}{e^{\beta(\epsilon_q - \mu)} - 1}, \tag{1.20}$$

where $\beta = 1/T$. As is evident from Eq. (1.20), μ must always be less then zero, otherwise the mean number of particles $N_q = [\exp \beta(\epsilon_q - \mu) - 1]^{-1}$ might become negative, which is physically senseless. With the decrease of temperature, the absolute value of μ will also decrease and, when we reach T_c, it becomes proportional to V^{-1}. That means that contribution to the density from the lowest energy level $\rho_0 = N_0/V$ has some finite value independent of the volume. Hence, this contribution should be separated from other levels with $q \neq 0$, for which $\rho_q \propto V^{-1}$. For these levels, summation in Eq. (1.20) can be replaced by integration, then

$$\rho = \rho_0 + \frac{(mT)^{3/2}}{2^{1/2}\pi^2} \int_0^\infty \frac{\sqrt{z}dz}{\exp^{(z-\nu)} - 1}, \tag{1.21}$$

where $\nu = \beta\mu$. The condition $\rho_0 = N_0/V$ defines ν, which is equal to $-(\rho_0 V)^{-1}$. As in a typical situation $\rho V = N \gg 1$, the value ν for $T < T_c$ is extremely small and it can be omitted in the integral of Eq. (1.21). The critical temperature is defined by the condition $\rho_0 = 0$ in the limit $V \to \infty$

and is equal to

$$T_c = \frac{3.31}{m}\rho^{3/2}. \tag{1.22}$$

The discussion presented up till now shows that the Bose-Einstein condensation only resembles a second-order phase transition, because the main feature of such transitions — broken symmetry — has not been established. The simplest way to reveal broken symmetry is to introduce an external field conjugated to the order parameter. Unfortunately, there is no real physical quantity that may play this role. This means that the only possible way in this direction is to find an order parameter apparently reflecting a symmetry breaking. Evidently, N_0 can be directly related to the order parameter, since it has a finite value at $T < T_c$ and is equal to zero at $T > T_c$. Let us consider $T = 0$, then all particles are condensed in the lowest energy level. The quantum mechanical number of particles operator has the form $\hat{N} = \sum_q a_q^+ a_q$, where a_q^+ and a_q are creation and annihilation operators in the state with momentum q. Taking into account the condensation in the level with $q = 0$, the expectation value of this operator can be written

$$N = \langle \hat{N} \rangle = \langle N \mid a_0^+ \mid N-1 \rangle \langle N-1 \mid a_0 \mid N \rangle. \tag{1.23}$$

The only nonzero matrix element of the operator a_0, as is seen from Eq. (1.23), is a large complex macroscopic quantity with the absolute value $\propto \sqrt{N}$. This quantity cannot change abruptly with the increase of temperature, but gradually decreases till at the critical temperature it becomes of the order of unity. This shows that the quantity

$$\Psi = \frac{\phi_1 + i\phi_2}{\sqrt{2}} = \frac{\langle N_0 - 1 \mid a_0 \mid N_0 \rangle}{\sqrt{V}} \tag{1.24}$$

is a good candidate for the order parameter of this phase transition. The Hamiltonian of noninteracting Bose particles

$$H = \sum_q \epsilon_q a_q^+ a_q$$

is invariant under the transformation $a_q \to e^{i\alpha} a_q$. The introduced order parameter Eq. (1.24) changes under this transformation acquiring the factor $e^{i\alpha}$. Now it is seen that Bose-Einstein condensation has all necessary attributes of the second-order phase transition: continuous transition, singular second derivatives, and the breaking of an inherent symmetry. By expanding Eq. (1.21) in the neighborhood of T_c one can see that the order parameter introduced in Eq. (1.24) varies as $\mid \Psi \mid \propto \mid \tau \mid^{\beta}$ with $\beta = 1/2$. To find other critical exponents we should consider our system as being subjected to the influence of an external complex field. This field, although a mathematical abstraction, helps us to find the exponents[7] $\delta = 3$ and $\gamma = 1$. The specific heat is continuous at $T = T_c$ corresponding to $\alpha = 0$, but it has a bend so that its derivative with respect to τ exhibits a jump.

Bose-Einstein condensation was the first example of an exactly solvable model exhibiting a second-order phase transition. Notwithstanding the coincidence of the values for critical exponents with the corresponding ones for the Landau approach, this model, nonetheless, adds to understanding of the phase transition problem. It demonstrates an additional way to define expectation quantities in the phase with a broken symmetry. According to this model the "dangerous" condensing mode should be excluded from any statistical (classical or quantum mechanical) averaging and treated like an external field with a value which is defined from some additional selfconsistent treatment. In the most general case, this additional condition is the requirement that a properly chosen thermodynamic potential is minimized. Unlike Bogoliubov's definition of expectation quantities, this way neither needs the introduction of an external field nor the consideration of limiting values. It is applicable even in systems having finite volume. The Bose-Einstein condensation will be further exploited in this book in Chapter 8, where some simple, but nontrivial, exactly solvable models are considered.

1.4 Ergodicity

Some additional physical insight on the problem of phase transition can be obtained from consideration of the order parameter relaxation. As mentioned in Sect. 1.3, some of the averaged quantities depend on the method of their definition (see Eq. (1.19)). Therefore, they reflect the prehistory of the states of the system, thus violating analiticity of thermodynamic functions with respect to the thermodynamic variables. This violation also implies the noncoincidence of time and ensemble averaged quantities, which corresponds to a loss of ergodicity in the system. The ergodic or, more strictly, quasi-ergodic hypothesis constitutes a mathematical basis of contemporary statistical physics. Hence, systems violating ergodicity should not be considered as traditional thermodynamic objects, unless some ways of decomposing the whole phase space on ergodic and nonergodic subspaces are found. In order to consider the problem of ergodicity in relation to phase transitions, we need to first find the reaction of the system to an external field (susceptibility) and then to study its relaxation after switching-off the field.

[7] To find δ one has to also introduce at least infinitesimally small interaction between Bose particles.

1.4.1 Susceptibility

In the nonordered phase the homogeneous susceptibility for the case of $\mathcal{O}(n)$ symmetry is a spherical tensor in the isotopic space $\chi_{ik} = \delta_{ik}\chi$. The quantity χ does not qualitatively differ from the corresponding value in the case $n = 1$. A substantial difference appears below the critical temperature, where the ordered phase is characterized by the appearance of an isotopic vector–order parameter. Therefore χ_{ik} can no longer be defined by one scalar quantity. Since in the plane perpendicular to the order parameter direction the symmetry is conserved, we will need to introduce $\chi_{\perp}$. Thus we describe a reaction of the system to an external field, transverse to the order parameter direction external field. Consequently, we will also need to know the longitudinal susceptibility $\chi_{\parallel}$. These two quantities will completely define the susceptibility tensor

$$\chi_{ik} = n_i n_k \chi_{\parallel} + (\delta_{ik} - n_i n_k)\chi_{\perp}. \tag{1.25}$$

The values for transverse and longitudinal susceptibilities can be simply obtained in the framework of Landau approximation by taking consecutive derivatives from Eq. (1.9). They give us

$$\begin{aligned} \chi_{\perp} &= \overline{\phi}/h; \\ \chi_{\parallel} &= \overline{\phi}/(h + 2b\overline{\phi}^3). \end{aligned} \tag{1.26}$$

From the discussion in this section, it seems quite natural that the transverse susceptibility should appear in addition to the longitudinal one which is present even in the case $n = 1$. The thing which may seem not so natural is divergence of $\chi_{\perp}$ when $h \to 0$. This fact is a mathematical consequence of the nonvanishing, in the limit $h \to 0$, of the order parameter at $T < T_c$. It could also be anticipated from a simple physical consideration. As the movement in the transverse direction involves no change in the free energy of the system (see Fig. 1.1) an infinitesimally small change of an external field should be immediately followed by the order parameter. On the contrary, in the longitudinal direction the value of the order parameter corresponds to the minimum of energy and any change would require some energy addition resulting in a finite value $\chi_{\parallel}$. Therefore, infinite value of $\chi_{\perp}$ is a direct consequence of the continuous degeneracy of our system.

1.4.2 The Ergodic Hypothesis

Let us suppose that we disturbed the equilibrium state of our system. In the neighborhood of a critical point all macroscopic processes are slowing down. Microscopic kinetics, scattering of particles, exchange of momentum and energy, and so on, are not affected by the slowing of the macroscopic behavior of the system. This leads to the situation where all other degrees of freedom can be considered as being in a local equilibrium state characterized by some slowly changing value of the order parameter $\phi(r,t)$. Such a quasi-equilibrium state allows one to consider the free energy functional $\mathcal{F}\{\phi\}$ at the given $\phi(r,t)$. If we have fixed the temperature and volume of the system, then, in the state of complete equilibrium, $\frac{\delta\mathcal{F}}{\delta\phi} = 0$. Under the influence of some deviation from the equilibrium state a process will arise returning ϕ to the equilibrium state. In accordance with the fundamental principles of nonequilibrium thermodynamics (Landau and Lifshits, 1976) the speed with which ϕ will relax to its equilibrium value is proportional to the conjugate thermodynamic force

$$-\frac{\partial\phi}{\partial t} = \Gamma\frac{\delta\mathcal{F}}{\delta\boldsymbol{\phi}(\boldsymbol{r},\boldsymbol{t})}, \tag{1.27}$$

where Γ is a kinetic factor which, according to the general ideas of the Landau approximation, can be treated in the neighborhood of T_c as a temperature independent quantity.

Let us consider only small homogeneous deviations from the equilibrium, then we can expand Eq. (1.27) and obtain

$$\begin{aligned} -\,\partial\delta\phi_{\|}/\partial t &= \Gamma\chi_{\|}^{-1}\overline{\phi}\,\delta\phi_{\|}; \\ -\,\partial\delta\phi_{\perp}/\partial t &= \Gamma\chi_{\perp}^{-1}\overline{\phi}\,\delta\phi_{\perp}. \end{aligned} \tag{1.28}$$

As was shown in Sect. 1.4.1 $\chi_{\perp} \to \infty$ when $h \to 0$. This means that transverse deviation will never return to its initial value. Hence, our system at $T < T_c$ is no longer an ergodic (or quasi-ergodic) system. Ergodicity is, usually, defined as a property of any phase trajectory on the surface with a fixed energy to cover this surface homogeneously. In other words, a chosen trajectory can come arbitrarily close to any point on the given energy surface. As a consequence of this property there is a coincidence of time and ensemble averaging. Eq. (1.28) shows that by fixing $\delta\ \phi_{\perp}$ we define a point in a phase space that can never be reached by phase orbits which have different order-parameter orientation. The ensemble averaging in this case would be an averaging over different orientations causing the loss of vital information about the ordered state. Time averaging, if one would be able to do it, will inevitably lead to the state with a fixed order parameter vector. The loss of ergodicity is another insight clarifying physical sense of the procedure described in Sect. 1.3.1. The exclusion of the condensing mode off the statistical averaging suggested there, from the point of view developed here, is simply an exclusion of a non-ergodic mode, and to some extent, this procedure can be seen as obligatory in the pursuit of a statistic mechanical approach.

Until now we have discussed the case with $n > 1$ and it might seem that nonergodicity is a consequence of the continuous degeneracy. This statement is, of course, not true. The right thing would be to say that any kind of degeneracy of the ground state leads to the appearance of a nonergodic mode. It is much more simple to demonstrate this property for the case $n > 1$, because the existence of an infinite relaxation time can be revealed by considering small deviations from the equilibrium. In the system with $n = 1$, small deviations will relax during a finite time interval, but strong enough values of the external field will change the sign of the order parameter and switching-off the field does not return the initial sign within a finite time. Here we have twofold degeneracy (see Fig. 1.1) and these degenerate states are divided by an infinite barrier in the limit $V \to \infty$. This means that all phase space is decomposed into two subspaces which can be classified by the order parameter sign. If one finds the system in the subspace with the positive sign, it will never change to the negative sign, unless an appropriately oriented external field is applied during some finite time interval. Of course, the difference between $n > 1$ and $n = 1$ is very crucial. Continuously degenerate systems lead to an infinite number of subsets in phase space and by some continuous change of $n - 1$ parameters one can cover all the subsets. This is not so in the case where $n = 1$, which is an example of a twofold separation of the phase space where there are no parameters, by continuous change of which, one could switch from one subset to the other. The distinction is even more clearly seen in the case with finite volume. This is, usually, the case realized in experiment. Now such terms as an ergodic or nonergodic system require more rigorous definition. All relaxation times are finite. If $n = 1$, then for the system to switch from, for instance, the positive sign of ϕ to a negative value the characteristic time is exponentially large $\propto \exp(V/T)$ while for $n > 1$ this time is proportional to some power of the volume.

1.5 Gaussian Approximation

Since in the Landau approximation Eq. (1.10) gives the most probable configuration $\boldsymbol{\phi}(\boldsymbol{r})$ and, therefore, takes care of the most important aspects of a phase transition, it would be quite natural to try to improve the approximation by considering small deviations from this configuration. Then the functional, Eq. (1.8) will be represented in the form

$$\mathcal{F}(T, \{\boldsymbol{h}\}, \{\boldsymbol{\phi}\}) = \mathcal{F}_{\mathcal{L}} + \frac{1}{2}\sum_{i,k}\int d\boldsymbol{r}d\boldsymbol{r}'\frac{\delta^2\mathcal{F}}{\delta\phi_i(\boldsymbol{r})\delta\phi_k(\boldsymbol{r}')}\delta\phi_i(\boldsymbol{r})\delta\phi_k(\boldsymbol{r}'), \quad (1.29)$$

where $\mathcal{F}_{\mathcal{L}}$ is the free energy in Landau approximation.

The second variational derivative in Eq. (1.29) should be taken at $\boldsymbol{\phi}(\boldsymbol{r})$ corresponding to Landau approximation and, hence, it can be expressed in terms of physically measurable quantities of this approximation. This can be proved with the help of the following procedure. Let us take the variational derivative from Eq. (1.9) with respect to $\boldsymbol{h}(\boldsymbol{r})$, then we obtain the equation

$$\sum_k\int d\bar{\boldsymbol{r}}\frac{\delta^2\mathcal{F}}{\delta\phi_i(\boldsymbol{r})\delta\phi_k(\bar{\boldsymbol{r}})}\chi_{kl}(\bar{\boldsymbol{r}}, \boldsymbol{r}') - \delta_{il}\delta(\boldsymbol{r}-\boldsymbol{r}') = 0, \quad (1.30)$$

where $\chi_{ik}(\boldsymbol{r}, \boldsymbol{r}')$ is a nonhomogeneous magnetic susceptibility, which in Landau approximation is defined by

$$\chi_{ik}(\boldsymbol{r}, \boldsymbol{r}') = \frac{\delta\phi_i(\boldsymbol{r})}{\delta h_k(\boldsymbol{r}')}. \quad (1.31)$$

From Eqs. (1.30) and (1.31) one can see that the searched-for second derivative is equal to the inverse susceptibility tensor.

1.5.1 Goldstone Branch of Excitations

Let us restrict consideration with the case $\boldsymbol{h} = const$. In this case the tensor of the inverse susceptibility in Eq. (1.29) depends only on the difference of the space coordinates. As a result, the quadratic form containing deviations $\delta\boldsymbol{\phi}(\boldsymbol{r})$ can be diagonalized with the help of the Fourier transformation Eq. (1.12). Then instead of Eq. (1.29) we obtain

$$\Delta\mathcal{F} = \mathcal{F} - \mathcal{F}_{\mathcal{L}} = \frac{1}{2}\sum_q\left\{\chi_{\parallel}^{-1}(q) \mid \phi_{\parallel q}\mid^2 + \chi_{\perp}^{-1}(q)\mid \boldsymbol{\phi}_{\perp q}\mid^2\right\}, \quad (1.32)$$

where the longitudinal and transverse inverse susceptibilities are defined by

$$\begin{aligned}\chi_{\parallel}^{-1}(q) &= cq^2 + h/\bar{\phi} + 2b\bar{\phi}^2; \\ \chi_{\perp}^{-1}(q) &= cq^2 + h/\bar{\phi}.\end{aligned} \quad (1.33)$$

As the Landau state is the state with the lowest energy of the system, the Gaussian additions represent the energy of excitations. A remarkable feature of the excitation energy $\epsilon_{\parallel,\perp}(q) \propto \chi^{-1}_{\parallel,\perp}(q)$ is again related to the case with $n > 1$. When $h = 0$ and $\tau < 0$, as has been mentioned before (see Sect. 1.3), we have continuous degeneracy and, therefore, motion in the transverse direction costs no energy to excite. This means that there must be a transverse homogeneous excitation with q and $\epsilon_\perp(q)$ equal to zero.[8] This branch of excitations having no gap (or mass) was at first studied by Goldstone, Salam, and Weinberg (1962) at zero temperature with a particular emphasis on a relativistic field theory application. It is now commonly accepted to call a zero boson mass excitation branch resulting from the symmetry breaking process *a Goldstone excitation mode.* In the case of $\mathcal{O}(n)$ symmetry we have exactly $(n-1)$ Goldstone modes.

1.5.2 Correlation Functions

In phase transition theory, n-point correlation functions are defined as average values of the type

$$G_{i_1\cdots i_n}(\boldsymbol{r}_1\cdots\boldsymbol{r}_n) = \langle\delta\phi_{i_1}(\boldsymbol{r}_1)\cdots\delta\phi_{i_n}(\boldsymbol{r}_n)\rangle. \tag{1.34}$$

For the case of Gaussian approximation only the correlation functions having an even number of points are different from zero. The procedure of averaging with the functional, Eq. (1.32) can be considered as a classical limit of averaging for a Bose field $\delta\boldsymbol{\phi}(\boldsymbol{r})$ and, as a consequence of this, Wick's theorem is applicable. Therefore, all higher order correlation functions can be expressed in terms of only two-point correlation functions, which are also called Green's functions or propagators. The two-point correlation functions can be easily obtained with the help of Eq. (1.32) in q-space

$$\begin{aligned} G_\parallel(q) &= \langle\phi_{\parallel q}\phi_{\parallel -q}\rangle = T\chi_\parallel(q); \\ G_\perp(q) &= \langle\phi_{i_\perp q}\phi_{i_\perp -q}\rangle = T\chi_\perp(q). \end{aligned} \tag{1.35}$$

Since Landau theory can only be justified as an expansion in the vicinity of critical temperature when $|\tau| \ll 1$ (see Sect. 1.2), T in the formulae of Eq. (1.35) is conventionally replaced by T_c. Let us now consider the behavior of correlation functions in r-space. If $\tau > 0$ and the external

[8] It is necessary to remember that the mode with q exactly equal to zero is a condensing mode and, according to Sect. 1.3, it must be excluded from the thermodynamic averaging as it has already been incorporated into $\mathcal{F}_\mathcal{L}$ in Eq. (1.29).

field is small enough to satisfy inequality $bh^2 \ll (\alpha\tau)^3$, the transverse and longitudinal correlation functions will coincide, and by making Fourier's transformation we obtain

$$G(r) = \frac{T_c}{4\pi c} \frac{\exp\{-r/\xi(\tau)\}}{r}, \tag{1.36}$$

where the characteristic length describing exponential decay of correlations is given by $\xi(\tau) = \sqrt{c/\alpha\tau}$.

In the case of negative τ and $h = 0$ the longitudinal and transverse correlation functions will no longer coincide.The difference between them can be, nonetheless, illustrated with the help of Eq. (1.36). This formula is to be formally considered as the one valid also at $\tau < 0$, but for the longitudinal correlations $\xi(\tau)$ should be replaced by $\xi_{\parallel}(\tau) = \sqrt{c/(2\alpha \mid \tau \mid)}$, while for the transverse ones replaced by $\xi_{\perp}(\tau) \to \infty$.

The latter is a direct consequence of the continuous degeneracy and the symmetry breaking. It can also be treated as a manifestation of the Goldstone excitation mode. The expression for the correlation function, given by Eq. (1.36), also holds for correlation functions in large external fields $bh^2 \gg (\alpha\tau)^3$. In this case we have the same expressions for correlation functions independent of the sign of τ. The longitudinal and transverse correlation lengths are now defined by $\xi_{\parallel}(h) = c^{1/2}/(\sqrt{3}b^{1/6}h^{1/3})$ and $\xi_{\perp}(h) = c^{1/2}/(b^{1/6}h^{1/3})$. Independently of the type of correlation functions, the correlation lengths diverge when τ and h tend to zero. This divergence again follows a power law behavior introducing two additional critical exponents. In the general case, the exponent reflecting the temperature dependance is denoted by ν ($\xi(\tau) \propto \mid \tau \mid^{-\nu}$, $h = 0$) and the one for the external field dependance by μ ($\xi(h) \propto h^{-\mu}$, $\tau = 0$). The Landau approximation gives the following values for these critical exponents, $\nu = 1/2$ and $\mu = 1/3$.

1.5.3 Microscopic Scales in Phase Transitions

Microscopic lengths are usually initially present in the problem under consideration. They may be lattice sizes, screening lengths, particular interaction lengths and so on. As will be seen in the next chapter, when the Ginzburg-Landau functional is being derived from a microscopic Hamiltonian of some system, the information about all these details of the microscopic arrangement are being lost. This means that spatial resolutions should not be considered at distances less than some length $r_0 \ll \xi(\tau)$. Consequently, momentum related values should be restricted by some cutoff momentum Λ of the order of r_0^{-1}. It is convenient to have an estimate for r_0 in terms of the constant quantities α, T_c, b which have already been introduced in Ginzburg-Landau functional Eq. (1.8). This can be done with the help of the following consideration (Patashinskii and Pokrovskii 1979). The energy of interactions is of the order of $Vb\bar{\phi}^4 \approx V\alpha^2/b$. On the other hand, this value has also the order of the number of degrees of freedom N by the energy related to one degree of freedom, which is of the order of T_c. The length obtained from V/N, which is a volume occupied by one degree of freedom, can yield a scale for the separation of the two nearest interacting degrees. Hence, for r_0 these considerations give an estimate

$$r_0 \approx \left(\frac{T_c b}{\alpha^2}\right)^{\frac{1}{3}}. \tag{1.37}$$

Formally, correlation functions defined in Sect. 1.5.2 diverge when any pair of spatial coordinates coincide. This divergence results from the neglect of microscopic scales, and it should be excluded by taking the smallest spatial separation to be of the order of r_0.

1.6 The Ginzburg Criterion

The Gaussian approximation allows one to estimate the region in which Landau theory is applicable. In order to do this, it is only necessary to compare any measurable quantity obtained in Landau approximation with the consequent addition due to the Gaussian integration in the partition function, Eq. (1.11). One of the simplest ways is to complete this procedure with the specific heat. The free energy $\mathcal{F}$ of the system, in the case when fluctuations are important, is defined by

$$\mathcal{F} = -T\ln Z = \mathcal{F}_{\mathcal{L}} - T\ln Z_G, \tag{1.38}$$

where Z_G is the Gaussian contribution to the partition function. Since we are only considering the region close to T_c, the addition to the density of entropy ΔS can be obtained from Eq. (1.38) as an expression

$$\Delta S = \frac{1}{V}\frac{\partial \ln Z_G}{\partial \tau} = -\frac{\alpha}{2T_c}\left\{G_{\parallel}(r_0) + (n-1)G_{\perp}(r_0)\right\}. \tag{1.39}$$

From this expression the fluctuational correction to the specific heat ΔC, is obtained as usual

$$\Delta C = \frac{\partial S}{\partial \tau} = \frac{\alpha^{3/2}}{16\pi T_c c^{3/2} \mid \tau \mid^{1/2}}. \tag{1.40}$$

Comparing this value with the specific heat jump in Landau approximation ($\delta C = -\alpha^2/2bT_c$) we arrive at the criterion of the applicability of Landau theory, which is traditionally called the Ginzburg criterion

$$\frac{b^2 T_c^2}{\alpha c^3} \ll \mid \tau \mid \ll 1. \tag{1.41}$$

This criterion was first found by Levaniuk in 1959 and by Ginzburg in 1960. The quantity on the left hand side of the inequality, Eq. (1.41) is a dimensionless parameter which is related only to characteristic values of the given system. This quantity is called the Ginzburg parameter and it can also be expressed as a ratio of the two characteristic lengths r_0 and $\xi_0 = \xi(1) = \sqrt{c/\alpha}$

$$G_i = \frac{b^2 T_c^2}{\alpha c^3} = \left(\frac{r_0}{\xi_0}\right)^6. \tag{1.42}$$

This parameter should be small to ensure the region for applicability of the Landau theory. As is seen from the inequality, Eq. (1.41), $\mid \tau \mid \gg G_i$ in the region of the mean field approximation. Since the parameter G_i is related only to two microscopic lengths the natural question arises: Are there any general considerations which could give us at least qualitative indication of its value? Or in other words: Why is the separation of interacting degrees of freedom different from the correlation length for these degrees far away from the transition point? There is no general answer to these questions, but for particular systems it is often possible to make quite reasonable predictions. For instance, if we deal with a system of nearest neighbor interacting magnetic moments located at lattice sites, then there is only one characteristic length in the problem. Therefore, one would expect G_i in this case to be of the order of unity and only some numerical reasons could lead to its smallness. Namely, due to the sixth power in Eq. (1.42), even in the case when $\xi_0 = 2r_0$ we have $G_i \approx 10^{-2}$. On the other hand, there are many physical systems having quite a number of characteristic lengths and without proper analysis it is impossible to anticipate the value for G_i. For example, in the case of a phase transition in a metal, an estimate of the correlation length ξ_0 can be $\hbar/p_0$, where p_0 is a characteristic momentum composed from the ordering energy per one degree of freedom T_c and Fermi velocity v_F of an electron $\xi_0 = \hbar v_F/T_c$. Consequently, it might seem to be quite plausible to take as a microscopic length r_0 a mean separation distance for electrons which is of the order of $\hbar/p_F$ where p_F is Fermi momentum of an electron. These two lengths would make an estimate $G_i \approx (T_c/\epsilon_F)^6$, which gives a very small value of G_i for any kind of phase transition in metals, because $T_c \ll \epsilon_F$ always. This estimate is not correct owing to a too rough approximation for r_0. Even in the case of superconductors, for which G_i in reality is a very small value, more rigorous evaluations should be made.

In this case ξ_0 is correlation radius of Cooper pairs at $T = 0$, which is $\hbar v_F/2\Delta = \hbar v_F/3,5T_c$, where we have taken BCS relation between energy gap Δ at $T = 0$ and T_c for weakly coupled superconductors. The density of the condensation energy $\alpha^2/4b$ at $T = 0$ can be related to the value of Δ and the density of electrons $n \approx (p_F/\hbar)^3$

$$\frac{\alpha^2}{b} \propto \Delta^2 n \epsilon_F^{-1}. \tag{1.43}$$

From this, equation with the help of Eq. (1.37), we find

$$r_0 \approx \frac{\hbar}{p_F}\left(\frac{\epsilon_F}{T_c}\right)^{1/3}. \tag{1.44}$$

For typical superconductors this value is at least one order of magnitude higher than $\hbar/p_F$ and for Ginzburg parameter we have

$$G_i \approx 10^2\left(\frac{T_c}{\epsilon_F}\right)^4. \tag{1.45}$$

In conventional superconductors the ratio $T_c/\epsilon_F \approx 10^{-3}$ and $G_i \approx 10^{-14}$. Hence, the fluctuation region is very narrow $\mid \tau \mid \ll 10^{-14}$ and is inaccessible for experimental observation. It can be substantially widened in the "dirty" superconductors due to essential decrease of the correlation length ξ_0 (Gorkov 1959). In high-T_c superconductors $T_c/\epsilon_F \approx 10^{-1}$, $G_i \approx 10^{-2}$, and this region has already been examined in a number of experimental studies (see, for example, Baun *et al* 1992). The example of superconductivity shows how fragile estimates of G_i can be. These estimates cannot be obtained *a priori* but only *a posteriori* after a derivation of the Ginzburg-Landau functional from a microscopic Hamiltonian (see next Chapter).

1.6.1 Critical Dimensions

The Landau theory of critical phenomena gives results which are independent of the dimension of space d. Gaussian corrections to this theory introduce such dependencies in it. This statement can be simply verified by considering a 2-point correlation function in a d-dimensional space. In q-space such correlation function is independent of d and is completely defined by Eq. (1.35), from which one can find its expression in r-space

$$G(r) = \frac{2T_c}{c} \cdot \frac{K_{\frac{d-2}{2}}(r/\xi)}{(2\pi r\xi)^{\frac{d-2}{2}}}, \tag{1.46}$$

where $K_\nu(z)$ is Macdonald's function of ν-th order.

As is known, the z-dependence of Macdonald's function is essentially influenced by its order. Namely, when $z \ll 1$, $K_\nu(z)$ has asymptotes

$$\begin{aligned} K_\nu(z) &\approx \tfrac{1}{2}\Gamma(\nu)\left(\tfrac{1}{2}z\right)^{-\nu}; \\ K_0(z) &\approx -\{\ln(z/2)+\gamma\}, \end{aligned} \tag{1.47}$$

where $\Gamma(\nu)$ is Gamma function and γ is Euler's number. For $z \gg 1$ the asymptotic behavior of K_ν is only slightly dependent on ν

$$K_\nu(z) \approx \sqrt{\frac{\pi}{2z}}\, e^{-z}\left\{1 + \frac{4\nu^2 - 1}{z} + \cdots\right\} \tag{1.48}$$

leading to exponential decay of the propagator, Eq. (1.46) $G(r) \propto r^{\frac{1-d}{2}}$ $\exp(-r/\xi)$ in the limit when $r \gg \xi$. In the opposite case $r \ll \xi$, $G(r) \propto r^{2-d}$.

Let us now study the intensity of fluctuations by considering the variance for $\delta\phi(\boldsymbol{r})$. This variance will be defined by the correlator, Eq. (1.46) with $r \approx r_0$. Strictly speaking, such small distances should not be considered since the Ginzburg-Landau functional can only be derived in the limit of slowly changing variables on the microscopic scales. Therefore, only such fluctuations of $\delta\phi(\boldsymbol{r})$ from which microscopic scale is excluded can be physically justified. Such an exclusion can be achieved by preliminary averaging of $\delta\phi$ over some finite volume $V \gg r_0^d$. Then the variance for $\delta\bar{\phi} = V^{-1}\int d\boldsymbol{r}\delta\phi(\boldsymbol{r})$ can be expressed in terms of propagators $G_\parallel$ and $G_\perp$

$$\langle(\delta\bar{\phi})^2\rangle \approx \frac{1}{V}\int d\boldsymbol{r}\left\{G_\parallel(\boldsymbol{r}) + (n-1)G_\perp(\boldsymbol{r})\right\}. \tag{1.49}$$

Now, with the help of Eq. (1.46), and taking into account Eq. (1.47) and Eq. (1.48) we find the estimate

$$\langle(\delta\bar{\phi})^2\rangle \approx \frac{T_c}{cR^d}\left\{\min(R^2, \xi_\parallel^2) + (n-1)\min(R^2, \xi_\perp^2)\right\}, \tag{1.50}$$

where $R \propto V^{1/d}$.

Landau theory is applicable until the variance of $\delta\bar{\phi}$, within a volume of linear size ξ, is much less than $\bar{\phi}^2(\tau < 0)$. For $d > 2$ the latter leads to inequality

$$\frac{b^2 T_c^2}{\alpha^\epsilon c^d} \ll \mid \tau \mid^\epsilon, \tag{1.51}$$

where $\epsilon = 4 - d$. This inequality, though it has been obtained from different starting points, generalizes the Ginzburg criterion, Eq. (1.41) for an arbitrary dimension d. Again, as has already been shown in Sect. 1.6 for $d = 3$, the Ginzburg parameter can be expressed as a ratio of the two characteristic lengths r_0 and ξ_0. The consideration, provided in Sect. 1.5.3, shows that $r_0 \approx (T_c b/\alpha^2)^{1/d}$. Taking into account that $\xi_0 = \sqrt{\alpha/c}$ independent of the value of d, we find

$$G_i = \left(\frac{r_0}{\xi_0}\right)^{2d} . \tag{1.52}$$

Similar to the inequality, Eq. (1.51), the Ginzburg criterion on the magnitudes of the external fields, which make the mean field theory applicable, can be obtained. Now, instead of the temperature dependent correlation length at $h = 0$, one must use ξ at $\tau = 0$, but $h \neq 0$. According to Sect. 1.5.2, this $\xi \propto c^{1/2}/b^{1/6}h^{1/3}$. The same procedure, which led to Eq. (1.51), yields

$$G_i^{3/2} \ll \left(\frac{h}{h_0}\right)^{\epsilon} , \tag{1.53}$$

where the characteristic field $h_0 = \alpha^{3/2}/b^{1/2}$ has been introduced.

The inequalities, Eq. (1.51) and Eq. (1.53) show that at $d \geq 4$ the Landau theory always has a domain of applicability in the vicinity of the critical point ($\tau = 0, \quad h = 0$). When $d < 4$ the fluctuation region appears immediately below $d = 4$ and it spreads with the decrease of d. Therefore, the dimension $d = 4$ plays a special role as a dimension above which the fluctuation influence on the phase transitions disappears. The other critical dimension is $d = 2$. A particular role of this dimension can be seen from Eq. (1.50). Let us suppose that $h \to 0$, then, according to Sect. 1.5.1, $\xi_\perp \to \infty$ and this leads to an infinite increase of fluctuations. This increase can be demonstrated with the help of the following reasoning. In the case where $d = 2$ the last term of Eq. (1.50) originating from the transverse propagator should be modified by taking into account the asymptotic behavior (see Eq. (1.47) for Macdonald's function with $\nu = (d-2)/2 = 0$). This asymptote will lead to the replacement of R^2 by $R^2 \ln R$ in the transverse term of Eq. (1.50) and, as a consequence of it, to the logarithmic increase of transverse fluctuations. On the other hand, $\delta\bar{\phi}$ must tend to zero when $R \to \infty$ simply due to the exclusion the condensing mode with $q = 0$. This contradiction can only be removed by taking $\xi_\perp$ as a finite value in the limit $h \to 0$. As is seen from Eq. (1.33) $\xi_\perp = (c\bar{\phi}/h)^{1/2}$, therefore $\bar{\phi} \to 0$, when $h \to 0$. This means that for planar systems ($d = 2$) in the presence of continuous degeneracy ($n > 1$) the mean value of the order parameter is always equal to zero. This statement was at first strictly proved by Mermin and Wagner (1966) for a plane Heisenberg Hamiltonian and by Hohenberg

(1967) for superconductivity and superfluidity. Nonetheless, the absence of the order parameter does not necessarily mean the absence of a phase transition. As it has been shown by Berezinsky (1970, 1971), in a system with a two-component order parameter there is a phase transition of a special kind leading to the appearance of rigidity in relation to the transverse fluctuations in the low temperature phase. This phase transition is very unusual and it happens to be in the system of long-wavelength, nonlinear, topological excitations, vortex excitations. These excitations are specific for the systems with $n = 2$ and they can also be found in the three dimensional space. In this case they lead to no essential physical effects[9] as their energy is proportional to the length of an individual vortex. Hence, they cannot be thermally excited and give no appreciable effect in thermodynamics. In the planar systems the energy of an individual vortex is finite[10] and they substantially influence thermodynamic behavior of the system at low temperatures. Kosterlitz and Thouless have shown (1973) that this transition manifests itself as a second-order phase transition. Physically, it represents transformation from a vortex plasma state with equal number of vortices having opposite vorticity to a molecular-like state with molecules composed from a pair of vortices with different polarities. In the literature the phase transition in the system with $d = 2$ and $n = 2$ is called a topological, or Kosterlitz-Thouless transition. In summary, we should say that $d = 2$ is a specific dimension at which existence of the phase transition in nature depends on the number of components of the supposed order parameter. If $n = 1$ we have a conventional phase transition (Ising model),[11] for $n = 2$ the mean value of the order parameter is absent, but a phase transition exists; while for $n > 2$ there is no phase transition.

From what has been said above, intuitively, one would expect absence of phase transitions at $d = 1$. The following simple physical reasoning shows that this is true (Landau and Lifshits 1976). Let us suppose that we have $n = 1$, and space dimension $d > 1$. Then, at temperatures below the critical one, the ordered state is doubly degenerate. Hence, there is an equal probability to find states with opposite magnetization (different signs of $\bar{\phi}$). If we have found a particular sing of magnetization, then, between regions of positive and negative magnetization, there is a state with changing configuration giving some positive addition to the free energy proportional to

[9] This statement is valid only in the absence of an external field. In the type II superconductors, magnetic fields induce the so called Abrikosov's phase (see Abrikosov 1957).

[10] Formally, when the width of a thin film tends to zero the energy of a vortex, being proportional to this width, tends to zero too.

[11] In this case there is no contribution to fluctuations from the last (transverse) term in Eq. (1.50).

L^{d-1}, where L is linear dimension of the sample. Therefore, if $L \to \infty$ the probability of finding the opposite configuration tends to zero. In the case $d = 1$ this probability is finite, independent of the size of the sample. In consequence, we will have an equal probability of finding opposite orientations of magnetization. As a result of this, the average magnetization must be zero at any finite temperature. The only temperature which could lead to a macroscopic magnetization of the sample is $T = 0$. In such a case zero temperature can be treated as a critical one, but there is no phase transition in the ordinary sense. We only have the state of matter above the critical temperature ($T = 0$). Nonetheless, when T tends to zero, fluctuations may substantially influence the behavior of some one-dimensional systems, for which $T = 0$ would be the critical point.

1.7 The Scaling Hypothesis

As we have seen in Sect. 1.5 Gaussian approximation describes small deviations around the most probable thermodynamic state. Gaussian probability distribution of fluctuations is applicable at any point of thermodynamic space where a thermodynamic description can be justified. It is a consequence of general thermodynamic principles (see, for example, Landau and Lifshits 1976). In the region of a phase transition, as has been shown in Sect. 1.4, special care should be taken in order to introduce the conventional statistical description. Below T_c, in the simplest case $n = 1$, the equilibrium state of the system is doubly degenerate. Hence, the probability distribution function can be represented as a sum of the two Gaussians centered around these degenerate states. As $| \tau |$ decreases and the equilibrium order parameter tends to zero, the Gaussians will tend to merge inevitably forming a non-Gaussian distribution in the vicinity of T_c, where $| \tau | \ll G_i$. Above T_c, where $\tau \gg G_i$, we again will have a Gaussian distribution but now, only one, as there is no degeneracy in the system. The deviation of the probability distribution function from Gaussian is closely related to the infinite increase of the correlation length in the critical point. The system can no longer be divided into statistically independent subsystems.[12] Therefore, in this region traditional methods starting from independent noninteracting quasiparticles (here fluctuations) will fail and nonperturbational approaches should be developed. An extremely important step in understanding the critical behavior in the fluctuation region was the scaling hypothesis. This hypothesis is a plausible anticipation which can be applied to any kind of Hamiltonian describing phase transitions. At first

[12]This condition is usually supposed to be obligatory in the statistical derivations of Gaussian distributions (Landau and Lifshits 1976).

the scaling hypothesis was advanced by Patashinskii and Pokrovskii (1964, 1966) as a supposition that all correlators in the critical region have power law behavior. They also found relations between the exponents of different correlators. In a more restrictive way this hypothesis was formulated by Kadanoff (1966). The basis for this hypothesis can be seen from the following consideration. The correlation length increasing with $\tau \to 0$ will inevitably exceed any characteristic scale in the fluctuating system. The fluctuations of the newly appearing phase become correlated in very large regions. It is quite natural now to suggest that the behavior of these fluctuations is completely unsensitive to the small-scale structure of the matter. The correlation length becomes the only characteristic scale for the fluctuating system and its divergence at the critical point leads to a singular behavior of all thermodynamic functions. In other words, the singular behavior of physically measurable quantities results from a single divergent quantity

$$\xi(\tau) \propto \begin{cases} \tau^{-\nu'} & \tau > 0 \quad ; \\ \mid \tau \mid^{-\nu} & \tau < 0 \quad , \end{cases} \tag{1.54}$$

where in the general case the exponent ν' could be different from ν.

1.7.1 Scaling Laws

Mathematical formulation of the scaling hypothesis leads to some relations between different critical exponents. These relations are traditionally called scaling laws. We need to find a way to express all measurable quantities in terms of the correlation length. After that, all singularities, according to the scaling hypothesis, should come from the only singularity in this length. First of all, it is necessary to mention that till now the correlation radius has only been defined by the Gaussian approximation. In the fluctuation domain this approximation is not applicable. Hence, it is necessary to introduce the correlation radius as a quantity not related to a particular approximation of the propagator. The Gaussian approximation shows that G in q-space is a Lorentzian having width $\Delta q \approx \xi^{-1}$. Intuitively it seems quite plausible that, independent of any particular approximation, the Green's function has a quite sharp maximum in the vicinity of $q = 0$ with the width defined by ξ^{-1}. As a consequence of this, we can define ξ^{-1} as the width of this maximum. A concrete definition can be chosen following Ma (1976) in the form

$$\xi^2 = -\frac{1}{2} G^{-1}(0) \frac{\partial^2 G(q)}{\partial q^2} \mid_{q=0} . \tag{1.55}$$

Since $k = q\xi$ is a dimensionless value, in the region where ξ essentially exceeds all microscopic lengths, we could write an expression for the propagator like

$$G(q) = \xi^y g(k). \tag{1.56}$$

When $k = 0$, the temperature behavior of G is described by the critical exponent γ (see Sect. 1.2.1) $G \propto | \tau |^{-\gamma}$ and from Eq. (1.56) we find

$$y = \begin{cases} \gamma'/\nu' & \tau > 0 \ ; \\ \gamma/\nu & \tau < 0 \ . \end{cases} \tag{1.57}$$

On the other hand, if we fix some finite value for k, then $\xi = k/q$ and from Eq. (1.56) it follows that $G(q) \propto q^{-y}$ and, therefore,

$$y = 2 - \eta, \tag{1.58}$$

where a new critical exponent η has been introduced. In the mean field approximation y is equal to 2. There is no reason for the mean value $y = 2$ to be conserved in the fluctuational domain.

What we have done until now can be described as follows. The propagator, which is a quantity having some unit of measure, has been expressed as the measure of some power of the correlation length. That has been done with the help of the dimensionless quantity k instead of q. As a result of such representation Eq. (1.57) and Eq. (1.58) have been obtained, which algebraically relates three critical exponents γ, ν, and η. Let us further continue with this procedure in considering other physically measurable quantities. The free energy divided by temperature is a dimensionless value defining the probability density Eq. (1.10). Consequently, it does not change owing to the introduction of dimensionless variables, but the density of the free energy F will change, as we now have to introduce a dimensionless space coordinate $x = r/\xi$. Since F has dimensions of the inverse volume we have

$$F \propto \xi^{-d} \propto | \tau |^{\nu d}, \tag{1.59}$$

from which the singular part of the specific heat is found as

$$C = -T\frac{\partial^2 F}{\partial T^2} \propto | \tau |^{\nu d - 2}, \tag{1.60}$$

or, in accordance with the definition of the critical exponent α ($C \propto | \tau |^{-\alpha}$), we have

$$\alpha = 2 - \nu d \, . \tag{1.61}$$

From Eq. (1.56) and the definition of propagators Eq. (1.35) we have the measure of the order parameter in q-space $\phi_{q=0} \propto \xi^{y/2}$. Since, according to the Fourier transformation, Eq. (1.12), the order parameter $\bar{\phi}$ is equal to $\phi_{q=0}/\sqrt{V}$, we obtain its measure in terms of ξ

$$\bar{\phi} \propto \xi^{\frac{y-d}{2}} \propto | \tau |^{\nu \frac{d-y}{2}}, \tag{1.62}$$

from which follows the relation for β

$$\beta = \nu \frac{d-y}{2}, \tag{1.63}$$

In order to find an expression for δ-exponent we need to find the measure for the external field h. This can be done with the help of the following thermodynamic identity

$$\bar{\phi} = -\frac{\partial F}{\partial h}, \tag{1.64}$$

This equation with the help of Eq. (1.59) and Eq. (1.62) can be transformed to

$$h \propto \xi^{-\frac{d+y}{2}}, \tag{1.65}$$

from where, taking into account Eq. (1.62) and the expression Eq. (1.58) for y, one will have

$$\delta = \frac{d+2-\eta}{d-2+\eta}. \tag{1.66}$$

Let us now consider the free energy density Eq. (1.59) in the presence of a finite external field h. In accordance with Eq. (1.65) we should have

$$F = \xi^{-d} f_{\pm}\left(h\xi^{\frac{d+y}{2}}\right), \tag{1.67}$$

where two different functions of the dimensionless quantity $a = h\xi^{\frac{d+y}{2}}$ has been introduced $f_+(\tau > 0)$ and $f_-(\tau < 0)$. The following consideration shows that the introduction of these different functions is inevitable. Let us transform Eq. (1.64) with the help of Eq. (1.67), then

$$\bar{\phi} = \xi^{\frac{y-d}{2}} f'_{\pm}(a) \propto \mid \tau \mid^{\beta} f'_{\pm}(a), \tag{1.68}$$

where $f'_{\pm}(a) = df_{\pm}(a)/da$. Since the order parameter is equal to zero when $h = 0$ and $\tau > 0$, but finite at $\tau < 0$ the derivative from f with respect to a has a jump at $a = 0$ depending on the sign of τ. Consequently the description of F with the help of the different functions $f_{\pm}$ is justified.

Additional relations between critical exponents can be obtained by the following deliberation. As the critical point is a point at which $h = 0$ and $\tau = 0$, it is natural to expect that in finite external fields the free energy density is a smooth function of τ and, hence, it can be expanded in a power series of τ

$$F = F_0(h) + F_1(h)\tau + \cdots \quad . \tag{1.69}$$

On the other hand, this expansion can be made out of the representation Eq. (1.67). The latter gives

$$F = \xi^{-d}\left\{A_+ a^{\kappa_1^+} + B_+ a^{\kappa_2^+}\right\} \quad \tau > 0;$$
$$F = \xi^{-d}\left\{A_- a^{\kappa_1^-} + B_- a^{\kappa_2^-}\right\} \quad \tau < 0. \tag{1.70}$$

Taking into account Eq. (1.54) and comparing these expressions with Eq. (1.69), we find

$$\kappa_1^+ = \tfrac{2d}{d+y} \;;\quad \kappa_2^+ = \tfrac{2}{\nu'} \cdot \tfrac{d\nu'-1}{d+y} \quad ;$$
$$\kappa_1^- = \tfrac{2d}{d+y} \;;\quad \kappa_2^- = \tfrac{2}{\nu} \cdot \tfrac{d\nu-1}{d+y} \quad . \tag{1.71}$$

At last, we must use the fact that functions $F_{0,1}(h)$ do not depend on the sign of τ. This immediately gives $\nu = \nu'$ and as a consequence (see Eqs. (1.57), (1.58), and Eq. (1.61)), $\gamma = \gamma'$, and $\alpha = \alpha'$.

In conclusion, we have found the following scaling laws (see Eqs. (1.57), (1.58), (1.61), and Eq. (1.63))

$$\begin{aligned} \gamma &= \gamma' = (2-\eta)\nu; \\ \alpha &= \alpha' = 2 - d\nu; \\ \alpha + 2\beta + \gamma &= 2; \\ \delta &= (d+2-\eta)/(d-2+\eta). \end{aligned} \tag{1.72}$$

Thus, the scaling hypothesis, though being only a reasonable guess, leads to very important consequences. It gives symmetry of the critical exponents in relation to the change of sign of τ. The universal equations of state, like the ones constituted by Eqs. (1.56), (1.67), and Eq. (1.68), follow from this hypothesis. And, lastly, it leads to scaling laws Eq. (1.72), simple algebraic relations between different critical exponents. All these consequences of the scaling hypothesis have been strictly proven only on the basis of the renormalization group theory approach, which will be considered in the following chapters.

Chapter 2

The Ginzburg-Landau Functional

2.1 Introduction

In this chapter we consider derivation of the Ginzburg-Landau functional (GLF) from different microscopic Hamiltonians. The words "microscopic Hamiltonian" mean that we start from some plausible representation of interactions which could take place in a particular physical system. These words do not necessarily mean that a Hamiltonian is obtained from the first principles, but rather that it somehow formulated on a microscopic scale as a model reflecting physical reality. There is no general recipe for the derivation of the GLF. Each kind of Hamiltonian requires, as a rule, a somewhat different approach. Nonetheless, there is quite general way of obtaining the GLF. This can be clearly seen from the examples given in this chapter and it should always contain the following steps:

1. Change from an operator representation of the partition function to a representation as an integration over continuous variables.

2. Redefinition of the physically measurable quantity having sense of the order parameter in terms of the new variables, thus defining an entity which could produce the Ginzburg-Landau field $\phi(\boldsymbol{r})$.

3. Finding "dangerous" modes and trial critical temperatures.

4. Expansion of the classical functional in the vicinity of the critical region.

5. Coarse graining by averaging over a microscopic spatial length scale and exclusion of nonessential modes resulting in a renormalization of trial parameters.

These are the necessary steps one has to take in order to get the Ginzburg-Landau functional. Some of these steps use traditional mathematical approaches independent of the particular physical problem. Namely, step one is conventionally performed with the help of the well-known Hubbard-Stratonovich transformation for any kind of the two-particle interaction Hamiltonian. In the identification of "dangerous" modes (step 3), the Gaussian expansion seems to be usually quite adequate to the problem. On the other hand, definition of the supposed Ginzburg-Landau field $\boldsymbol{\phi}(\boldsymbol{r})$ is a process substantially affected by a concrete microscopic model, as well as the procedure of coarsening and exclusion of nonessential modes.

2.2 Classical Systems

In this chapter we classify different Hamiltonians by their variables. If the variables are commuting quantities, they may be considered as classical variables and the consequent Hamiltonian is treated as a classical one. Noncommuting variables produce quantum Hamiltonians considered in Sect. 2.3. In this section, the Ising model is considered along with the classical Heisenberg magnetic and interacting particles Hamiltonians. The commutability of the hamiltonian variables substantially simplifies the problem of the GFL derivation independently of their nature (discrete Sect. 2.2.1, vector on some fixed surface Sect. 2.2.2, or traditional position vectors of particles Sect. 2.2.3).

2.2.1 The Ising Model

In 1944 Onsager published a manuscript the results of which were an unexpected contradiction with the Landau theory. He found the exact solution for the Ising model on the planar square lattice described by the Hamiltonian

$$H = -\frac{1}{2}\sum_{l,l'} m_l J_{ll'} m_{l'} - h\sum_{l} m_l, \tag{2.1}$$

where m_l can have only ± 1 values, l are vectors numbering lattice sites, $J_{ll'}$ is some interaction energy between two m placed at l and l' sites. In Onsager's and some following works[1] it was found that critical exponents of

[1]Onsager considered the case with $h = 0$ and only nearest neighbor interaction $J_{ll'}$.

the two dimensional Ising model were substantially different from the corresponding ones of Landau theory. The most striking result was that the specific heat had a logarithmic singularity instead of a jump (see Sect. 1.2.1) and, therefore, the partition function is a nonanalytic function in the vicinity of the transition point. This result questioned the main supposition of applicability of the expansions (see Eqs. (1.1), (1.4)) in the critical region.

It is convenient to use the notation J and h for J/T and h/T. This notation cannot lead to any misunderstanding, since in the case when the evident dependence on T is essential it can be simply extracted by the inverse substitution. The partition function for the Ising model can be written

$$Z = \mathrm{Tr}\exp\left\{\frac{1}{2}\sum_{l,l'} m_l J_{ll'} m_{l'} + \sum_l h_l m_l\right\}, \tag{2.2}$$

where the site dependent field h_l is introduced for further convenience. We consider a simple hypercubic d-dimensional lattice with a total number of lattice cells equal to N. As is conventionally accepted, we deal with periodic boundary conditions in all d dimensions.

Now let us carry out the first step discussed in Sect. 2.1 by introducing ordinary continuous variables η_l instead of discrete quantities m_l. This can be done by the use of Hubbard-Stratonovich identity

$$\exp\left\{\frac{1}{2}\sum_{l,l'} m_l J_{ll'} m_{l'}\right\} = \frac{1}{\sqrt{\det 2\pi J}}\prod_l \int_{-\infty}^{\infty} d\eta_l$$

$$\times \exp\left\{-\frac{1}{2}\sum_{l,l'} \eta_l J^{-1}_{ll'} \eta_{l'} + \sum_l \eta_l m_l\right\}. \tag{2.3}$$

With the help of this formula the partition function Eq. (2.2) can be reduced to

$$Z = \sqrt{\det\left(\frac{2}{\pi J}\right)}\prod_l \int_{-\infty}^{\infty} d\eta_l \exp\{-H\{\eta_l\}\}, \tag{2.4}$$

where

$$H\{\eta_l\} = \frac{1}{2}\sum_{l,l'} \eta_l J^{-1}_{ll'} \eta_{l'} - \sum_l \ln\cosh(\eta_l + h_l). \tag{2.5}$$

In order to examine the physical sense of the new variables η_l let us find the magnetization $\langle m_l\rangle$ from the representation of Z given by Eq. (2.4)

$$\langle m_l\rangle = \frac{\partial \ln Z}{\partial h_l} = \langle \tanh(\eta_l + h_l)\rangle, \tag{2.6}$$

where in the right-hand side of this equation the averaging is performed with the probability function

$$\rho\{\eta_l\} = \frac{1}{Z} \exp\{-H\{\eta_l\}\}. \tag{2.7}$$

According to the physical sense of the order parameter for Ising model, to the nonlinear function of $(\eta_l + h_l)$ in Eq. (2.6) should be ascribed the meaning of the Landau field ϕ. Unfortunately, the nonlinear relation is not convenient for practical purposes. The nature of the transformation Eq. (2.3) allows for a linear relation. This relation can be obtained with the help of the following consideration. By definition one has

$$\prod_k \int d\eta_k \frac{\partial}{\partial \eta_l} \rho\{\eta_k\} = 0.$$

On the other hand, after carrying out the derivative with respect to η_l, we arrive at the identity

$$\sum_k J_{lk}^{-1} \langle \eta_k \rangle - \langle \tanh(\eta_l + h_l) \rangle = 0. \tag{2.8}$$

This identity shows that the magnetization Eq. (2.6) is linearly related to the mean value of η. In the case of the ferromagnetic ordering[2] η, after some additional coarsening, can be taken as the Landau smooth field variable ϕ. The procedure of coarsening should be done in a way so as not to change the partition function. Therefore, we have to perform integration in Eq. (2.4) over rapidly varying modes which can be identified with the help of the Fourier transformation

$$\eta_l = \frac{1}{\sqrt{N}} \sum_q \eta(\boldsymbol{q}) e^{i\boldsymbol{q} \cdot \boldsymbol{l}}. \tag{2.9}$$

At first, let us restrict our consideration at the stage of the Gaussian approximation with $h = 0$. In this approximation the Hamiltonian Eq. (2.5) with the use of the transformation Eq. (2.9) can be reduced to

$$H = \frac{1}{2} \sum_q \{J^{-1}(\boldsymbol{q}) - 1\} \mid \eta(\boldsymbol{q}) \mid^2, \tag{2.10}$$

where the Fourier transform $J(\boldsymbol{q})$ is defined by

$$J_{ll'} = \frac{1}{N} \sum_q J(\boldsymbol{q}) e^{i\boldsymbol{q} \cdot (\boldsymbol{l} - \boldsymbol{l}')}. \tag{2.11}$$

[2]The term "ferromagnetic" in relation to the Ising model means that at $T = 0$ and $h = 0$ the ground state of the Hamiltonian, Eq. (2.1) corresponds to all $m_l = +1$, or $m_l = -1$. Antiferromagnetic ordering corresponds to two sublattices with opposite m.

In obtaining this relation, the property $\eta(\boldsymbol{q}) = \eta^*(-\boldsymbol{q})$ has been used. This property is a direct consequence of the fact that the variables η_l have real values.

At this moment, is necessary to recall that $J(\boldsymbol{q})$ is a temperature dependent quantity and to restore the initial J defined by the Hamiltonian Eq. (2.1). Then, taking also into account definitions for the Gaussian propagator (see Sect. 1.5), we find

$$G(\boldsymbol{q}) = \langle | \, \eta(\boldsymbol{q}) \, |^2 \rangle = \frac{J(\boldsymbol{q})}{T - J(\boldsymbol{q})}. \tag{2.12}$$

As is shown in Chapter 1 the appearance of "dangerous" modes can be traced as a divergence of the propagator. The quantity $J(\boldsymbol{q})$ is a function restricted to the range of $\boldsymbol{q}$ in the first Brillouin zone (see, for instance, Ziman 1972). In the case of the ferromagnetic ordering this function has absolute maximum at $\boldsymbol{q} = 0$. With decreasing of temperatures the function G will first diverge at $q = 0$ at $T = T_{co} = J(0)$. Therefore, in this case the "dangerous" mode is the mode with $q = 0$. This mode will condense below the genuine critical temperature and, hence, it must be excluded from the integration in the partition function Eq. (2.4). Those modes which have small q values are important modes. They are responsible for the critical singularities in physically measurable quantities. All other modes can be excluded from consideration by carrying out integration over them in Gaussian approximation. This integration procedure will renormalize trial parameters of the Ginzburg-Landau functional which now could be obtained by simple expansion of the Hamiltonian Eq. (2.5) in powers of $\eta(\boldsymbol{q})$ having small values of $\boldsymbol{q}$. Some of the trial parameters can be obtained even without this expansion. They are already present in an indirect form in the propagator Eq. (2.12). Let us expand $J(q)$ in the vicinity $q = 0$. Then we arrive at the expression of the correlator which has already been found in Chapter 1

$$G(q) = \frac{1}{\xi_t^2 q^2 + \tau_t}, \tag{2.13}$$

where

$$\xi_t^2 = \frac{1}{2dJ(0)} \frac{\partial^2 \ln J(\boldsymbol{q})}{\partial \boldsymbol{q}^2}\bigg|_{q=0} \quad ; \quad \tau_t = \frac{T - J(0)}{J(0)} \quad .$$

Now let us define in a more rigorous manner what should be considered as important modes. Landau theory is applicable when $\tau \ll 1$. Therefore, only those modes should be retained which give the same order of magnitude contribution into the inverse Green's function (see Eq. (2.13)). For these modes we have $q \leq q_0 \ll \xi_t^{-2}$.

At last, it is necessary to perform the fourth and fifth steps in the

GLF derivation. These two steps are interchangeable. It is possible to realize the expansion in the Hamiltonian Eq. (2.5) in powers of η_l and then perform the coarse graining, leaving only spatial resolutions exceeding the microscopic scale of the lattice. The other way would be coarsening before the expansion is made. Both approaches lead to the same result and which one to choose depends on the particular model under consideration and is a question of convenience. Here we pursue the first approach. Let us introduce notations $\eta_l = \eta_l^< + \eta_l^>$, where in $\eta_l^<$ only $q < q_0$ are retained in the Fourier transformation Eq. (2.9), while $\eta_l^>$ contains all $|\, \boldsymbol{q} \,|$ exceeding q_0. Integration over the modes with $q > q_0$ in the partition function expressed in Eq. (2.4) leads to the following expansion for the renormalized Hamiltonian

$$H_R = \frac{1}{2} \sum_{q<q_0} \{\xi_t^2 q^2 + \tau U\} \mid \eta_q \mid^2 + \frac{W}{4} \sum_l \left(\eta_l^<\right)^4, \tag{2.14}$$

where τ is defined as usual $\tau = (T - T_c)/T_c$ with T_c obtained from the condition

$$\frac{T_c}{T_{co}} - 1 + a_1(T_c) = 0. \tag{2.15}$$

Factors U and W are defined by

$$\begin{aligned} U &= \tfrac{T_c}{T_{co}} + T_c \tfrac{\partial a_1(T_c)}{\partial T_c}, \\ W &= \tfrac{1}{3} - \tfrac{4}{3} a_1(T_c) - a_2(T_c), \end{aligned} \tag{2.16}$$

where $a_{1,2}(T)$

$$a_n(T) = \frac{1}{N} \sum_{q>q_0} \left(\frac{J(q)}{T - J(q)} \right)^n$$

In Eq. (2.14) the summation over $\boldsymbol{l}$ on the discrete lattice can be replaced by the integration over the continuous variable $\boldsymbol{r}$. This replacement is possible due to the restriction $q < q_0$ which means that η_l does not change at the spacing $\Delta l \approx a$ but only when $\Delta l \approx q_0^{-1} \gg a$. To complete the derivation of the GLF, it is necessary to define a proportionality factor which will relate $\eta(\boldsymbol{r})$ to $\phi(\boldsymbol{r})$. This factor can be found with the help of the external field h. Now, it is convenient to make a change of variables in the partition function Eq. (2.4). Let us shift η_l to $(\eta_l + h_l)$, then the term ΔH containing the external field in the Hamiltonian Eq. (2.5) can be written as

$$\Delta H = -\sum_{l,l'} J_{ll'}^{-1} \eta_{l'}^< h_l = -\frac{1}{\Omega_0 J(0)} \int d\boldsymbol{r} \eta(\boldsymbol{r}) h(\boldsymbol{r}), \tag{2.17}$$

where $\Omega_0 = a^d$ is the volume of the lattice cell, it has also been assumed that h_l is a slowly varying value on the microscopic scale a. From this equation one can find the scaling factor as

$$\phi(\boldsymbol{r}) = \frac{T_c}{T_{co}\,\Omega_0}\eta(\boldsymbol{r}). \tag{2.18}$$

Finally we arrive at the traditional form of the Ginzburg-Landau functional (see Eq. (1.8)) with

$$\begin{aligned} c &= \Omega_0 \tfrac{T_{co}^2}{T_c}\xi_t^2, & \alpha &= \Omega_0 \tfrac{T_{co}^2}{T_c} U, \\ \xi_0^2 &= \tfrac{\xi_t^2}{U}, & b &= \Omega_0^3 \tfrac{T_{co}^4}{T_c^3} W. \end{aligned} \tag{2.19}$$

It is interesting to check the qualitative estimate for the microscopic scale r_0 found in Chapter 1. According to Sect. 1.5.3, $r_0 = (T_c b/\alpha^2)^{1/d}$, on the other hand, using the parameters obtained in Eq. (2.19), we arrive at

$$r_0 = a\left(\frac{W}{U^2}\right)^{\frac{1}{d}},$$

from which it is seen that the estimate for r_0 corresponds to the genuine microscopic scale a, since usually the factor $(W/U^2)^{1/d}$ is of the order of unity.

2.2.2 The Heisenberg Model

The classical Heisenberg Hamiltonian has the same form as for the Ising model, Eq. (2.1)

$$H = -\frac{1}{2}\sum_{l,l'} \boldsymbol{m}_l J_{ll'} \boldsymbol{m}_{l'} - \boldsymbol{h}\sum_l \boldsymbol{m}_l. \tag{2.20}$$

The only difference is that, instead of the discrete values m_l, in Eq. (2.20) the n-component classical vectors $\boldsymbol{m}_l$ stand. These vectors have a fixed value of magnitude $|\,\boldsymbol{m}_l\,| = const.$ We choose the magnitude of the vectors to be equal to 1, then the case $n = 1$ corresponds to the Ising model. In the partition function Eq. (2.2), instead of the operation of taking trace we have the multiple integration

$$Z = \prod_l \int d\boldsymbol{m}_l \exp(-H). \tag{2.21}$$

As the Hamiltonian Eq. (2.20) is a bilinear form in the variables $\boldsymbol{m}_l$ it may seem that Eq. (2.21) contains a Gaussian integral. In some sense this is a true statement, but the variables are defined in the $(n-1)$-dimensional non-Euclidian space. They are defined on the sphere enclosed by the isotopic n-dimensional space with the unit radius. To introduce Euclidian variables, it is necessary to use the Hubbard-Stratonovich identity in the same manner as it was done for the Ising model (see Eq. (2.3)). Due to the vector nature of the Heisenberg variables, the new variables $\boldsymbol{\eta}_l$ are also vectors. As the whole procedure is identical with the one considered in Sect. 2.2.1 we will only show some intermediate results which may yield additional physical information. The Hamiltonian analogous to Eq. (2.5) has the form

$$H = \frac{1}{2} \sum_{l,l'} \boldsymbol{\eta}_l J^{-1}_{ll'} \boldsymbol{\eta}_{l'}$$

$$- \sum_l \ln \left[\Gamma(\frac{n}{2}) \left(\frac{2}{\mid \boldsymbol{\eta}_l + \boldsymbol{h}_l \mid} \right)^{\frac{n}{2}-1} I_{\frac{n}{2}-1} (\mid \boldsymbol{\eta}_l + \boldsymbol{h}_l \mid) \right], \qquad (2.22)$$

where $\Gamma(\nu)$ is gamma function, I_ν is Bessel's function of an imaginary variable.

The zero order Gaussian propagator, which results from the Hamiltonian Eq. (2.22), is

$$G(\boldsymbol{q}) = \langle \mid \eta_i(\boldsymbol{q}) \mid^2 \rangle = \frac{J(\boldsymbol{q})}{T - J(\boldsymbol{q})/n}. \qquad (2.23)$$

Again, as shown in the previous section, "dangerous" modes are defined by the divergence of this propagator. In order to have some additional diversity, here we consider a more general case. For the ferromagnetic case the trial critical temperature is defined by $T_c = J(0)/n$. As was first shown by Ivanchenko and Fillipov (1984), more complicated magnetic structure must arise when the absolute maximum of $J(\boldsymbol{q})$ is realized at some $\boldsymbol{q} = \boldsymbol{q}_1 \neq 0$. Since the function $J(\boldsymbol{q})$ is a Fourier transform of a function J_l defined on the crystal lattice, the appearance of a maximum at $\boldsymbol{q} = \boldsymbol{q}_1$ means that there are k equivalent maxima defined by the point symmetry group of the crystal (where k is the number of the symmetry elements).

Antiferromagnetic and Incommensurate Structures

To understand what kind of ordering may occur below the transition temperature in this case, let us disregard all possible renormalization effects and construct the magnetization $\boldsymbol{m}_l$ of the system. In the general case, this magnetization should consist of the contributions of all equivalent maxima

$$\boldsymbol{m}_l = \frac{1}{\sqrt{N}} J(\boldsymbol{q}_1) \sum_{j=1}^{k} e^{i\boldsymbol{q}_j \cdot \boldsymbol{l}} \boldsymbol{\eta}(\boldsymbol{q}_j), \tag{2.24}$$

where we used the symmetry property $J(\boldsymbol{q}_j) = J(\boldsymbol{q}_1)$. Eq. (2.24) shows that the magnetization is an oscillating function of $\boldsymbol{l}$. This means that appearance of "dangerous" modes with $\boldsymbol{q} \neq 0$ not breaks only the inherent symmetry in the isotopic space, but also the translational symmetry of the system. A new period, the period of the magnetization oscillations, enters. The simplest case would be if the maxima of $J(\boldsymbol{q})$ are reached at the symmetrical points on the boundary of the Brillouin zone. For instance, let us look at the case of the simple cubic Brillouin zone in which the function $J(\boldsymbol{q})$ has maxima at the corners of the hypercube. In this example all $\boldsymbol{q}_j$ points are equivalent, because they differ from each other by a reciprocal lattice vector. As a consequence, all the amplitudes $\boldsymbol{\eta}(\boldsymbol{q}_j)$ are also equal ($\boldsymbol{h} = 0$). Therefore, we have

$$\boldsymbol{m}_l = 2^d J(\boldsymbol{q}_1) \boldsymbol{\eta}(\boldsymbol{q}_1) \prod_{i=1}^{d} \cos(\pi n_i), \tag{2.25}$$

where n_i are integers numerating lattice sites in different directions ($\boldsymbol{l} = a \sum \boldsymbol{e}_i n_i$, $\boldsymbol{e}_i$ are unit vectors along the cube ribs). As is seen from Eq. (2.25) the magnetization corresponds to the simplest case of antiferromagnetic ordering of equal magnetic moments in two sublattices with opposite orientations. The Landau field for such an ordering can be constructed from a properly coarsened difference of the macroscopic magnetizations of these sublattices. Sometimes, especially in the vicinity of the critical point when $\boldsymbol{h} \neq 0$, it is necessary to introduce two interacting fields representing the magnetizations of the different sublattices. In a general case, when $\boldsymbol{q}_j$ are neither on the boundary of the Brillouin zone nor on a line of a symmetry direction, only those amplitudes $\boldsymbol{\phi}(\boldsymbol{q}_j)$ which corresponds to the time reversal symmetry points ($\boldsymbol{q}_j = -\boldsymbol{q}_i$) are equal. Therefore, the periodicity of the magnetization is not the same as that of the uderlying crystal, and incommensurate structure will be established. This structure, in the general case should also be a non-collinear one. There is also a qualitative feature distinguishing this structure from the case when maxima are reached at the symmetry points on the boundary of the Brillouin zone. The manifestation of extrema at the symmetry points on the zone boundary surface is a consequence of the translational crystallographic symmetry, while the appearance of a maximum in a general point may happen only accidentally, due to a particular choice of the dependence $J(\boldsymbol{l})$. The location of such maxima should be strongly influenced by the renormalization effects, which are temperature-dependent. Therefore, spatial periods of magnetic structures in this case, are dependent on both the temperature and the external field. Such complicated structures have not yet been fully studied theoretically and experimentally.

Heisenberg Magnets with Two Exchange Mechanisms

Let us consider the case when the function $J(\boldsymbol{q})$ has two sets of maxima. The simplest situation for this example corresponds to the symmetry related maxima $J(\boldsymbol{q}_j)$ at the corners of the Brillouin zone (antiferromagnetic) and the maximum $J(0)$ at the center of the zone (ferromagnetic). Following the method discussed in Sect. 2.2.1 we must find the region of important modes in the vicinity of the absolute maximum and carry out the integration over the other modes. In the case when there is more than one set of maxima, two qualitatively different possibilities exist. The first one is realized when the absolute maximum (let us assume for definiteness $J(\boldsymbol{q}_1)$) has a substantially higher magnitude than the other. Then the value of function $J(\boldsymbol{q})$ on the boundary surface restricting the important modes $J(\ \boldsymbol{q}) = J_0$ exceeds the other maximum. In such a case the presence of the additional maximum $J(0)$ cannot be revealed in the behavior of the system. This maximum can only contribute to the renormalization of the trial characteristic values of the critical behavior. Qualitatively this case corresponds to the problem with one kind of Landau field $\boldsymbol{\phi}_a$. In the case where $J(0) \approx J_0$ or even $J(0) > J_0$ both sets of maxima will contribute into the critical behavior and to obtain a complete description of the system we need to introduce two Landau fields $\boldsymbol{\phi}_a$ and $\boldsymbol{\phi}_f$. The procedure of the GLF derivation in this case is qualitatively the same as for the case with one field. It is necessary to find regions of important modes in the vicinity of each maxima. All the maxima corresponding to antiferromagnetic modes are equivalent, therefore, it is convenient to consider them as one maximum divided from the ferromagnetic maximum in q-space on the vector $\boldsymbol{q}_1 = \pi \sum \boldsymbol{e}_i / a$. The expansion of the function $J(\boldsymbol{q})$ in the vicinity of each maxima leads to the following two trial propagators

$$G(q) = \frac{n}{\xi_{tf}^2 q^2 + \tau_{tf}} \quad ; \quad G(k) = \frac{n}{\xi_{ta}^2 k^2 + \tau_{ta}} \ , \qquad (2.26)$$

where $\boldsymbol{k} = \boldsymbol{q}_1 - \boldsymbol{q}$,

$$\xi^2_{tf} = \frac{1}{2dJ(0)} \frac{\partial^2 \ln J(\boldsymbol{q})}{\partial \boldsymbol{q}^2}\bigg|_{q=0}; \qquad \tau_{tf} = \frac{nT - J(0)}{J(0)};$$

$$\xi^2_{ta} = \frac{1}{2dJ(\boldsymbol{q}_1)} \frac{\partial^2 \ln J(\boldsymbol{q})}{\partial \boldsymbol{q}^2}\bigg|_{q=q_1}; \quad \tau_{ta} = \frac{nT - J(\boldsymbol{q}_1)}{J(\boldsymbol{q}_1)}.$$

The following procedure is exactly the same as has already been developed for the Ising model (see Sect. 2.2.1). The only difference, which will influence the final results, is the presence in $\eta^<$ of the two regions in $\boldsymbol{q}$-space separated by the vector $\boldsymbol{q}_1$. This difference leads to the following representation, evidently demarcating the different physical contributions

$$\eta_l^< = \eta_{lf} + e^{i\, \boldsymbol{q}_1 \boldsymbol{l}} \eta_{la}, \tag{2.27}$$

It is also worth mentioning that for the antiferromagnetic ordering there are no physical fields which could define a proportionality factor and, hence, the normalization of the $\phi_a(\boldsymbol{r})$ originating from η_{la} is arbitrary. The ferromagnetic Landau field normalization is defined by Eq. (2.18) with T_{tf} replacing T_{co} and T_a replacing T_c.[3] If we preserve the same factor for the antiferromagnetic Landau field, the resulting Ginzburg-Landau functional will have only seven parameters $T_{a,f}$, $\alpha_{a,f}$, $c_{a,f}$, and b, instead of nine (including b_a, b_f, and b_{af}). These parameters can be obtained with the help of equations defining $T_{a,f}$

$$\frac{nT_a}{J(\boldsymbol{q}_1)} - 1 + \frac{a_1(T_a)}{n} = 0,$$
$$\frac{nT_f}{J(0)} - 1 + \frac{a_1(T_f)}{n} = 0, \tag{2.28}$$

and the following expressions for the renormalization factors $U_{a,f}$

$$U_a = \frac{nT_a}{J(\boldsymbol{q}_1)} + \frac{T_a}{n}\frac{\partial a_1(T_a)}{\partial T_a},$$
$$U_f = \frac{nT_f}{J(0)} + \frac{T_f}{n}\frac{\partial a_1(T_f)}{\partial T_f}. \tag{2.29}$$

Now the Ginzburg-Landau functional for the two interacting fields can be represented by the form

$$\mathcal{F} = \mathcal{F}_a + \mathcal{F}_f + \mathcal{F}_{int} - \int d\boldsymbol{r} \boldsymbol{h}(\boldsymbol{r}) \cdot \boldsymbol{\phi}_f(\boldsymbol{r}), \tag{2.30}$$

[3] In this case we assume the critical temperature to be equal to $T_a (> T_f)$. As shown in Chapter 5, in the critical region the system may order ferromagnetically notwithstanding the inequality $T_a > T_f$.

where

$$\begin{aligned} \mathcal{F}_a &= \int dr \left\{ \tfrac{\alpha_a \tau_a}{2} \phi_a^2(\boldsymbol{r}) + \tfrac{c_a}{2} \left(\nabla \phi_a(\boldsymbol{r})\right)^2 + \tfrac{b}{4} \left(\phi_a^2(\boldsymbol{r})\right)^2 \right\}, \\ \mathcal{F}_f &= \int dr \left\{ \tfrac{\alpha_f \tau_f}{2} \phi_f^2(\boldsymbol{r}) + \tfrac{c_f}{2} \left(\nabla \phi_f(\boldsymbol{r})\right)^2 + \tfrac{b}{4} \left(\phi_f^2(\boldsymbol{r})\right)^2 \right\}. \end{aligned} \tag{2.31}$$

The part of the functional Eq. (2.30) describing the interaction between the ϕ_a and ϕ_f fields contains two possible invariants of the $\mathcal{O}(n)$ symmetry group

$$\mathcal{F}_{int} = b \int dr \left\{ \frac{1}{2} \phi_a^2(\boldsymbol{r}) \phi_f^2(\boldsymbol{r}) + \left(\phi_a(\boldsymbol{r}) \phi_f(\boldsymbol{r})\right)^2 \right\}. \tag{2.32}$$

In the formulae Eqs. (2.31) and (2.32) the parameters α, c, b are defined by (compare with Eq. (2.19))

$$\begin{aligned} c_a &= \Omega_0 \frac{n T_{tf}^2}{T_a} \xi_{ta}^2, \quad & \alpha_a &= \Omega_0 \frac{n T_{tf}^2}{T_a} U_a, \quad & b &= \Omega_0^3 \frac{T_{tf}^4}{T_a^3} W_n, \\ c_f &= \Omega_0 \frac{n T_{tf}^2}{T_a} \xi_{tf}^2, \quad & \alpha_f &= \Omega_0 \frac{n T_{tf}^2}{T_a} U_f, \quad \xi_{a,f}^2 & &= \frac{\xi_{ta,f}^2}{U_{a,f}}, \end{aligned} \tag{2.33}$$

where

$$W_n = \frac{1}{n^2(n+2)} \left\{ 1 - \frac{4a_1(T_a)}{n} - \frac{a_2(T_a)(n+8)}{n^2(n+2)^2} \right\}$$

Note, that the chosen normalization of the field variables $\phi_{a,f}$ is also convenient for describing the real magnetizations $(\boldsymbol{m}_{1,2})$ of the different sublattices, which can be expressed as

$$\boldsymbol{m}_1 = \frac{\phi_f + \phi_a}{2} \quad , \quad \boldsymbol{m}_2 = \frac{\phi_f - \phi_a}{2} \quad .$$

2.2.3 Interacting Particles

The case of interacting classical particles has a direct relation to the first phase transition experimentally studied by Andrews (1869). The description of liquid–vapor critical behavior always requires the introduction of at least two different kinds of interactions: short-range repulsion and long-range attraction. Namely, the competition of these two interactions defines all the properties of the liquid state of matter. The presence of only one kind of the interactions would lead either to instability (in the case of attractive forces) or to the loss of liquid features.[4]

[4] Short-range repulsive forces cannot lead to the liquid–vapor transition in the classical case.

From the mathematical point of view the study of the liquid state is a very complicated problem. The theory has no small parameter which allows for a perturbational approach to thermodynamics. Nonetheless, the case of the gas with short-range repulsive interactions has been studied quite thoroughly.[5] Thermodynamic functions for this system were calculated with high accuracy. Starting from these functions, the long-range potential can be considered as a perturbation. Such an approach was first provided by Zwanzig (1954) and later developed by Barkev and Henderson (1967), Storer (1969), and Weeks *et al.* (1971). Of course, the perturbation parameter used in these works (long-range potential) could not be considered as small for real physical systems. Below, in this section we discuss the approach developed by Ivanchenko and Lisyansky (1983, 1984). This approach uses as a small parameter the ratio of two characteristic radii: the radius of the short-range repulsive potential r_+ and the radius of the long-range attractive potential r_-. For a number of physical systems the value r_+/r_- is really a small parameter.

The Hamiltonian of a classical gas having N particles and subjected to the influence of an external field with a potential $U(\boldsymbol{r})$ can be written

$$H = H_0 + H_U + H_I, \tag{2.34}$$

where

$$\begin{aligned} H_0 &= \textstyle\sum_{i=1}^{N} \frac{\boldsymbol{p}_i^2}{2m}, \\ H_U &= \textstyle\sum_{i=1}^{N} U(\boldsymbol{r}_i), \\ H_I &= \textstyle\frac{1}{2}\sum_{i\neq j} V(\mid \boldsymbol{r}_i - \boldsymbol{r}_j \mid). \end{aligned} \tag{2.35}$$

We assume that the interaction potential $V(\boldsymbol{r})$ can be divided into two parts — repulsive and attractive. In performing such a division there is some liberty, as the constraints on the short-range repulsive $V_+(\boldsymbol{r})$ and long-range attractive $V_-(\boldsymbol{r})$ potentials are not strict. We assume that the Fourier transform of V_+ is positive and V_- is negative ($\pm V_\pm(\boldsymbol{q}) > 0$); next, the potential $V_-(\boldsymbol{r})$ is a smooth function of $\boldsymbol{r}$, and, at last, the radius of the attractive forces essentially exceeds the radius of the repulsive forces. The way to divide a given physical potential into two potentials satisfying these three conditions is not unique. In principle, this property can be utilized in deriving diverse approximation schemes for obtaining thermodynamic functions of a gas with short-range repulsive interaction. Such a problem

[5]See Guinta *et al.* (1985), Tarazona (1985), and references therein.

has been extensively considered and is treated as a solved problem. Let us introduce functions $L_\pm(\boldsymbol{r})$ defined by

$$\int d\boldsymbol{r} L_\pm(\boldsymbol{r}' - \boldsymbol{r}) L_\pm(\boldsymbol{r} - \boldsymbol{r}'') = \pm V(\boldsymbol{r}' - \boldsymbol{r}'').$$

The functions $L_\pm$ are real, as a direct consequence of the inequalities $\pm V_\pm(\boldsymbol{q}) > 0$. With the help of these functions the Hamiltonian H_I can be reduced to

$$H_I = -\frac{NV(0)}{2} + \frac{1}{2}\int d\boldsymbol{x} \left[\sum_{i=1}^{N} L_+(\boldsymbol{r}_i - \boldsymbol{x})\right]^2$$

$$-\frac{1}{2}\int d\boldsymbol{x} \left[\sum_{i=1}^{N} L_-(\boldsymbol{r}_i - \boldsymbol{x})\right]^2, \tag{2.36}$$

where $V(0) = V(\boldsymbol{r} = 0)$ has been assumed to be finite. At the end of the derivation this assumption appears as nonessential and $V(0)$ can be set to infinity. Eq. (2.36) enables one to use the Hubbard-Stratonovich identity (see, for instance, Eq. (2.3)) which helps to reduce the expression for the grand partition function to the continual integration over two real functions $\eta_+(\boldsymbol{x})$ and $\eta_-(\boldsymbol{x})$

$$Z = \sum_N^\infty \frac{\xi^n}{N!} \prod_{i=1}^N \int d\boldsymbol{r}_i \int \frac{D\eta_+(\boldsymbol{x}) D\eta_-(\boldsymbol{x})}{A_+ A_-}$$

$$\times \exp\left\{-\frac{T}{2}\int d\boldsymbol{x}\left[\eta_+^2(\boldsymbol{x}) + \eta_-^2(\boldsymbol{x})\right] - \frac{1}{T}\sum_{i=1}^N U(\boldsymbol{r}_i)\right.$$

$$\left. + i\sum_{i=1}^N \int d\boldsymbol{x} L_+(\boldsymbol{r}_i - \boldsymbol{x})\eta_+(\boldsymbol{x}) - \sum_{i=1}^N \int d\boldsymbol{x} L_-(\boldsymbol{r}_i - \boldsymbol{x})\eta_-(\boldsymbol{x})\right\}, \tag{2.37}$$

where

$$\xi = \zeta \exp(V(0)/2T), \; \zeta = (mT/2\pi\hbar^2)^{3/2} \exp(\mu/T),$$

and μ is the chemical potential. The factor ζ results from the integration over the momentum variables $\boldsymbol{p}_i$. At last, the two constants $A_\pm$ are defined by

$$A_\pm = \int D\eta_\pm(\boldsymbol{x}) \exp\left[-T\int d\boldsymbol{x}\eta_\pm^2(\boldsymbol{x})/2\right].$$

By carrying out the summation in Eq. (2.37) and after the change of variables

$$\Psi(\boldsymbol{x}) = T\int d\boldsymbol{y}L_+(\boldsymbol{x}-\boldsymbol{y})\eta_+(\boldsymbol{y}),$$

$$\Theta(\boldsymbol{x}) = -T\int d\boldsymbol{y}L_-(\boldsymbol{x}-\boldsymbol{y})\eta_-(\boldsymbol{y}),$$

we arrive at the expression

$$Z = \int \frac{D\Psi(\boldsymbol{x})D\Theta(\boldsymbol{x})}{A_+A_-}\exp\left\{-\frac{1}{2T}\int d\boldsymbol{x}d\boldsymbol{x}'\Psi(\boldsymbol{x})K_+(\boldsymbol{x}-\boldsymbol{x}')\Psi(\boldsymbol{x}')\right.$$
$$-\frac{1}{2T}\int d\boldsymbol{x}d\boldsymbol{x}'\Theta(\boldsymbol{x})K_-(\boldsymbol{x}-\boldsymbol{x}')\Theta(\boldsymbol{x}')$$
$$\left.+\xi\int d\boldsymbol{x}\exp\left[\frac{i\Psi(\boldsymbol{x})+\Theta(\boldsymbol{x})-U(\boldsymbol{x})}{T}\right]\right\}, \tag{2.38}$$

where the constants $A_\pm$ are again defined as the Gaussian measure integrals on the new variables Ψ, Θ, operators $K_\pm(\boldsymbol{x}-\boldsymbol{x}')$ are defined by

$$\pm\int d\boldsymbol{y}K_\pm(\boldsymbol{x}-\boldsymbol{y})V_\pm(\boldsymbol{y}-\boldsymbol{x}') = \delta(\boldsymbol{x}-\boldsymbol{x}')$$

In Eq. (2.38) the order of integration is very important. The functional integral over the variable Θ is formally divergent due to the term $\exp(\Theta(\boldsymbol{x})/T)$ in the curly brackets of Eq. (2.38).[6] Nonetheless, the oscillating character of the factor $\exp(i\Psi(\boldsymbol{x})/T)$ makes the integral convergent, if one carries out the integration over the Ψ variable first. Therefore, we have to represent the partition function in the form

$$Z = \int \frac{D\Theta(\boldsymbol{x})}{A_-}\exp\left\{-\frac{1}{2T}\int d\boldsymbol{x}d\boldsymbol{x}'\Theta(\boldsymbol{x})K_-(\boldsymbol{x}-\boldsymbol{x}')\Theta(\boldsymbol{x}')\right\}Z_+\{\Theta\}, \tag{2.39}$$

where Z_+ is a partition function of the gas with the short range repulsive interactions subjected to the influence of the external field potential $U(x)$ and an effective field $-\Theta(\boldsymbol{x})$ produced by the attractive interactions

$$Z_+\{\Theta\} = \int \frac{D\Psi(\boldsymbol{x})}{A_+}\exp\left\{-\frac{1}{2T}\int d\boldsymbol{x}d\boldsymbol{x}'\Psi(\boldsymbol{x})K_+(\boldsymbol{x}-\boldsymbol{x}')\Psi(\boldsymbol{x}')\right.$$
$$\left.+\xi\int d\boldsymbol{x}\exp\frac{i\Psi(\boldsymbol{x})+\Theta(\boldsymbol{x})-U(\boldsymbol{x})}{T}\right\} = \exp\sum_{m=1}^{\infty}\frac{\xi^m}{m!}u_m\{\Theta\}. \tag{2.40}$$

[6]This is a consequence of the instability in systems with purely attractive forces.

The quantities u_m in Eq. (2.40) are formally defined by

$$u_m\{\Theta\} = \frac{\partial^m}{\partial \xi^m} \ln Z_+\{\Theta\}\bigg|_{\xi=0}$$

and they can be calculated with the help of the following procedure. Let us consider

$$u_1\{\Theta\} = \left\{\int d\boldsymbol{x} \exp\left[\frac{\Theta(\boldsymbol{x}) - U(\boldsymbol{x})}{T}\right] \int D\Psi(\boldsymbol{x}) \exp\left[-\frac{1}{2T}\int d\boldsymbol{x}_1 d\boldsymbol{x}_2 \right.\right.$$
$$\left.\left. \times\ \Psi(\boldsymbol{x}_1) K_+(\boldsymbol{x}_1 - \boldsymbol{x}_2)\Psi(\boldsymbol{x}_2) + i\frac{\Psi(\boldsymbol{x})}{T}\right]\right\}$$
$$\times \left\{\int D\Psi(\boldsymbol{x}) \exp\left[-\frac{1}{2T}\int d\boldsymbol{x}_1 d\boldsymbol{x}_2 \Psi(\boldsymbol{x}_1) K_+(\boldsymbol{x}_1 - \boldsymbol{x}_2)\Psi(\boldsymbol{x}_2)\right]\right\}^{-1}.$$

Applying the shift of variables $\Psi(\boldsymbol{x}_i)$ to $\Psi(\boldsymbol{x}_i) + iV_+(\boldsymbol{x}_i - \boldsymbol{x})$ one can cancel the linear in Ψ terms in the exponential function of the nominator, and as a result we get

$$u_1 = \int d\boldsymbol{x} \exp\left\{\frac{\Theta(x) - U(x) - V_+(0)/2}{T}\right\}$$

The same procedure leads to the following representation for the general term of the series Eq. (2.40)

$$u_m\{\Theta\} = \exp\left\{-\frac{mV_+(0)}{2T}\right\} \int d\boldsymbol{x}_1 d\boldsymbol{x}_2 \dots d\boldsymbol{x}_m B(\boldsymbol{x}_1, \dots, \boldsymbol{x}_m)$$
$$\times \exp\left\{\frac{\Theta(\boldsymbol{x}_1) + \dots + \Theta(\boldsymbol{x}_m) - U(\boldsymbol{x}_1) - \dots - U(\boldsymbol{x}_m)}{T}\right\},$$

where the functions $B(\boldsymbol{x}_1, \dots, \boldsymbol{x}_m)$ are combinations of Mayer's functions

$$f(\boldsymbol{x}_1, \boldsymbol{x}_2) = \exp[-V(\boldsymbol{x}_1 - \boldsymbol{x}_2)/T] - 1.$$

These combinations are constructed as follows

$$\begin{aligned} B(\boldsymbol{x}_1) &= B_1 = 1,\ B_{12} = f(\boldsymbol{x}_1, \boldsymbol{x}_2) \equiv f_{12}, \\ B_{123} &= f_{12}f_{13} + f_{12}f_{23} + f_{13}f_{23} + f_{12}f_{13}f_{23}, \quad \dots \end{aligned} \tag{2.41}$$

Until this moment, all transformations have been exact and none of the particular properties of the potentials $V_\pm$, mentioned above, have been exploited. Now we are going to make use of the assumed differences in the r-dependencies of the potentials. The assumption that $V_+(\boldsymbol{x})$ is a short range potential and that $V_-(\boldsymbol{x})$ along with $U(\boldsymbol{x})$ are smooth, on the scale of r_+, functions of $\boldsymbol{x}$, leads to the following approximation

$$Z_+\{\Theta\} = \exp \sum_{m=1}^{\infty} \frac{b_m}{m!} \zeta^m e^{-\frac{mV_-(0)}{2T}} \int d\boldsymbol{x} \exp\left\{ m \frac{\Theta(\boldsymbol{x}) - U(\boldsymbol{x})}{T} \right\}$$

$$= \exp\left\{ \frac{1}{T} \int d\boldsymbol{x} P_+(\mu + \Theta(\boldsymbol{x}) - U(\boldsymbol{x}), T) \right\}, \tag{2.42}$$

where $P_+(\mu, T)$ is the pressure in the system of particles with the short-range interaction potential V_+, the quantities b_m are the so-called group integrals

$$b_m = \int d\boldsymbol{x}_2 \ldots d\boldsymbol{x}_m B(\boldsymbol{x}_1, \ldots \boldsymbol{x}_m).$$

In the derivation of Eq. (2.42) the following property of the $B_{1,\ldots,m}$ functions has been used. These functions are different from zero only when $\boldsymbol{x}_1, \ldots, \boldsymbol{x}_m$ are close to each other. It is possible to estimate the accuracy of the above procedure. Under the derivation of the expression Eq. (2.42) the terms having the order of magnitude $r_+^2[\nabla\Theta(\boldsymbol{x})]^2/\Theta^2(\boldsymbol{x})$, or smaller, have been neglected. In the functional Eq. (2.40), the main contribution comes from the functions $\Theta(\boldsymbol{q}) \propto [K_-(\boldsymbol{q})]^{1/2}$. As the function $V_-(\boldsymbol{x})$ is a smooth long-range potential, the Fourier spectrum for the $K_-(\boldsymbol{x})$ has been considered as restricted from above by the value $q_0 \propto 1/r_-$. Therefore, $|\nabla\Theta| \propto \Theta/r_-$ and the obtained result, Eq. (2.42), is a zero order approximation in the power series over the parameter $(r_+/r_-)^2$. Using Eqs. (2.39) and (2.42) one can write the expression for the thermodynamic function Ω in the form

$$\Omega = -T \ln \int \frac{D\Theta(\boldsymbol{x})}{A_-} \exp\left\{ -\frac{\Phi\{\Theta\}}{T} \right\}, \tag{2.43}$$

where

$$\Phi\{\Theta\} = \int \frac{d\boldsymbol{x} d\boldsymbol{x}'}{2} \Theta(\boldsymbol{x}) K_-(\boldsymbol{x} - \boldsymbol{x}') \Theta(\boldsymbol{x}') - \int d\boldsymbol{x} P_+(\mu + \Theta(\boldsymbol{x}) - U(\boldsymbol{x}), T). \tag{2.44}$$

Mean Field Approximation

Let us apply the steepest descent method to Eq. (2.43), from which we obtain the equation defining $\Theta_0(\boldsymbol{x})$ corresponding to the minimum of the functional Φ as

$$n_+(\mu + \Theta_0(\boldsymbol{x}) - U(\boldsymbol{x}), T) = \int d\boldsymbol{x} K_-(\boldsymbol{x} - \boldsymbol{x}') \Theta_0(\boldsymbol{x}'), \tag{2.45}$$

where $n_+(\mu, T)$ is the particle density in the system with the short range repulsion potential defined from the usual thermodynamic relation $n_+ = \partial P_+/\partial\mu|_T$.[7] The restriction of the whole procedure with the case of a homogeneous solution of Eq. (2.44) ($U(\boldsymbol{x}) = 0$) gives

$$\Theta_0 = an_+(\mu + \Theta_0), \quad a = \int d\boldsymbol{x} V_-(\boldsymbol{x}) > 0, \tag{2.46}$$

The approximation $\Omega_0 = \Phi[\Theta_0]$ corresponds to the consideration of the attractive part of the interaction potential in the mean field approach. Let us also notice that in this approximation the density of particles in the system can be expressed as

$$n(\mu, T) = -\frac{1}{V}\frac{\partial\Omega_0}{\partial\mu}\bigg|_T = -\frac{1}{V}\frac{\partial\Omega_0}{\partial\Theta_0}\bigg|_{T,\mu}\frac{\partial\Theta_0}{\partial\mu}\bigg|_T$$

$$+ \frac{\partial P_+(\mu + \Theta_0, T)}{\partial\mu}\bigg|_{\Theta_0,T} = n_+(\mu + \Theta_0). \tag{2.47}$$

With the help of Eqs. (2.45) and (2.47) the pressure in the system can be obtained in the form

$$P = P_+(\mu + \Theta_0) - \frac{an_+^2(\mu + \Theta_0)}{2}. \tag{2.48}$$

The set of Eqs. (2.45), (2.46), and (2.47) define, in the mean field approximation, the equation of state in a parametric form. The parameter is the chemical potential μ. With the help of Eq. (2.46) it is possible to eliminate the parameter μ and to reduce the equation of state to traditional form

$$P(n, T) = P_+(n, T) - \frac{an^2}{2}. \tag{2.49}$$

As follows from numerical calculations, at low densities $P_+(n, T)$ practically equal to the pressure of the ideal gas $P_+ = nT$.[8] If the mean spacing between particles becomes of the order of the repulsion radius r_+, i.e. $n \approx r_+^{-d}$, then P_+ would have a stronger dependence on n. The simplest approximation, describing such behavior of P_+, would be $P_+ = nT/(1 - nr_+^d)$. In this case Eq. (2.49) reduces to the traditional van der Waals equation.

Formally exact definitions of the quantities $P_+(\mu, T)$ and $n_+(\mu, T)$ can

[7] The functions $P_+(\mu, T)$ and $n_+(\mu, T)$ are assumed to be known.

[8] Densities satisfying inequality $n \ll r_+^{-d}$.

be made with the help of the virial expansion (see Landau and Lifshits 1976), which can be obtained from Eq. (2.40)

$$P_+(\mu,T) = T\sum_{m=1}^{\infty}\frac{b_m}{m!}\left\{\zeta e^{-\frac{V_-(0)}{2T}}\right\}^m ,$$

$$n_+(\mu,T) = T\sum_{m=1}^{\infty}\frac{b_m}{(m-1)!}\left\{\zeta e^{-\frac{V_-(0)}{2T}}\right\}^m .$$

It is interesting to mention that during this derivation the value $V(0)$ has been considered as a finite quantity. Nonetheless, in the final formulae for the pressure in the system and density of particles, $V(0)$ is not present, but instead the value $V_-(0) = V(0) - V_+(0)$ appears. Therefore, in the final formulae we can put $V(0)$ to infinity, insofar as the value $V_-(0)$, renormalizing the chemical potential, is finite.

Ginzburg-Landau Functional

The self-consistent approximation Eq. (2.44) is not a good approximation in the vicinity of the critical point. Let us choose $\Theta(\boldsymbol{x}) = \Theta_0(\mu,T)+\eta(\boldsymbol{x})$, where the function $\eta(\boldsymbol{x})$ is a small deviation from the steepest descent value of Θ, and let us also expand the functional $\Phi\{\Theta\}$ into the functional Taylor series

$$\Phi\{\Theta\} = \Phi\{\Theta_0\} + \Delta\Phi\{\eta\} ,$$

where

$$\Delta\Phi\{\eta\} = \frac{1}{2}\int d\boldsymbol{x}d\,\boldsymbol{x}'\eta(\boldsymbol{x})K_-(\boldsymbol{x}-\boldsymbol{x}')\eta(\boldsymbol{x}')$$
$$-\int d\boldsymbol{x}\left\{\Delta\mu\left(\frac{\eta(\boldsymbol{x})}{a}+n\right)+\frac{1}{2}n_+^{(1)}(\mu_0+\Theta_0)\eta^2(\boldsymbol{x})\right.$$
$$\left.+\frac{1}{3!}n_+^{(2)}(\mu_0+\Theta_0)\eta^3(\boldsymbol{x})+\frac{1}{4!}n_+^{(3)}(\mu_0+\Theta_0)\eta^4(\boldsymbol{x})\right\}, \qquad (2.50)$$

$\Delta\mu = \mu - \mu_0$, the function $\mu_0(n,T)$ is defined by Eqs. (2.46) and (2.47) (now n is an independent thermodynamic parameter),

$$n_+^{(k)}(\mu+\Theta_0) = \left.\frac{\partial^{k+1}P_+(\mu+\Theta_0)}{\partial\mu^{k+1}}\right|_{\Theta_0} .$$

The factors before the powers of η in Eq. (2.49) can be obtained after differentiation in a form such as

$$n_+^{(1)} = \frac{n_\mu}{1+an_\mu}, \quad n_+^{(2)} = \frac{n_{\mu\mu}}{(1+an_\mu)^3},$$
$$n_+^{(3)} = \frac{n_{\mu\mu\mu}}{(1+an_\mu)^4} - 3a\frac{n_{\mu\mu}^2}{(1+an_\mu)^5}.$$

The derivatives with respect to μ can be expressed with the help of the function $f(n,T) = (\partial P/\partial n)_T$ (where P is the pressure defined by Eq. (2.49)), as

$$n_+^{(1)} = \frac{n}{f+an}\,, \quad n_+^{(2)} = n\frac{f-nf_n}{(f+an)^3}\,,$$

$$n_+^{(3)} = \frac{n}{(f+an)^4}\left\{\frac{3(3f^2-2nff_n+n^2f_n^2)}{f+an} + 2nf_n - n^2 f_{nn} - 2f\right\}, \tag{2.51}$$

The final aim of this deliberation is the derivation of the Ginzburg-Landau functional in the critical region, which means that we need to find all factors in the power series of the Landau field ϕ in the region of temperature and densities $\tau,\ v \ll 1$, where in addition to τ we have a new thermodynamic parameter $v = (n-n_c)/n_c$. Therefore, it is necessary to find two conditions, instead of one (see Sects. 2.2.1 and 2.2.2) defining the critical temperature and critical density of particles n_c. These conditions are

$$\left.\frac{\partial P}{\partial n}\right|_T = 0\,, \quad \left.\frac{\partial^2 P}{\partial n^2}\right|_T . \tag{2.52}$$

The expression for the function $f(\tau,v)$ in the critical region, follows from these conditions

$$f(\tau,v) = \frac{h(\tau,v)}{n_c} = \frac{\alpha\tau + \beta v^2 + \gamma\tau v}{n_c}, \tag{2.53}$$

where the factors α, β, γ are defined by

$$\begin{aligned}
\alpha &= T_c n_c \tfrac{\partial}{\partial T}\left(\tfrac{\partial P}{\partial n}\right)_T\Big|_{T=T_c,n=n_c}, \\
\beta &= \tfrac{1}{2}n_c^3\left(\tfrac{\partial^3 P}{\partial n^3}\right)_T\Big|_{T=T_c,n=n_c}, \\
\gamma &= T_c n_c^2 \tfrac{\partial}{\partial T}\left(\tfrac{\partial^2 P}{\partial n^2}\right)_T\Big|_{T=T_c,n=n_c}.
\end{aligned} \tag{2.54}$$

Note, that though the last term in Eq. (2.53) is less than the first one, it should be retained due to the finite contribution to the factor at the η^3 term.

Finally, introducing the notation $\phi(\boldsymbol{x}) = \eta(\boldsymbol{x})/an_c$ and making an expansion in the vicinity of the critical point we obtain the searched-for

Ginzburg-Landau functional

$$\mathcal{F} = \Delta\Phi\{\phi\} = \int d\boldsymbol{x} \left\{ \frac{c}{2} \left[\nabla\phi(\boldsymbol{x})\right]^2 + \frac{h}{2}\phi(\boldsymbol{x})^2 \right.$$

$$\left. - \frac{h - h_v}{3!}\phi^3(\boldsymbol{x}) + \frac{1}{4!}h_{vv}\phi^4(\boldsymbol{x}) - \Delta\mu n_c\phi(\boldsymbol{x}) \right\} - \Delta\mu n_c V, \qquad (2.55)$$

where $c = a^2 n_c \int d\, \boldsymbol{x}\boldsymbol{x}^2 K_-(\boldsymbol{x})$.

In order to find the physical sense of the order parameter ϕ, let us consider the mean value of the density of particles in the system

$$\overline{n(\boldsymbol{r})} = -\frac{\delta \ln Z}{\delta U(\boldsymbol{r})} = \int \frac{D\Theta(\boldsymbol{x})}{ZA_-} \exp\left\{ -\frac{\Phi\{\Theta\}}{T} \right\} n_+[\mu + \Theta(\boldsymbol{r}) - U(\boldsymbol{r})].$$

Let us again expand the integral expression the way it has been done earlier (see Eq. (2.50)) in the vicinity of the Θ_0

$$\overline{n(\boldsymbol{r})} = n + \int \frac{D\phi(\boldsymbol{x})}{Z_1} \exp\left\{ -\frac{\Delta\Phi\{\phi\}}{T} \right\} n_+^{(1)}(\mu_0 + \Theta_0)\phi(\boldsymbol{r}), \qquad (2.56)$$

where

$$Z_1 = \int D\phi(\boldsymbol{x}) \exp\left\{ -\frac{\Delta\Phi\{\phi\}}{T} \right\}.$$

In the neighborhood of the critical point this expression can be reduced to

$$\frac{\overline{n(\boldsymbol{r})} - n}{n_c} = \int \frac{D\phi(\boldsymbol{x})}{Z_1} \exp\left\{ -\frac{\Delta\Phi\{\phi\}}{T} \right\} \phi(\boldsymbol{r}) = \overline{\phi(\boldsymbol{r})}.$$

Thus, the value of the parameter ϕ averaged over the thermodynamic ensemble is proportional to the deviation of the mean value of density from its steepest descent quantity n. When n coincides with n_c, which along with P_c and T_c is defined by the set of Eqs. (2.46), (2.47), and (2.52), the sense of the order parameter becomes quite simple: $\overline{\phi(\boldsymbol{r})} = (\overline{n(\,\boldsymbol{r})} - n_c)/n_c$. The Ginzburg-Landau functional at $n = n_c$ has the form

$$\mathcal{F} = \Delta\Phi\{\phi\} = \int d\boldsymbol{x} \left\{ \frac{c}{2} \left[\nabla\phi(\boldsymbol{x})\right]^2 + \frac{\alpha\tau}{2}\phi(\boldsymbol{x})^2 \right.$$

$$\left. - \frac{\alpha - \gamma}{3!}\tau\phi^3(\boldsymbol{x}) + \frac{2\beta}{4!}\phi^4(\boldsymbol{x}) - \Delta\mu n_c\phi(\boldsymbol{x}) \right\} - \Delta\mu n_c V. \qquad (2.57)$$

The density correlation function can be obtained following the same way as for the order parameter

$$G(\boldsymbol{r},\ \boldsymbol{r}') = \frac{1}{\overline{n}}\overline{(n(\boldsymbol{r}) - \overline{n})(n(\boldsymbol{r}') - \overline{n})} - \overline{n}\delta(\boldsymbol{r} - \boldsymbol{r}')$$

$$= \frac{T^2}{n_c}\frac{\delta^2 \ln Z}{\delta U(\boldsymbol{r})\delta U(\boldsymbol{r}')} - \overline{n}\delta(\ \boldsymbol{r} - \boldsymbol{r}') = \overline{\phi(\ \boldsymbol{r})\phi(\boldsymbol{r}')} - \overline{\phi(\boldsymbol{r})}\ \overline{\phi(\boldsymbol{r}')}.$$

Let us find the equation of state in the mean field approximation using the functional Eq. (2.57):

$$P - P_c = \frac{1}{V}T \ln \int \frac{D\phi}{A_-} \exp\left\{-\frac{\Delta\Phi\{\phi\}}{T}\right\}.$$

Excluding $\Delta\mu$ with the help of the equation $\delta\Phi/\delta\phi = 0$, we find

$$P - P_c = \alpha\tau v + \frac{1}{3}\beta v^3 + \frac{1}{2}\gamma\tau v^2. \tag{2.58}$$

From this equation we can get Eq. (2.52), which proves the self-consistence of the procedure used. Eq. (2.58) contains the term τv^2, which may seem to be smaller than the first two terms. Nonetheless, if one omits this term, the essential feature of the vapor-liquid coexistence curve vanishes. The curve is symmetrical, i.e. $v_l = -v_g = \sqrt{-\alpha\tau/\beta}$. The so-called diameter of this curve becomes temperature independent. Note that from the Van der Vaals equation equality $\alpha = \gamma$ emerges. In this case we would have the factor at ϕ^3 term equal to zero. Now one can see, that it is necessary to keep the last term in Eq. (2.58). Otherwise, the main components in the sum $v_l + v_g$ are compensated. In order to find the real diameter of the coexistence curve, we need to take into account the terms in the equation of state, of higher order than τv and v^3:

$$\Delta P = \alpha\tau v + \frac{1}{3}\beta v^3 + \frac{1}{2}\gamma\tau v^2 + \frac{1}{4}\beta v^4$$

then one can find

$$v_l + v_g = \frac{1}{\beta}\left(\frac{9}{5}\alpha - \gamma\right)\tau. \tag{2.59}$$

This equation states a well-known experimental "law of rectilinear diameter" (see Greer and Moldover 1981). This law has long been considered as resulting either from the critical fluctuations (Patashinskii and Pokrovskii 1979) or due to the higher order terms in Landau expansion (Vause and Sak 1980, Nicoll 1981). From the present derivation it is seen that this law is an intrinsic property of interacting particle systems and it can be revealed in the framework of conventional approximations within the mean field theory approach. Naturally, in the very close vicinity of the critical point, where $\tau \ll G_i$, Eq. (2.59) no longer holds. In the scaling region the sum $v_l + v_g$ is not proportional to τ but to $\tau^{1-\alpha}$ (see for instance Fomichev and Khohlachev 1974). It is interesting to estimate the region where one can apply the mean field approximation. As has been shown in Chapter 1, the mean field theory is applicable when $G_i \ll \tau \ll 1$. The Ginzburg parameter can be estimated with the help of Van der Waals equation and the definition Eq. (1.42). As a result we have $G_i \propto (r_+/r_-)^6$. In the considered model we have assumed that $(r_+/r_-)^2 \ll 1$. Therefore, we always have a substantial region for the applicability of the mean field theory. As the result obtained in Eq. (2.59) describes experimental data for the dependance of the diameter of the coexistence curve on temperature, it is clear that this region is present in real vapor-liquid phase transitions.

The derivation of GLF presented in this section may create the impression that some of the steps discussed in Sect. 2.1 are missing. In fact, they are all present but some of them are disguised due to the specific nature of the problem. The explicit appearance of the procedure of coarsening and the exclusion of the nonessential modes has not been present in this derivation. Nonetheless, this procedure has been included obliquely due to the expansion over the parameter $(r_+/r_-)^2$. From the example of this derivation one can see that there is more than one way to follow all the steps leading to a particular Ginzburg-Landau Functional.

2.3 Quantum Systems

Quantum systems add a new feature to the GLF derivation. Noncommuting variables do not permit the Hubbard-Stratonovich identity to be applied in the same way possible for classical Hamiltonians. The main obstacle, which one now has to overcome, is a problem of representation of quantum Hamiltonians in terms of commuting variables. Such a problem sometimes can be solved for a given Hamiltonian with a particular structure. After that, the procedure goes along the lines described in Sect. 2.2. The possibility of a particular solution for a chosen Hamiltonian should not create the impression that there is no general approach to the problem in the quantum case. In fact, this approach exists and it also sheds some additional light on the question of the lower critical dimensionality. The way applicable to any kind of Hamiltonian insists upon the introduction of an additional dimension. As is well known (see for instance Abrikosov *et al.* 1963) the partition function for any system can be written as

$$Z = \mathrm{Tr}e^{-\beta H} = Z_0 \left\langle T_\tau \exp\left\{ -\int_0^\beta d\tau H_I(\tau) \right\} \right\rangle, \tag{2.60}$$

where $\beta = 1/T$, T_τ is the time ordering operator on the scale of imaginary time, $0 \leq \tau \leq \beta$, $\langle \cdots \rangle$ denotes ensemble averaging with the zero order Hamiltonian H_0. The interaction Hamiltonian in Eq. (2.60) can now be treated as a classical quantity because the operation of time ordering takes care of all the effects of noncommutability. The additional dimension introduced in this equation cannot change the dimensionality of the system as a whole, since we deal with a finite slab in the direction of τ. Nonetheless, when $T \to 0$, $\beta \to \infty$ the effective dimension of the system changes. This accounts for the fact that at $d = 1$ in a quantum system we may have a phase transition with $T_c = 0$ (see Sect. 1.6.1).

2.3.1 The Heisenberg Hamiltonian

The quantum Heisenberg Hamiltonian formally looks like the Hamiltonian Eq. (2.20) for the classical model

$$H = -\frac{1}{2}\sum_{l,l'} \boldsymbol{S}_l J_{ll'} \boldsymbol{S}_{l'} - \boldsymbol{h}\sum_l \boldsymbol{S}_l, \tag{2.61}$$

where the spin operators $\boldsymbol{S}_l$ satisfy the commutation relations

$$[S_l^\alpha S_{l'}^\beta]_- = i\delta_{l,l'} e_{\alpha\beta\gamma} S_l^\gamma, \tag{2.62}$$

and have $\boldsymbol{S}^2 = S(S+1)$.

The application of the Hubbard-Stratonovich identity to the time ordered average in Eq. (2.60) leads to

$$\frac{Z}{Z_0} = \prod_l \int \frac{D\boldsymbol{\eta}_l(\tau)}{A} \exp\left\{ -\frac{1}{2\beta^2}\int_0^\beta d\tau \sum_{l,l'} \boldsymbol{\eta}_l(\tau) J_{ll'}^{-1} \boldsymbol{\eta}_{l'}(\tau) - \beta\Omega\{\boldsymbol{\eta}_l(\tau)\} \right\}, \tag{2.63}$$

where Ω is given by

$$\Omega\{\boldsymbol{\eta}_l(\tau)\} = -T\ln\left\langle T_\tau \exp\left\{ \frac{1}{\beta}\sum_l \int_0^\beta d\tau \boldsymbol{\eta}_l(\tau)\boldsymbol{S}_l(\tau) \right\} \right\rangle. \tag{2.64}$$

The dangerous modes now can be revealed with the help of the expansion of Eq. (2.64) in the power series of η when $\boldsymbol{h} = 0$. The latter leads to the following Gaussian propagator

$$G(\boldsymbol{q},\omega_n) = \langle|\boldsymbol{\eta}_n(\boldsymbol{q})|^2\rangle = \frac{J(\boldsymbol{q})}{T - \delta_{n,0}J(\boldsymbol{q})S(S+1)/3}, \tag{2.65}$$

where $\omega_n = 2\pi nT$, and the Gaussian Hamiltonian is defined by

$$H = \sum_{n,q} G^{-1}(\boldsymbol{q},\omega_n)|\boldsymbol{\eta}_n(\boldsymbol{q})|^2$$

with

$$\boldsymbol{\eta}_n(\boldsymbol{q}) = T\int_0^\beta d\tau \boldsymbol{\eta}(\boldsymbol{q},\tau)e^{i\omega_n\tau}$$

From Eq. (2.65) one can see that the dangerous mode is defined by the conditions $T = J(\boldsymbol{q})S(S+1)/3$. Modes with $n \neq 0$ are unimportant and they should be excluded from consideration with the help of Gaussian integration as well as the modes $\eta_0(\boldsymbol{q})$ having $q \geq q_0$. This procedure does not differ from the one used in Sect. 2.2. As a consequence, one can obtain all the results previously found for the classical case. The parameters of the Ginzburg-Landau functional are different in values from their classical analogues but they have the same orders of magnitudes.

2.3.2 Bose Gas

The case of a nonideal Bose gas to some extent resembles the classical system of interacting particles. The Hamiltonian of this system has a one-to-one correspondence to the Hamiltonian Eq. (2.34) with

$$\begin{aligned} H_0 &= \int d\boldsymbol{x}\hat{\Psi}^+(\boldsymbol{x})\left(-\tfrac{\nabla^2}{2m}\right)\hat{\Psi}(\boldsymbol{x}) \\ H_I &= \tfrac{1}{2}\int d\boldsymbol{x}d\boldsymbol{x}'\hat{\rho}(\boldsymbol{x})V(\boldsymbol{x}-\boldsymbol{x}')\hat{\rho}(\boldsymbol{x}'), \end{aligned} \tag{2.66}$$

where $\hat{\Psi}^+$ and $\hat{\Psi}$ are Bose field operators, $\hat{\rho} = \hat{\Psi}^+\hat{\Psi}$ is the particle density operator. The qualitative difference between the classical and quantum cases lies in the following. In the classical case we need to have an attractive part in addition to the repulsion in the interaction potential $V(\boldsymbol{x})$ for a phase transition to happen. The quantum case does not need an introduction of any attraction as the quantum statistics by itself ensures such attraction. Therefore, for the sake of simplicity, we can regard the potential $V(\boldsymbol{r})$ as purely repulsive.

As in the classical case we need to consider the partition function for the grand canonical ensemble. Hence, in Eq. (2.60) the ensemble average should be understood as the average with the effective Hamiltonian $\overline{H}_0 = H_0 - \mu\hat{N}$ with the operator $\hat{N}$ for the total number of particles in the system equal to $\int d\,\boldsymbol{x}\hat{\rho}(\boldsymbol{x})$. After the application of the Hubbard-Stratonovich identity we arrive at the following representation for the partition function

$$\frac{Z}{Z_0} = \int \frac{D\Theta(\boldsymbol{x},\tau)}{A} \exp\left\{-\frac{\Phi\{\Theta(\boldsymbol{x},\tau)\}}{T}\right\}, \tag{2.67}$$

where

$$\Phi\{\Theta\} = \frac{T}{2}\int_0^\beta d\tau \int d\boldsymbol{x}d\boldsymbol{x}'\Theta(\boldsymbol{x},\tau)K(\,\boldsymbol{x}-\boldsymbol{x}')\Theta(\boldsymbol{x}',\tau) - T\ln\langle T_\tau\sigma(\beta)\rangle$$

with

$$\langle T_\tau\sigma(\beta)\rangle = \left\langle T_\tau \exp\left\{-i\int_0^\beta d\tau \int d\boldsymbol{x}\Theta(\boldsymbol{x},\tau)\hat{\rho}(\boldsymbol{x},\tau)\right\}\right\rangle, \tag{2.68}$$

all other notations can be understood by a comparison of this expression with Eq. (2.39). The formula Eq. (2.68) can be physically interpreted as follows. The expression in the functional integral Eq. (2.67) contains the trace of the operator

$$e^{-\beta\overline{H}_0}T_\tau \exp\left\{-i\int_0^\beta d\tau \int d\boldsymbol{x}\Theta(\boldsymbol{x},\tau)\hat{\rho}(\boldsymbol{x},\tau)\right\}, \tag{2.69}$$

which could be considered as an equilibrium statistical operator for the system of noninteracting electrons in the imaginary field Θ, provided $\Theta(\boldsymbol{x})$ is a τ-independent value. In such a case we could say that this statistical operator corresponds to the Hamiltonian

$$\overline{H}_0 + i\int d\,\boldsymbol{x}\Theta(\boldsymbol{x})\rho(\boldsymbol{x}).$$

In reality, however, Θ depends on the imaginary time τ, and we do not have such a description. Nonetheless, it is possible to interpret the trace of the operator Eq. (2.69) as a partition function considered as a functional of the imaginary field $\Theta(\boldsymbol{x},\tau)$.

The following procedure is analogous to the one used in Sect. 2.2.3. Let us apply the steepest descent method to Eq. (2.67) to define the equation of the mean field approximation

$$\int d\boldsymbol{x}'K(\boldsymbol{x}-\boldsymbol{x}')\Theta_0(\boldsymbol{x}') = -i\frac{\langle T_\tau\sigma(\beta)\hat{\rho}(\boldsymbol{x},\tau)\rangle}{\langle T_\tau\sigma(\beta)\rangle}. \tag{2.70}$$

The restriction of this consideration to the time-independent and space-homogeneous solution leads to

$$\Theta_0 = -ia\frac{\langle T_\tau \sigma(\beta)\hat{\rho}\rangle}{\langle T_\tau \sigma(\beta)\rangle} = -ian(\mu + i\Theta_0, T). \tag{2.71}$$

In the same way as was done in Sect. 2.2.3 one can find expression for the pressure in this approximation

$$P(\mu, T) = P_0(\mu + i\Theta_0) + \frac{1}{2}an^2(\mu + i\Theta_0), \tag{2.72}$$

where $P_0(\mu + i\Theta_0)$ is the pressure for an ideal Bose gas with the chemical potential $\mu + i\Theta_0$. Since our aim is to construct the Ginzburg-Landau functional, let us set $\Theta(\boldsymbol{x}, \tau) = \Theta_0 + \eta(\boldsymbol{x}, \tau)$ and expand the functional $\Phi\{\Theta\}$ in Eq. (2.67) in a power series of η

$$\Phi\{\Theta\} = \Phi\{\Theta_0\} + \Delta\Phi\{\eta\}$$

where the Gaussian part of $\Delta\Phi_G$ can be written as

$$\Delta\Phi_G\{\eta\} = \frac{T}{2}\int d\tau d\tau' \int d\boldsymbol{x} d\boldsymbol{x}' \eta(\boldsymbol{x}, \tau) \{\delta(\tau - \tau')K(\boldsymbol{x} - \boldsymbol{x}')$$
$$+ \langle T_\tau \hat{\rho}(\boldsymbol{x}, \tau)\hat{\rho}(\boldsymbol{x}', \tau')\rangle_{oc}\} \eta(\boldsymbol{x}', \tau'), \tag{2.73}$$

where $\langle\cdots\rangle_{oc}$ means connected average with the Hamiltonian $\overline{H}_0 + i\Theta_0\hat{N}$.

The Fourier transform of the expression in the curly brackets in Eq. (2.73) determines the inverse trial propagator. This propagator can be represented in the form

$$G(\boldsymbol{q}, \omega_n) = \frac{V(\boldsymbol{q})}{1 + V(\boldsymbol{q})F(\boldsymbol{q}, \omega_n)}. \tag{2.74}$$

The definition for the function F follows from Eq. (2.73)

$$F(\boldsymbol{q}, \omega_n) = T\sum_{n'} \int \frac{d\boldsymbol{q}'}{(2\pi)^3} \bar{G}_0(\boldsymbol{q}', \omega'_n)\bar{G}_0(\boldsymbol{q}' - \boldsymbol{q}, \omega'_n - \omega_n), \tag{2.75}$$

where $\bar{G}_0$ is a one particle Matsubara Green's function defined by

$$\bar{G}_0 = \int_0^\beta d\tau \int d\boldsymbol{x} e^{i(\omega\tau - \boldsymbol{q}\boldsymbol{x})} \langle T_\tau \hat{\Psi}(\boldsymbol{x}, \tau)\hat{\Psi}^+(0, 0)\rangle_0 = \frac{1}{-i\omega + \epsilon(\boldsymbol{q})}, \tag{2.76}$$

with $\epsilon(\boldsymbol{q}) = \boldsymbol{q}^2/2m - \mu + i\Theta_0$.

Applying the standard procedure of summation over the Matsubara frequencies to Eq. (2.76) we obtain

$$F(\boldsymbol{q}, \omega_n) = \int \frac{d\boldsymbol{q}'}{(2\pi)^3} \frac{\nu(\epsilon(\boldsymbol{q}')) - \nu(\epsilon(\boldsymbol{q}' - \boldsymbol{q}))}{i\omega_n - \epsilon(\boldsymbol{q}') + \epsilon(\boldsymbol{q}' - \boldsymbol{q})}, \tag{2.77}$$

where $\nu(\epsilon)$ is the Bose distribution function. As we are interested here only in the case of the vapor-liquid transition, it is possible to restrict consideration with the limit of the nondegenerate Bose gas. This assumption simplifies calculations in Eq. (2.77) due to the replacement of $\nu(\epsilon)$ with the Boltzmann distribution function. As a result, Eq. (2.77) can be reduced in the limit of small q to

$$F(\boldsymbol{q}, \omega_n) = \frac{\Lambda_T^2}{\Gamma} \left\{ i\omega_n + \left(\frac{\boldsymbol{q}}{\Lambda_T} \right)^2 - 2T \right\}, \tag{2.78}$$

where $\Lambda_T = \sqrt{2mT}$ is the thermal characteristic momentum, $\Gamma^{-1} = \Lambda_T e^{\beta(\mu + i\Theta_0)}/6\pi^{3/2}T^2$.

Expanding $V(\boldsymbol{q})$ in the vicinity $\boldsymbol{q} = 0$ and replacing $\eta(\boldsymbol{q}, \omega)$ in Eq. (2.73) by $\eta(\boldsymbol{q}, \omega)\sqrt{V(0)}$ we arrive at

$$\Delta\Phi_G\{\eta\} = \frac{T}{2} \sum_n \int \frac{d\boldsymbol{q}}{(2\pi)^3} |\eta(\boldsymbol{q}, \omega_n)|^2 \left\{ \tau + cq^2 + i\frac{\omega_n \Lambda_T^2}{\Gamma V(0)} \right\}, \tag{2.79}$$

where τ is defined as usual ($\tau = (T - T_{ct})/T_{ct}$), with the trial critical temperature given by $T_{ct} = \Gamma V(0)/2\Lambda_T^2$, and c is defined by the expression

$$c = \frac{1}{V(0)} \left(\frac{1}{6} \frac{\partial^2 V}{\partial \boldsymbol{q}^2} \bigg|_{q=0} + \frac{\Lambda_T T_{ct}}{\Gamma} \right)$$

Now the problem is almost completely reduced to the classical one. The only difference is the presence of the discrete Matsubara frequency dependence in the Gaussian Hamiltonian Eq. (2.79). This dependence can be used for the developing of the dynamical analysis (with the help of the analytical continuation procedure) in the vicinity of a critical point (see Ivanchenko *et al.* 1990). In the thermodynamic case considered here, this dependence results in the following two qualitatively different possibilities. The first one is realized, when $2\pi T_{ct}\Lambda_T^2 \gg \Gamma V(0)$. In this case, all modes with $n > 0$ are unimportant and they should be excluded with the help of the Gaussian integration, as well as the modes $\eta^>(\boldsymbol{q}, 0)$ having $q > q_0$. If $2\pi T_{ct}\Lambda_T^2 \leq \Gamma V(0)$, the situation does not correspond to the one considered in Sect. 2.2, because now we must retain some number of modes with $n < n_{max}$, where n_{max} is of the order of $\Gamma V(0)/2\pi T_{ct}\Lambda_T^2$. In this case we will have n interacting fields, which could also be treated as an n-component order parameter with a strong anisotropy in the isotopic space.

It is necessary to make the following comment about Eq. (2.79). This equation has been obtained in the system of units with $\hbar = 1$. In order to show that in the system of Bose particles even the ordinary vapor-liquid transition can be a quantum related effect one has to restore $\hbar$ in the final equation. This can be done by the formal replacement of m onto $m/\hbar^2$. As a result of this replacement T_{ct} turns out to be proportional to $\hbar$ and if $\hbar \to 0$ the phase transition vanishes. The latter is a consequence of the assumption made, at the beginning of the derivation, namely the absence of the attractive part in the potential $V(\boldsymbol{q})$.

2.3.3 Bose-Einstein Condensation

Until now we have not touched the case of Bose-Einstein condensation. This phase transition is more simple in its mathematical treatment. As is shown in Sect. 1.3.1, the order parameter can be composed from a properly averaged value of $\hat{\Psi}$-Bose operators. This shows that classical variables should be directly related to $\hat{\Psi}^+$, $\hat{\Psi}$ operators. The relation can be found with the help of the following consideration. As is well known, a general term of the expansion Eq. (2.60) in the power series over the interaction Hamiltonian H_I, can be decomposed with the help of Wick's theorem onto 2^n (n is power of H_I) multiplications of single particle Green's functions integrated over all possible space and time coordinates. The same decomposition takes place in the situation of a classical averaging with the Gaussian Hamiltonian, because Wick's theorem is also applicable here. Therefore, if we appropriately choose a zero order propagator we will have the same power series. By this method one can solve the problem of obtaining a representation of the quantum Hamiltonian in terms of commuting variables. Applying this program to the Hamiltonian Eq. (2.66) we arrive at

$$\frac{Z}{Z_0} = \int \frac{D\Psi^*(\boldsymbol{x},\tau)D\Psi(\boldsymbol{x},\tau)}{A} \exp\left\{-\frac{\Phi\{\Psi^*(\boldsymbol{x},\tau);\Psi(\boldsymbol{x},\tau)\}}{T}\right\}$$

where

$$\Phi = T\int d\tau d\tau' \int d\boldsymbol{x}d\boldsymbol{x}' \left\{\Psi^*(\boldsymbol{x},\tau)\bar{G}_0^{-1}(\boldsymbol{x}-\boldsymbol{x}',\tau-\tau')\Psi(\boldsymbol{x},\tau)\right.$$

$$\left.+\delta(\tau-\tau')\Psi^*(\boldsymbol{x},\tau)\Psi(\boldsymbol{x},\tau)V(\boldsymbol{x}-\boldsymbol{x}')\Psi^*(\boldsymbol{x}',\tau')\Psi(\boldsymbol{x}',\tau')\right\}. \qquad (2.80)$$

In this equation the zero order propagator $\bar{G}_0$ has the definition resulting from Eq. (2.76) with $\Theta_0 = 0$. Practically, the representation Eq. (2.80) has solved the problem of the GLF derivation for the case of the Bose-Einstein condensation. The work, which should be done to complete this derivation is the integration over the unimportant modes. Here we will only consider this procedure in a quite general way. The region of essential modes in this problem corresponds to $\omega_n = 0$, $q \leq \Lambda_T$. The division of the function

$\Psi(\boldsymbol{x},\tau)$ into important and unimportant contributions should be made as follows

$$\Psi(\boldsymbol{x},\tau) = \Psi^{<}(\boldsymbol{x}) + \Psi^{>}(\boldsymbol{x},\tau), \tag{2.81}$$

where

$$\Psi^{<}(\boldsymbol{x}) \int \frac{d\boldsymbol{x}}{(2\pi)^3}\Theta(\Lambda_T - q)\Psi(\boldsymbol{q},\omega_n = 0)e^{i\boldsymbol{q}\boldsymbol{x}},$$

the function Θ here is Heaviside's step function.

Corresponding to the separation defined in Eq. (2.81) one can also divide the Hamiltonian $H = \Phi/T$ into parts $H^{<} + H^{>} + H_{int}$, where H_{int} contains both kinds of fields $\Psi^{<}$ and $\Psi^{>}$. The expression for the partition function now can be rewritten as

$$\frac{Z}{Z_0 Z^{>}} = \int \frac{D\Psi^{*<}(\boldsymbol{x})D\Psi^{<}(\boldsymbol{x})}{A^{<}} \exp\left\{-(H^{<} + \Delta H^{<})\right\}, \tag{2.82}$$

where $\Delta H^{<}$ results from the integration over the unimportant modes, i.e.

$$\Delta H\{\Psi^{*<},\Psi^{<}\} = -\ln\langle e^{-H_{int}}\rangle_{>}.$$

The notation $\langle\cdots\rangle_{>}$ means a value averaged over the ensemble defined by the Hamiltonian $H^{>}$.

From Eq. (2.82) the Gaussian part H_G of the Hamiltonian $H^{<} + \Delta H^{<}$ can be obtained in the form

$$H_G = \int_{q<\Lambda_T} \frac{d\boldsymbol{q}}{(2\pi)^3}|\Psi^{<}(\boldsymbol{q})|^2 \left\{\frac{q^2}{2m} - \mu + \Sigma(q,\mu,T)\right\}, \tag{2.83}$$

where the self energy part $\Sigma(q,\mu,T)$ is defined as a Fourier transform of the expression

$$\Sigma(\boldsymbol{x}-\boldsymbol{x}',\mu,T) = \left.\frac{\delta^2\Delta H^{<}\{\Psi^{*<},\Psi^{<}\}}{\delta\Psi^{*<}(\boldsymbol{x})\delta\Psi^{<}(\boldsymbol{x}')}\right|_{\Psi^{<}=0}.$$

A trial critical point $T = T_{c0}$, corresponds to the first occurence of a zero energy mode at $q = 0$ in Eq. (2.83), which gives the condition

$$\mu = \Sigma(q=0,\mu,T)|_{T=T_{c0}}. \tag{2.84}$$

This equation defines the chemical potential at the critical point as a function of the critical temperature. Therefore, we need an additional independent condition which could help in the definition of the critical point. This condition is exactly the same as the one used for the case of the noninteracting Bose particles (see Sect. 1.3.1), i.e. the constant density of matter ρ, which formally defines μ as a function of T through the equation

$$\rho = -T\left.\frac{\partial \ln Z}{\partial\mu}\right|_T = \rho(\mu,T). \tag{2.85}$$

This equation and Eq. (2.84) define the trial values of the chemical potential $\mu = \mu_{c0}$ and temperature $T = T_{c0}$ at the critical point. After expanding the Hamiltonian $H = H^< + \Delta H^<$ in the power series of Ψ^*, Ψ, and τ we arrive at the following expression[9]

$$H = \int d\,\boldsymbol{x} \left\{ \frac{1}{2m^*} |\boldsymbol{\nabla}\Psi(\,\boldsymbol{x})|^2 + \alpha\tau |\Psi(\boldsymbol{x})|^2 + \frac{b}{4} |\Psi(\,\boldsymbol{x})|^2 \right\}, \tag{2.86}$$

where m^*, α, and b are defined as follows

$$\begin{aligned} \frac{1}{m^*} &= \frac{1}{m} + \left.\frac{\partial^2 \Sigma}{\partial q2}\right|_{q=0,T=T_{c0},\mu=\mu_{c0}}; \\ \frac{\alpha}{T} &= \left.\left[\left.\frac{\partial \Sigma}{\partial \mu}\right|_T - 1 \right] \left.\frac{\partial \mu}{\partial T}\right|_\rho + \left.\frac{\partial \Sigma}{\partial T}\right|_\mu \right|_{q=0,T=T_{c0},\mu=\mu_{c0}}; \\ b &= 2\int d\boldsymbol{x} V(\boldsymbol{x}) + \int d\boldsymbol{x} d\boldsymbol{x}_1 d\boldsymbol{x}_2 \left.\frac{\delta^4 \Delta H^<}{\delta\Psi^*(\boldsymbol{x}_1)\delta\Psi^*(\boldsymbol{x}_2)\delta\Psi(\boldsymbol{x}_3)\delta\Psi(\boldsymbol{x})}\right|_{\Psi=0}. \end{aligned}$$

In the derivation of Eq. (2.86) it has been assumed that the interaction potential has a finite radius $r_0 \ll \Lambda_T^{-1}$. It is also necessary to mention that in the case of an ideal gas Bose-Einstein condensation (see Sect. 1.3.1), the derivative $\partial\mu/\partial T|_\rho$ has happened equal to zero at the critical point. In the situation of the interacting particles this derivative is not equal to zero and in the case of the liquid Helium λ-transition it ranges from 14 to 40, depending on the pressure in the system (see Buckingham and Fairbank (1961), Kierstead (1967)). This means, that in H_e^4 the interaction of particles is quite strong.

2.3.4 Fermi Gas

The vapor-liquid phase transition in the system of Fermi particles can only happen if there is a substantial attractive part in the interaction potential. The procedure of the GLF derivation for such a transition repeats the steps considered in Sects. 2.2.3 and 2.3.2, differing only in nonessential details. In this section we will treat a case which so far has not been considered — superconductivity transition. This phenomenon, although a direct analogue of the Bose-Einstein condensation (superfluidity), differs from it in two essential aspects. First, superconductivity represents the condensation not of individual particles but of strongly correlated pairs. Second, the pairs are formed from electrons which are subjects of electromagnetic interactions. As the Fermi statistics provide an inherent repulsion, to form a Cooper pair one needs to introduce an attraction into the interaction Hamiltonian. Here

[9] Here and below we omit the superscript $<$ in the notations for Ψ^*, Ψ functions.

we will consider only the case of a weak interaction with the Hamiltonian considered by Gorkov (1958)

$$H = \sum_{\sigma} \int d\boldsymbol{x} \hat{\Psi}_{\sigma}^{+}(\boldsymbol{x}) \frac{\hat{\boldsymbol{p}}^2}{2m} \hat{\Psi}_{\sigma}(\boldsymbol{x}) - \lambda \int d\boldsymbol{x} \hat{\Psi}_{\uparrow}^{+}(\boldsymbol{x}) \hat{\Psi}_{\downarrow}^{+}(\boldsymbol{x}) \hat{\Psi}_{\downarrow}(\boldsymbol{x}) \hat{\Psi}_{\uparrow}(\boldsymbol{x}), \quad (2.87)$$

where the operator of momentum in the general case contains the vector potential contribution ($\hat{\boldsymbol{p}} = -i\boldsymbol{\nabla} - e\boldsymbol{A}(\boldsymbol{x})$) and the negative sign before the interaction Hamiltonian is chosen to have λ positive.

In order to find classical representation for this problem we need to introduce a continuous integration over the variables, which will evidently relate to the strong pair correlations in the system. This can be done if one chooses the following form for Hubbard-Stratonovich depiction

$$\frac{Z}{Z_0} = \int \frac{D\eta^*(\boldsymbol{x},\tau) D\eta(\boldsymbol{x},\tau)}{A} \exp\left\{ - \frac{\Phi\{\eta^*(\boldsymbol{x},\tau), \eta(\boldsymbol{x},\tau), \boldsymbol{A}(\boldsymbol{x})\}}{T} \right\}, \quad (2.88)$$

where

$$\Phi\{\eta^*, \eta, \boldsymbol{A}\} = T \int_0^{\beta} d\tau \int d\boldsymbol{x} \frac{|\eta(\boldsymbol{x},\tau)|^2}{\lambda} - T \ln \langle T_\tau \sigma(\beta) \rangle$$

with

$$\langle T_\tau \sigma(\beta) \rangle = \left\langle T_\tau \exp\left\{ - \int_0^{\beta} d\tau \int d\boldsymbol{x}\, \eta^*(\boldsymbol{x},\tau) \hat{\Psi}_{\downarrow}(\boldsymbol{x},\tau) \hat{\Psi}_{\uparrow}(\boldsymbol{x},\tau) + h.c. \right\} \right\rangle . \quad (2.89)$$

This expression can be physically interpreted the same way as was done with Eq. (2.68). However, in Eq. (2.89), instead of the τ-dependent field Θ, we have pair sources η^* and η. And there is also some additional difference which should be taken into account. In the case of the vapor-liquid transition we had to make expansion over the small values of η, τ and v (see Sect. 2.2.3) and above the critical point an averaged value of η is not zero. In this case the situation is more simple, because only one thermodynamic parameter τ is present and the absence of the mean values of η^*, η above the critical point. Therefore, to define critical trial parameters we need to expand the functional $\Phi\{\eta^*, \eta\}$ in the power series of η^* and η. The inverse Gaussian propagator resulting from this expansion is

$$G^{-1}(\boldsymbol{x}, \boldsymbol{x}', \tau - \tau') = \frac{\delta(\tau - \tau')\delta(\boldsymbol{x} - \boldsymbol{x}')}{\lambda} - \langle T_\tau R^+(\boldsymbol{x},\tau) R(\boldsymbol{x}',\tau') \rangle, \quad (2.90)$$

where R^+, R are creation and annihilation pair operators ($R = \hat{\Psi}_{\downarrow}\hat{\Psi}_{\uparrow}$).

The correlation function for the pair operators in Eq. (2.90) can be expressed with the help of single particle Fermi propagators like this

$$\langle R^{+}(\boldsymbol{x},\tau)R(\boldsymbol{x}',\tau')\rangle = G^{2}(\boldsymbol{x}',\boldsymbol{x},\tau'-\tau), \tag{2.91}$$

where the single particle propagator satisfies the equations (see Abrikosov *et al* 1963)

$$\begin{aligned}\left\{i\omega - \frac{(i\boldsymbol{\nabla} + e\boldsymbol{A}(\boldsymbol{x}))^2}{2m} + \mu\right\} G(\boldsymbol{x},\boldsymbol{x}',\omega) = \delta(\boldsymbol{x}-\boldsymbol{x}'),\\ \left\{i\omega - \frac{(i\boldsymbol{\nabla}' - e\boldsymbol{A}(\boldsymbol{x}'))^2}{2m} + \mu\right\} G(\boldsymbol{x},\boldsymbol{x}',\omega) = \delta(\boldsymbol{x}-\boldsymbol{x}'),\end{aligned} \tag{2.92}$$

where ω are the odd Matsubara frequencies ($\omega_n = \pi(2n+1)T$).

The solution of these equations can be simply obtained for the space-homogeneous case ($\boldsymbol{A}(\boldsymbol{x}) = 0$).This case can be solved as usual with the help of the Fourier transformation leading to $G_0(\ \boldsymbol{q},\omega) = i\omega - \epsilon(\boldsymbol{q}) + \mu$. The presence of an external magnetic field complicates calculations. This case can be reduced to homogeneous situation for fields slowly varying, in comparison with the correlation length ξ_0. In such a situation the solution of Eq. (2.92) can be represented as follows (see Gorkov 1959)

$$G(\boldsymbol{x},\boldsymbol{x}',\omega) = e^{i\phi(\boldsymbol{x},\boldsymbol{x}')}G_0(\boldsymbol{x}-\boldsymbol{x}',\omega), \tag{2.93}$$

where

$$\phi(\boldsymbol{x},\boldsymbol{x}') = e\left\{\boldsymbol{x}\cdot\boldsymbol{A}(\boldsymbol{x}) - \boldsymbol{x}'\cdot\boldsymbol{A}(\boldsymbol{x}')\right\}$$

Using this expression, after some standard calculations,[10] we obtain the Gaussian part of the Ginzburg-Landau functional in the form

$$\Delta\Phi_G = \int d\boldsymbol{x}\left\{\frac{1}{4m}\left|\left(\frac{\partial}{\partial\boldsymbol{x}} - 2ie\boldsymbol{A}(\boldsymbol{x})\right)\Psi(\boldsymbol{x})\right|^2 + \alpha\tau\left|\Psi(\boldsymbol{x})\right|^2\right\}, \tag{2.94}$$

where α and Ψ are defined by

$$\alpha = \frac{6\pi^2T_c^2}{7\zeta(3)\mu}\,,\ \Psi(\boldsymbol{x}) = \sqrt{\frac{7\zeta(3)\rho}{8(\pi T_c)^2}}\eta(\boldsymbol{x},\omega=0)\,,$$

where ρ is the electron density in the system, ζ is the Riemann's function, and the critical temperature T_c is defined by the equation (see Eq. (2.90))

$$\frac{1}{\lambda} - \int_0^{\beta_c} d\tau \int d\boldsymbol{x} G_0^2(\boldsymbol{x},\tau) = 0.$$

[10]Details of these calculations can be found in the original paper by Gorkov (1959) and also in the later version of the GLF derivation by Svidzinsky (1982). The latter is much closer to the derivation accepted in this chapter.

This equation can be reduced to the traditional BCS relation for the critical temperature with the help of the Fourier transformation and a conventional summation over the discrete frequencies

$$1 = \frac{\lambda m p_F}{2\pi^2} \int_0^{\omega_D} d\xi \frac{\tanh(\beta_c \xi/2)}{\xi},$$

where p_F is Fermi momentum and ω_D is the Debye cut-off energy. This cut-off reflects the fact that the attractive interaction results due to the exchange of phonons between electrons.

The step of coarsening or exclusion of unimportant modes in the case of weak interaction between electrons does not add anything new to the GLF derivation, because all renormalization contributions are very small. Therefore, for this case it is now possible to fulfil the final step of the derivation, i.e. to expand the functional Eq. (2.89) till the fourth power of $|\eta(\boldsymbol{x}, \omega_n)|$ leaving in the expansion only the terms with $n = 0$. As a result, of such an expansion we obtain the following addition to the Gaussian part Eq. (2.94)

$$\Delta\Phi = \int d\boldsymbol{x} \frac{\alpha}{2\rho} |\Psi(\boldsymbol{x})|^4. \tag{2.95}$$

Until now we have not considered the expression for the electric current in the system. This expression can be obtained with the help of the general definition

$$\boldsymbol{J}(x) = -T \frac{\delta \ln Z}{\delta \boldsymbol{A}(\boldsymbol{x})},$$

from which follows

$$\boldsymbol{J}(\boldsymbol{x}) = -T \frac{\delta \ln Z_0}{\delta \boldsymbol{A}(\boldsymbol{x})} - \left\langle \frac{ie}{m} \left(\Psi^* \frac{\partial \Psi}{\partial \boldsymbol{x}} - \Psi \frac{\partial \Psi^*}{\partial \boldsymbol{x}} \right) + \frac{2e^2}{m} \boldsymbol{A}(\boldsymbol{x}) |\Psi|^2 \right\rangle. \tag{2.96}$$

The first term in this equation represents the current in the system of noninteracting electrons in the presence of an external magnetic field $\boldsymbol{H} = curl \boldsymbol{A}$. This term should be equal to zero due to the absence of persistent currents in normal metals. As a result, from Eqs. (2.94), (2.95), and Eq. (2.96) we have the Ginzburg-Landau functional

$$\Phi = \int d\boldsymbol{x} \left\{ \frac{1}{4m} \left| \left(\frac{\partial}{\partial \boldsymbol{x}} - 2ie\boldsymbol{A}(\boldsymbol{x}) \right) \Psi(\boldsymbol{x}) \right|^2 + \alpha\tau |\Psi(\boldsymbol{x})|^2 + \frac{b}{2} |\Psi(\boldsymbol{x})|^4 \right\}$$

and the expression for the operator of the electric current defined by

$$\hat{\boldsymbol{J}}(\boldsymbol{x}) = -\frac{\delta\Phi}{\delta \boldsymbol{A}(\boldsymbol{x})}.$$

In this derivation we have only considered the limit of pure superconductors. This limit corresponds to the large value of the electron mean free path l in the comparison with the correlation length ξ_0. In the situation when the impurity scattering is essential ($l \leq \xi_0$), one should perform an appropriate averaging over the random impurity distribution. Such averaging must be done directly on the thermodynamic potentials, as the partition function is not a self-averaging value. Therefore, the effective partition function Z_e should be represented in the form

$$Z_e = \exp(\overline{\ln Z}),$$

where the bar over $\ln Z$ denotes an impurity averaged quantity. The procedure of the averaging is usually made with the help of the so-called "replica" method. The method reduces the problem of the logarithm averaging to a convenient deliberation over a traditional system with k equivalent interacting fields. This method is based on the following representation for the logarithm

$$\overline{\ln Z} = \frac{d}{dk}\overline{Z^k}\bigg|_{k=0} .$$

As is seen from this equation, such a representation requires analytical continuation from the discrete values of k (number of replica) to $k \to 0$, after the completion of the calculational procedure. If one restricts the whole consideration with the limit $p_F l \gg 1$, then the interaction between different replicas becomes unimportant and the problem will be reduced to the one previously considered with the obvious replacement in Eq. (2.91) of the single particle Green's function by the corresponding function averaged over the impurity distribution. As a result, such an approximation in the case $l \ll \xi_0$ leads to the replacements $\alpha \to \alpha\gamma$, $b \to b\gamma^2$, and $\Psi \to \Psi/\sqrt{\gamma}$, where $\gamma \approx 0.019\xi_0/l$.

Chapter 3

Wilson's Renormalization Scheme

3.1 Introduction

At present there is a diversity of formulations for the Renormalization Group (RG) approach to the theory of critical behavior. Analysis of different methods can be found in the review articles by Fisher (1974), Bresin *et al* (1976), Di Castro and Jona-Lasinio (1976), Wegner (1976) and in the books by Ma (1976), Patashinskii and Pokrovskii (1979), Amit (1984), Baker (1990), Zinn-Justin (1989), and others. From the variety of different formulations one needs to conceive only two disparate approaches in order to be able to grasp any other approach. The first approach, which has been successfully applied to the theory of phase transitions in the critical region, is Wilson's RG formulation (1971). In this approach the scaling hypothesis in Kadanoff's formulation (see Chapter 1) has been exploited to the fullest extent. Kadanoff (1966) substantially developed the intuitive idea of the scaling invariance of thermodynamic functions in the vicinity of the critical point. He showed that this invariance can be explained on the basis of a simple assumption. The essence of this assumption can be formulated as follows. The experimentally-seen scaling invariance is a direct consequence of some coarsened "dynamical" invariance of the Ginzburg-Landau functional in the critical region. The latter in turn is an intuitively-logical consequence of the existence of a single characteristic length — correlation radius ξ. In reality, this is a very restrictive condition and this condition is the origin of invariance of the Ginzburg-Landau functional under very unusual transformations at the critical point. In the following discussion we will call these transformations Kadanoff transformations and the consequent invariance of the Ginzburg-Landau functional, Kadanoff's invariance.

The second approach is based on the formal analogy between the following two problems. The first problem is calculation of the mean values in the thermodynamics, and the second is the consequent calculation in quantum field theory. Divergences, arising in the mean values at the critical point, are related to the vanishing of the characteristic length. These divergences resemble infra-red singularities in field theory in their mathematical nature. However, many different endeavors of applying Feynmann's graph technique to the problem of the critical behavior (see for example Di Castro and Jona-Lasinio (1969)) only led to success after Wilson's RG formulation. Wilson has supplied a strict mathematical content to the intuitive idea inherent in Kadanoff's transformations. Wilson and Fisher (1972) applied the Feynmann graph technique (which is not a must in this approach), found a new small expansion parameter $\epsilon = 4 - d$, and, most importantly, they introduced the idea that all critical exponents are smooth functions of the space dimensionality d. All these helped them to find the first nontrivial approximations in the critical region (1972). Notwithstanding the fact that Wilson's works were in the main based on the first approach, they showed proper ways to apply the field theoretical RG to the theory of critical phenomena.

At first the existence of the specific invariance in the so-called renormalizable quantum field theories was found by Stueckelberg and Petermann (1953). By itself, the fact of the renormalization invariance would not be a very interesting phenomenon, from the calculational point of view, if not for the work by Gell-Mann and Low in 1954. In this work the Green's functions in quantum electrodynamics were studied at small distances. Gell-Mann and Low derived quite general functional relations for the analysis of the whole sum of perturbation series. And though these relations could not have been considered as a new calculational method, nonetheless, they had introduced, with the help of simple mathematical hypotheses, a new kind of analysis for the asymptotic behavior of the Green's functions in the ultraviolet region. As a result, simple scaling properties had been discovered in the quantum field theory at small distances. Later, Bogoluybov and Shirkov (1955, 1956) reformulated renormalization functional and differential equations in the language of renormalized momenta. After that the group transformations were not related directly to the cut-off momentum Λ and therefore they allowed analysis of not only ultraviolet but also infrared singularities . The field theoretical RG is discussed in detail in the book by Bogoluybov and Shirkov (1984).

The field theoretical RG was first successfully applied to the theory of critical behavior by Di Castro in 1972 (see also the review by Di Castro and Jona-Lasinio (1976)) and later in a somewhat different formulation it was very effectively used by Bresin *et al* (1976).

The so-called infra-red problem, in its first applications, was formulated in field theory as a problem of asymptotic behavior of mean values at fixed momenta when $\Lambda \to \infty$. In the problems of condensed matter physics there is always a natural cut-off momentum. It can be, for instance, momentum at the boundary of the Brillouin zone, size of a molecule, or the inverse spacing between nearest sites in a lattice. Therefore, in the critical point the cut-off Λ is fixed, but divergences result in the limit when the momentum $q \to 0$. It is easy to see that in field theory and in the theory of the critical behavior the problem is of exactly the same nature. It is the problem of describing a system in the limit $q/\Lambda \to 0$. Hence, there is a deep lateral analogy in the mathematical formulations of these theories which contributed to the success of the field theoretical analysis of critical behavior.

Two questions naturally arise from the above:

1. As the two approaches are different in their mathematical and ideological essence, are they really equivalent in their physical content?

2. To what extent could the regions of applicability of these methods overlap?

The first question has been discussed by Di Castro and Jona-Lasinio (1976). They have shown that for the simplest models (for instance ϕ^4, $\mathcal{O}(n)$-symmetrical model), at least in the lowest orders in the power series of ϵ, results coincide. The second question is also possible to reformulate in such a way. As we now already know that in many cases the two approaches lead to the same (though approximate) results, to what kind of problems is this coincidence is restricted? The answer to this question will be given (to some extent) in this and the following chapters.

3.2 Kadanoff's Invariance

The correlation function (see Sect. 1.5.2) away from the critical region Eq. (1.46) for $r \leq \xi$ can be approximated by

$$G(r) \approx \frac{\exp(-r/\xi)}{r^{d-2}}. \tag{3.1}$$

This equation results from the mean field approach, but nonetheless it shows that interaction of the fluctuations is essential in the volume of the order of ξ^d. Even in the mean field approximation, correlation length increases infinitely as $\tau \to 0$. Therefore, the region of effective contributions to the interaction of fluctuations also increases and at the critical point one has to take into account all interactions of all magnetic moments in the volume of the sample. The solution of such a problem, from the point of view of traditional approximations, is completely hopeless. The conventional methods become useless even in the region where the correlation radius exceeds characteristic microscopic lengths.

Kadanoff in 1966 proposed a very original way to treat the problem. His method dealt with the problem of accounting for long-range correlations by solving the problem for only nearest neighbor interactions. The quintessence of the method can be (as it has been made in the original work by Kadanoff) demonstrated in the simplest way with the help of the Ising model (see Sect. 2.2.1). This model has only one microscopic characteristic length — lattice cell size a, and only one dimensionless interaction constant $J/T = K$.

Let us consider some quite small, but arbitrary volume Ω_0. If $\Omega_0 \ll \xi^d$, then all magnetic moments (spins) in this volume will be correlated. Their behavior correlates to such an extent, that if in Ω_0 they number n^d, the magnetization of the volume will have only one of the two values $\pm n^d$. The next step is to divide the whole space into regions which would contain n^d spins. The shape of the regions' boundary surfaces should be chosen in a way such as to allow such a division. Now we may normalize the magnetizations by n^d and introduce a new effective interaction K' between adjacent regions. Of course, K' should be some function of the original interaction K, i.e. $K' = f(K)$. As a result we have again obtained the Ising model, but with the new lattice constant $a' = na$ and the new interaction constant K'. Let us contract the lattice to have the initial lattice cell. After that we arrive at the initial problem on the same lattice but with the interaction constant K'. The correlation radius resulting form the contraction becomes n times smaller. Hence, the equation

$$\xi(K') \equiv \xi[f(K)] = \xi(K)/n \tag{3.2}$$

should be satisfied. When the system is in the critical point $K = K_c = J/T_c$ and the correlation radius $\xi \to \infty$. The latter means, that if $\xi(K)$ is not a multiply defined function, then the equality

$$K_c = f(K_c), \tag{3.3}$$

follows from Eq. (3.2), which is a mathematical expression for Kadanoff's invariance. Therefore, the Ising model Hamiltonian in the critical point is invariant with respect to the set of transformations considered. Evidently, to describe asymptotic behavior one has to break the symmetry expressed by Eq. (3.3). This symmetry should now be treated as a zero order approximation. As Kadanoff did not know a rigorous practical way to find $f(K)$, he proceeded with further assumptions. Let us suppose that $f(K)$ is an analytical function even in the vicinity of the critical point $K = K_c$ i.e.

$$f(K) = f(K_c) + \lambda(K - K_c), \tag{3.4}$$

where $\lambda = df/dK$ at $K = K_c$. So far as $K \propto T^{-1}$, one should write for K close to K_c, that $\xi(K) \propto |K - K_c|^{-\nu}$ from which it follows

$$\frac{\xi[f(K)]}{\xi(K)} = \left\{ \frac{f(K) - K_c}{K - K_c} \right\}^{-\nu}, \tag{3.5}$$

or, taking into account Eqs. (3.2), (3.3), and (3.4) we arrive at

$$n = \lambda^{\nu} \ , \ \nu = \frac{\ln n}{\ln \lambda}. \tag{3.6}$$

Thus Kadanoff had shown that there is a way the theory could lead to the description of the nonanalytic behavior of the physically measurable quantities in the vicinity of the critical point. This is completely based on the calculus of analytical functions. The constructive nature of this approach can be seen if one recalls that conventional methods (perturbation theory and Feynmann graph technique) are well adapted only for obtaining analytical expansions. Nonetheless after the appearance of Kadanoff's work, about five years lapsed before the next essential step in the phase transition theory occurred. Such a long period between Kadanoff's (1966) and Wilson's (1971) work can only be explained by an obstacle of principal importance. In order to calculate the function $f(K)$, it was necessary to have understood that as a result of the Kadanoff's transformations the system could not be returned to the initial one, with only one interaction constant. These transformations convert the initial system into a system described by the Hamiltonian of a more general kind. From the class of such Hamiltonians it is necessary to seek the point having Kadanoff symmetry.

3.3 Wilson's Theory

At the beginning of the 1970's Wilson's RG formulation on the basis of Kadanoff's program was frequently called the "new RG". This term underlined the distinction between the proposed renormalization scheme and the former field theoretical formulation. Now, after quite some time has passed, it is no longer reasonable to maintain this terminology. This is why in what follows, in describing this renormalization scheme we will call it Wilson's theory, Wilson's equation, and so on.

3.3.1 Derivation of the RG Equation

Before considering the derivation of the Wilson's equation, let us ponder the main steps of Kadanoff's transformation. The first step is uniting spins to create new block spins and the normalization which reduces their values to the initial ones. The second step is the change of scale by the contraction of the lattice, reducing the new lattice to the initial one. From the physical point of view, the first step coarsens the description. Instead of considering details in the behavior of individual spins, new normalized entities of spins are introduced, and they are considered as individual spins. Different variants of methods for introducing rigorously mathematical coarsened spins are described in the book by Ma (1976). For the following it is quite enough to consider only the case of coarsening with the help of a partial integration in q-space. This integration should be carried out under the condition that physically measurable quantities do not change. Below, for convenience, we will use the word "Hamiltonian" for the Ginzburg-Landau functional normalized by temperature ($H = \mathcal{F}/T$.) As usual (see Eq. (1.11)) the partition function can be represented by

$$Z = \prod_{q<\Lambda} \int \frac{d\,\phi_q}{\sqrt{2\pi V}} \exp[-H\{\phi_q\}], \tag{3.7}$$

where $\phi_q \equiv \phi(\boldsymbol{q})$. If one first carries out integration in Eq. (3.7) over the narrow band in q-space where $\Lambda(1-\delta) < q < \Lambda$, then it is possible to introduce a new Hamiltonian $H'\{\phi_q\}$, by rewriting the partition function as

$$Z = \prod_{q<\Lambda(1-\delta)} \int \frac{d\phi_q}{\sqrt{2\pi V}} \exp[-H'\{\phi_q\}], \tag{3.8}$$

where $H'\{\phi_q\}$ is defined by

$$H'\{\phi_q\}] = -\ln\left\{\prod_{\Lambda(1-\delta)<q<\Lambda} \int \frac{d\phi_q}{\sqrt{2\pi V}} \exp[-H\{\phi_q\}]\right\}.$$

The Hamiltonian defined by this equation does not contain fields with $q > \Lambda(1-\delta)$. Therefore, part of the information about the spacial resolution of the field $\phi(\boldsymbol{x})$ at small distances is lost. Thus, the definition of the Hamiltonian with the help of Eq. (3.8) constitutes the first step of Kadanoff's transformation. We have used this example as a simple demonstration of the essence of the coarsening procedure. Below, such a particular form of this procedure will not be used. Our purpose is to find an exact RG equation in differential form. Therefore, we need to perform an infinitesimally small coarsening transformation, i.e. we should let δ approach zero.

Unfortunately, the direct use of Eq. (3.8) in this limit leads to some mathematical troubles. These were discussed in details by Wegner and Houghton (1973) and by Wilson and Kogut (1974). The reason for these troubles is completely defined by the introduction of sharp boundaries in the momentum space. Such boundaries lead to the appearance of essentially nonlocal interactions in configurational space. Hence, one has to avoid the manifestation of the sharp boundaries between integrated and nonintegrated modes of the field ϕ_q. In order to succeed in this direction, Wilson (see Wilson and Kogut (1974)) introduced a special procedure of "incomplete integration".

Let us explicitly define dependence of the Hamiltonian H on some particular mode ϕ_q considering, for simplicity, real field ϕ_q. Let us also introduce notation $\rho(\phi_q, l) = \exp[-H(\phi_q, l)]$, where l continuously denotes the moment of coarsening. Suppose that $\rho(\phi_q, l)$ satisfies the differential equation

$$\frac{\partial \rho(\phi_q, l)}{\partial l} = h(\boldsymbol{q}) \frac{\partial}{\partial \phi_q} \left(\frac{\partial}{\partial \phi_q} + \phi_q \right) \rho(\phi_q, l). \tag{3.9}$$

The solution of this equation with the initial condition $H(\phi_q, 0) = H$ (here H is the initial physical Hamiltonian) will define the coarsened Hamiltonian $H(\phi_q, l)$. As is seen, Eq. (3.9) satisfies the main physical condition of the RG transformation: the conservation of measurable values. This condition can be formulated as the independence of mean values of the stage of the coarsening, i.e. of l. Really, by carrying out the integration in Eq. (3.9) over ϕ_q we obtain

$$\frac{dZ}{dl} = 0 \, , \; Z = \int_{-\infty}^{\infty} d\phi_q \rho(\, \phi_q, l). \tag{3.10}$$

In order to understand the behavior of the solution of Eq. (3.9) at an arbitrary value of l, it is convenient to find this solution with the help of the Green's function $\hat{G}$ as

$$\rho(\phi_q{}', l) = \int d\phi_q \hat{G}(\phi_q{}', \phi_q, l) \rho(\phi_q, 0), \tag{3.11}$$

where

$$\hat{G}(\phi_q{}', \phi_q, l) = \frac{1}{\sqrt{2\pi(1 - \exp[-h(q)l])}} \exp\left\{ -\frac{[\phi_q{}' - \phi_q e^{-h(q)l}]^2}{2(1 - e^{-h(q)l})} \right\}. \tag{3.12}$$

If $l \to 0$ the Green's function $\hat{G}$ turns into $\delta(\phi_q{}' - \phi_q)$, thus we have an initial Hamiltonian bearing no tokens of coarsening. When l is finite, the resolution of the initial structure decreases as the function $\hat{G}$ gradually weakens its dependence on ϕ_q with the increase of l. The information about the initial Hamiltonian is also being lost, as the ϕ_q in Eq. (3.12) contains the factor $\exp[-h(q)l]$. This constitutes Wilson's idea of the "incomplete integration". Evidently, when $l \to \infty$ the integration procedure is complete. The information about the initial distribution is completely lost

$$\rho(\phi_q, \infty) = \frac{Z}{\sqrt{2\pi}} \exp\left\{-\frac{[\phi_q]^2}{2}\right\}, \tag{3.13}$$

and, therefore, any kind of initial distribution will be reduced to a Gaussian distribution.

Let us now compare the two procedures of coarsening given by Eqs. (3.8) and (3.9). In Eq. (3.8) the complete integration has been carried out over the modes with large q. With the increase of l the number of the integrated modes increases. At every step of such a coarsening we would obtain a new Hamiltonian expressed in the initial variables ϕ_q but with $q < \Lambda \exp(-l)$. In Eq. (3.9) we obtain a new Hamiltonian in new variables but the modes with the fixed q do not vanish. Only the role of these modes changes in the new distribution. It is quite clear that in order to diminish the influence of the modes with high momenta, we need to have the function $h(\boldsymbol{q})$ being an increasing function with the increase of q. Wilson and Kogut (1974) assumed the following expression for $h(q)$ leading to the suppression of the higher momenta modes

$$h(q) = c + 2(q/\Lambda)^2. \tag{3.14}$$

Wilson's idea of the "incomplete integration" has been demonstrated in Eq. (3.9) only for a fixed real mode ϕ_q. In reality, however, the function $\phi(\boldsymbol{x})$ is real and hence $\phi_q^* = \phi_{-q}$. This compels us to consider an equation with at least two variables $\boldsymbol{\alpha}(\boldsymbol{q})$ and $\boldsymbol{\beta}(\boldsymbol{q})$ for every fixed mode $\phi_q = \boldsymbol{\alpha}(\boldsymbol{q}) + i\boldsymbol{\beta}(\boldsymbol{q})$ with $\boldsymbol{\alpha}(\boldsymbol{q}) = \boldsymbol{\alpha}(-\boldsymbol{q})$ and $\boldsymbol{\beta}(\boldsymbol{q}) = -\boldsymbol{\beta}(-\boldsymbol{q})$. It is also necessary to generalize the consideration by introducing, instead of the equation with some fixed $\boldsymbol{q}$, an equation for a whole set of different $\boldsymbol{q}$. This generalization should be provided with the necessary summation over $\boldsymbol{q}$ in the half space by choosing one of the $\boldsymbol{q}$ components positive in order to make $\boldsymbol{\alpha}(\boldsymbol{q})$ and $\boldsymbol{\alpha}(-\boldsymbol{q})$, as well as $\boldsymbol{\beta}(\boldsymbol{q})$ and $\boldsymbol{\beta}(-\boldsymbol{q})$, independent variables. The consequent generalization requires rewriting Eq. (3.9) in terms of all $\boldsymbol{\alpha}(\boldsymbol{q})$ and $\boldsymbol{\beta}(\boldsymbol{q})$, and the summation at the right hand side over all discrete values of $\boldsymbol{q}$. After the change from discrete to continuous values of $\boldsymbol{q}$, one can obtain the following equation for ρ

$$\left(\frac{\partial \rho}{\partial l}\right)_c = \int_q h(\boldsymbol{q}) \left\{\phi_q \cdot \frac{\delta}{\delta \phi_q} + \frac{\delta^2}{\delta \phi_q \delta \phi_{-q}} + nV\right\} \rho,$$

where we introduce the notation

$$\int_q \equiv \int \left(\frac{d\boldsymbol{q}}{2\pi}\right)^d .$$

Taking into account that $\rho = \exp(-H)$ one can obtain equation for H

$$\left(\frac{\partial H}{\partial l}\right)_c = \int_q h(\boldsymbol{q}) \left\{ \phi_q \frac{\delta H}{\delta \phi_q} + \frac{\delta^2 H}{\delta \phi_q \delta \phi_{-q}} - \frac{\delta H}{\delta \phi_q} \frac{\delta H}{\delta \phi_{-q}} - nV \right\}, \quad (3.15)$$

where the subscript c at the derivative with respect to l means that this is only part of the RG transformation arising due to the procedure of coarsening.

The procedure of the "incomplete integration" can also be considered as a transformation of variables. According to Wegner (1976) this transformation can be represented in an infinitesimal form as

$$\phi_q = \phi'_q + \delta l \boldsymbol{\Psi}_q \left\{ \phi'_q \right\} . \quad (3.16)$$

In order to obtain Eq. (3.15) with the help of Eq. (3.16) one should choose the function $\boldsymbol{\Psi}_q$ in the form

$$\boldsymbol{\Psi}_q = h(\boldsymbol{q}) \left\{ \phi_q - \frac{\delta H}{\delta \phi_{-q}} \right\} . \quad (3.17)$$

The derivation of Eq. (3.15) with the use of Eqs. (3.16) and (3.17) is given in Wegner's review (1976). Here we want to mention that the equivalency of the procedures of the "incomplete integration" and the transformation of variables should not create an impression of indefiniteness of the RG transformation. After a brief glance, it may seem that the only restriction on the transformation Eq. (3.16) is the condition of the increase of suppression of modes with increasing of q. More than that, the freedom in the choice of the function Ψ_q could encourage one to construct different approximation schemes. Unfortunately, this is not the case. Notwithstanding the indefiniteness in the definition of the function Ψ_q in the transformation Eq. (3.16), there is an additional restriction, which so far has not been taken into account. This restriction occurs due to the inherent physical nature of the RG transformation sought for. The transformation should lead to the points of Kadanoff's invariance. Unluckily, this restriction, however strong it is, only manifests itself at the final stage of an RG transformation and in Wilson's scheme it cannot be enforced at the initial stage of the definition. As will be seen later, even the choice of the function $h(\boldsymbol{q})$ in the form of Eq. (3.14) leads to reasonable physical approximations solely due to the liberty left in the definition of the constant c.[1] For a number of years, the absence of a general method for the definition of Ψ_q was a very serious obstacle restricting the use of Wilson's equation in concrete calculations. An essential step forward was made in 1985 by Riedel *et al.* In this work a new procedure for eliminating the most important redundant eigenvector was introduced.[2] This redundant eigenvector reflects the indefiniteness in the scale transformation of the variables $\phi_q \to \phi_q/\sigma$.

Let us now return to the last step of Kadanoff's transformation — contraction of the system in the coordinate space or dilation in the momentum space. This step can also be considered as a transformation of variables. Unlike the previous transformation (see Eq. (3.16)) suppressing some particular modes, this transformation changes the labels of the modes, i.e. wave vectors $\boldsymbol{q}$. As we have already conserved the measurable values at the first stage, so we should preserve this condition at the final stage. In addition, one is compelled to watch for the conservation of the number of the degrees of freedom. This means that if $\boldsymbol{q}'$ is increased by λ times, the volume of the system should be decreased by λ^d times as

$$q_i' = \frac{2\pi n_i}{L_i'} = \lambda q_i = \frac{2\pi n_i \lambda}{L_i}\,; \left\{-\frac{N^{1/d}}{2} \le n_i \le \frac{N^{1/d}}{2}\right\},$$

from where $\lambda L_i' = L_i$. It is convenient to have the Jacobian of the transformation to be equal to unity. This gives the additional condition $\phi_{q'}/\sqrt{V'} = \phi_q/\sqrt{V}$, from which for the infinitesimal change of the scale $\lambda = 1 + \delta l$ one can obtain

$$\begin{aligned} V &= (1+\delta l d)V', \\ \boldsymbol{q}' &= (1+\delta l)\boldsymbol{q}, \\ \phi_q &= \left(1+\delta l \tfrac{d}{2}\right)\phi'_{q'} \\ &= \left(1+\delta l \tfrac{d}{2}\right)\phi'_q + \delta l \boldsymbol{q}\frac{\partial \phi'_q}{\partial \boldsymbol{q}}. \end{aligned} \tag{3.18}$$

[1] From the requirement of Kadanoff's invariance this constant has been obtained by Wegner (1976) in ϵ-expansion till the second order of ϵ.

[2] Redundant eigenvectors appear (see Sect. 3.3.3) due to an incompleteness of the RG transformation in this form. The procedure of their elimination would eventually complete the definition of the RG transformation.

Using Eq. (3.18) one can find the following equation describing the transformation of the Hamiltonian under the action of the dilation

$$\left(\frac{\partial H}{\partial l}\right)_d = dV\frac{\partial H}{\partial V} + \int_q \left(\frac{d}{2}\phi_q + \boldsymbol{q}\frac{\partial \phi_q}{\partial \boldsymbol{q}}\right)\cdot\frac{\delta H}{\delta \phi_q}, \tag{3.19}$$

where the subscript d at the derivative with respect to l shows that this part of the RG transformation appears due to the dilation transformation in $\boldsymbol{q}$-space. The complete Wilson's RG equation results from the sum of the transformation Eqs. (3.15) and (3.19)

$$\frac{\partial H}{\partial l} = \hat{R}\{H\} = dV\frac{\partial H}{\partial V} + \int_q \left[\left(\frac{d}{2} + h(\boldsymbol{q})\right)\phi_q + \boldsymbol{q}\frac{\partial \phi_q}{\partial \boldsymbol{q}}\right]\cdot\frac{\delta H}{\delta \phi_q}$$
$$+ \int_q h(\boldsymbol{q})\left[\frac{\delta^2 H}{\delta \phi_q \delta \phi_{-q}} - \frac{\delta H}{\delta \phi_q}\cdot\frac{\delta H}{\delta \phi_{-q}} - nV\right]. \tag{3.20}$$

The dilation transformation may initiate some questions. One of them is: why is the factor at the δl in the transformation of momenta chosen to be unity but not any other number, for instance b? The obvious answer that l is a "thing by itself, the sense of which can only be defined at the points: $l = 0$ (exact description) and $l \to \infty$ (completely coarsened system)" cannot be considered as satisfactory. The dilation procedure was done after the coarsening procedure. Therefore, at the first stage, the scale of l has been defined. The Kadanoff transformation requires a linear contraction of the lattice to be equal to the number of spins entered into a block spin along the linear size ($n^{1/d}$). This demand translated into $\boldsymbol{q}$-space means the restoration of the initial cut-off parameter Λ, which is being decreased as a result of the "incomplete integration". The transformation Eq. (3.8) shows that the cut-off parameter has decreased on the value $\delta\Lambda$, but is it possible to identify the decrease of this parameter from Eq. (3.15)? The answer to this question requires more careful consideration of the coarsening transformation. Let us simplify the problem again by restricting the field ϕ_q to be real. It is also convenient to measure momenta in the units of the cutoff parameter Λ. We are going now to find the change of the simple Hamiltonian of noninteracting fluctuations with the structure $H_0 = g_2(\boldsymbol{q}){\phi_q}^2/2 + g_0$. First of all it is necessary to notice that there are two ways which describe equivalently the evolution of the Hamiltonian undergoing the RG transformation. The first is demonstrated in Eq. (3.11), where the transformed Hamiltonian is expressed in new variables. The other way is to fix variables ϕ_q and let the vertexes of the Hamiltonian evolve with l. Thus, the evolution of the Hamiltonian is defined by the change of vertexes. Below we will use this approach and, therefore for the given example one has only two vertexes changing with l — $g_2(q, l)$ and

$g_0(l)$. For practical purposes the change of the constant g_0 is not important and the value of interest is only the part of the Hamiltonian H_0 dependent on ϕ_q. In the Hamiltonian of the general kind for the Ising model, the operator part of the noninteracting spins ($H_0 \propto \sum_i m_i^2$) is always equal to zero due to the condition $m_i^2 = 1$. Therefore, this part of the Hamiltonian does not change under the influence of Kadanoff's transformation. If one introduces a space constituted by the vertexes describing interactions of different pairs of spins, then the noninteracting Hamiltonian for the Ising model represents a fixed projection under the Kadanoff transformation. In the case of Ginzburg-Landau functional of a general kind only the quadratic part can serve as a consequent analog of the fixed projection. Let us find what kind of functions can be used for $g_2(q)$ in such a case.[3] Instead of carrying out the integration in Eq. (3.11) with the use of Eq. (3.12), one can find the variation $(\delta g_2)_c$ under the influence of the infinitesimal increase of δl from Eq. (3.9) directly. This variation should be compensated by the change of the cutoff parameter, thus realizing the second stage of the RG transformation. Hence, this change can be obtained by exchanging $\boldsymbol{q}$ with $\boldsymbol{q}/(1 - \delta l b)$ which gives $(\delta g_2)_\Lambda$. By putting $(\delta g_2)_c = (\delta g_2)_\Lambda$ one can obtain the equation

$$h(q)g_2(q)[1 - g_2(q)] = bq^2 \frac{dg_2}{dq^2}. \tag{3.21}$$

The general integral of this equation is

$$g_2(q) = \frac{Aq^{2c/b} \exp(2q^{2c/b})}{1 + Aq^{2c/b} \exp(q^{2c/b})}, \tag{3.22}$$

where A is the constant of integration.

Analysis of the solution Eq. (3.22) shows that the choice of the function $h(q)$ and the scale (factor b) defines the dependence of the vertex g_2 on momenta. Reasonable expressions for the function $g_2(q)$ can only be obtained when the ratio $c/b = k$ is an integer. The case with $k = 0$ corresponds to the short range correlation function $G(r)$ and therefore has nothing to do with the critical behavior.[4] The conventional critical behavior corresponding to the Landau theory can be seen only when $k = 1$ ($g_q \propto q^2$ when $q^2 \to 0$).

Up to now this analysis has been confined to the case of only one mode with a fixed value of q. Due to such extreme simplification it has been possible to compensate the coarsening change of $g_2(\boldsymbol{q})$ by the consequent

[3] For noninteracting fluctuations fixed projection of the RG transformations generates the so-called fixed Hamiltonian, as a consequence of the absence of the higher order terms. The value g_0 in this case is uniquely defined by $g_2(q)$(see below).

[4] As usual $G(q) \propto g_2^{-1}(q)$.

variation of the scale of q. Strictly speaking, such a consideration cannot be considered as applicable to real physical situations. In a more or less practical case the Hamiltonian H_0 should be given not as $g_2(\boldsymbol{q}){\phi_q}^2/2$ but as an integral over all q ($\int_q g_2(\boldsymbol{q}){\phi_q}^2/2$). The change of the scale in such a case will lead to the variation of the integration element. This variation compels one to introduce the change of the scale for the field ϕ_q. Generally speaking, the choice of the variation of ϕ_q governed by Eq. (3.18) is quite arbitrary. The condition for the conservation of the Jacobian of the transformation is not a physical necessity. The conservation of measurable quantities allows for the appearance of a factor in the partition function. Nonetheless, the variation of the scale for ϕ_q will be determined by the change of the q-scale and by the choice of the function $h(\boldsymbol{q})$. One can see how this happens by repeating the derivation which has led to Eq. (3.22). In this more general case, it is necessary to carry out the integration over all the modes and to take into account that the appearance of the factor $(1 + \delta lb)$ with $\boldsymbol{q}$ generates some scale change of the field ϕ_q, which can be described by the factor $(1 + \delta la)$. These changes lead to the following replacement of Eq. (3.21) by

$$h(q)g_2(q)[1 - g_2(q)] = bq^2\frac{dg_2}{dq^2} + \frac{bd}{2} - a.$$

In order to reveal critical behavior, one could confine this analysis to the limit $q \to 0$. Only those solutions are critical which vanish in this limit. Therefore, the solution

$$g_2(q) = \frac{bd}{2} - \frac{a}{c} + Aq^{2c/b}, \tag{3.23}$$

which valid at small q ($q \ll 1$) can only be critical if $c \neq 0$. The critical solution corresponds to the choice $a = bd/2$. An additional condition $c/b = 1$ also follows from Eq. (3.23) if one chooses the Landau model of critical behavior. Hence, the choice of $b = 1$, as in Eq. (3.18), defines $a = d/2$. In turn, this gives $h(0) = c = 1$

This deliberation has a quite restricted domain of applicability. We have not taken into account higher vertexes of the GLF, which, naturally, would add to the definition of the ratio $h(0)/b$. It is also necessary to choose the function $g_2(q)$ as an invariant of the RG transformation.[5] The latter can also be considered as requiring the constancy of all the factors in the power series in $\boldsymbol{q}^2$ for the function g_2. Wilson assumed the choice $g_2(q) = q^2$ and the constant $h(0)$ being a function of l. This function can only be defined

[5]This means that the RG transformation and the function $g_2(q)$ could not be independently defined.

from the invariancy condition for the function $g_2(q)$. A practical realization of such a program leads to a very complicated problem, the full solution of which is given in the Chapter 5 on the basis of a different RG formulation.

At last we have to understand what kind of Hamiltonians may satisfy the RG equation. The initial Hamiltonian in the Landau model has a nonlocal vertex $g_2(q)$ in the second power of ϕ_q and a local at the fourth. Here we say "the vertex is local" if it does not depend on momenta. This is the necessary condition to have a local contribution to the Hamiltonian in x-space. The RG transformation Eq. (3.20) generates higher order vertexes as well as nonlocality in the Hamiltonian. If one would suppose that the RG equation does not contain nonlinear terms, then the highest power in ϕ_q in the Hamiltonian is conserved and only nonlocality is generated, due to the presence of the function $h(q)$. The nonlinear term in Eq. (3.20) contributes to the generation of all powers in ϕ_q, as well as to the nonlocality of vertexes. Therefore, in the general case the Hamiltonian should have the structure

$$H = Vg_0 + \sum_k \frac{1}{2^{k-1}} \left\{ \sum_{\alpha_1} \int_{q_1} \cdots \sum_{\alpha_k} \int_{q_k} g_k^{\alpha_1 \cdots \alpha_k}(\boldsymbol{q}_1 \cdots \boldsymbol{q}_k) \right.$$

$$\left. \times\, \delta\left(\sum_{i=1}^{k} \boldsymbol{q}_i \right) \prod_{i=1}^{k} \phi_{q_i}^{\alpha_i} \right\}, \tag{3.24}$$

where the δ-functions reflect the supposed spacial homogeneity of the Hamiltonian. As is seen from Eq. (3.24) the k-order vertex is a k-rank tensor in isotopic space. For particular systems the structure of these tensors should be specified. According to the derivation of the RG equation neither of the RG operations (coarsening or dilation) affects the inherent symmetry of the system in the isotopic space. Therefore, the Hamiltonian Eq. (3.24) must conserve the symmetry of the starting Hamiltonian. In addition to this, the RG transformation conserves parity with respect to ϕ_q.

3.3.2 Linearized RG Equation

The RG transformation defined by Eq. (3.20) enables one to study the point of the Kadanoff invariance for the generalized GLF. Namely this point defines, according to Eq. (3.5), the singular behavior of thermodynamic functions. This means that at $T = T_c$ ($l \to \infty$) Eq. (3.20) should have the solution H^* defined by

$$\hat{R}\{H^*\} = 0. \tag{3.25}$$

The Hamiltonian satisfying this equation is called a "fixed point" Hamiltonian or a "fixed point" solution of the RG transformation.

In reality, the reduction of temperature down to the critical one will lead to an essential renormalization of the trial critical temperature T_{c0}. Hence, one cannot fix $T = T_c$ (unless experimentally). This means that between the solutions satisfying Eq. (3.25) at least one of the Hamiltonians should correspond to the renormalized Hamiltonian at the critical point. The particular role of the critical point and the sense of Eq. (3.25) can be cleared by the formal analysis of the dependence of H on l. When $l = 0$ there is only the initial Hamiltonian with fixed vertexes. When $l \to \infty$ there are the following possibilities:

1. The Hamiltonian approaches H^* and remains constant.

2. The Hamiltonian has at least one of vortex with modulus increasing with l.

3. All the vertexes of the Hamiltonian are finite but at least one of them does not have a definite limit when $l \to \infty$.

The last possibility is not considered in this book, because such a behavior, though very interesting from both mathematical and physical points of view, does not seem to be related to the critical fluctuations. The examples of the realization of the second possibility will be given in the following chapters.[6] Here we will only consider the first case with the genuine fixed point. This case is characterized by a large probability of the complete loss of memory in H^* about the initial distribution, i.e. about the initial vertexes at $l = 0$. The latter means, that in the range of the given symmetry of the initial Hamiltonian one can expect the universal critical behavior for a very wide class of systems. All of these systems should have the same fixed-point Hamiltonian H^*. It is also quite clear that under the influence of small deviations from the given fixed point the system's behavior also should not be affected by the initial Hamiltonian.[7] Of course, these intuitive deliberations require a more strict mathematical foundation. To provide this, one is compelled to consider the evolution of the Hamiltonian after a small deviation from the fixed point. Such considerations should also enable one to define the critical behavior in the system in the manner proposed by Kadanoff (see Eq. (3.6)). Let us write the Hamiltonian H in the form $H^* + \Delta H$. The use of the form of the operator $\hat{R}$ defined by Eq. (3.20) gives

$$\frac{\partial \Delta H}{\partial l} = \hat{L}\Delta H + \hat{L}_n \Delta H, \tag{3.26}$$

[6]They include the first-order phase transitions or "runaway" behavior.
[7]Provided this fixed point is a stable one.

where $\hat{L}$ is a linear operator, the action of which is defined as follows

$$\hat{L}\Delta H = \int_q \left[\left(\frac{d}{2} + h(\boldsymbol{q})\right)\phi_q + \boldsymbol{q}\frac{\partial \phi_q}{\partial \boldsymbol{q}} - 2h(q)\frac{\delta H^*}{\delta \phi_{-q}}\right]\frac{\delta \Delta H}{\delta \phi_q} + \int_q h(q)\frac{\delta^2 \Delta H}{\delta \phi_q \delta \phi_{-q}} + dV\frac{\partial \Delta H}{\partial V}, \qquad (3.27)$$

and $\hat{L}_n$ is a nonlinear operator defined by

$$\hat{L}_n \Delta H = -\int_q h(q)\frac{\delta \Delta H}{\delta \phi_q}\frac{\delta \Delta H}{\delta \phi_{-q}}. \qquad (3.28)$$

When the value ΔH is quite small, the nonlinear term Eq. (3.28) can be neglected in Eq. (3.26). The left linear part of Eq. (3.26) allows for the application of the traditional analysis. The deviation ΔH from the Hamiltonian H^* can be realized differently. First of all it is quite natural to study the deviation from the fixed Hamiltonian confined to the critical surface. This surface is considered in the space constituted by the whole set of vertexes g_n of the Hamiltonian and is defined by the two conditions: $T = T_c$ and the absence of the external field ($h = 0$). On the critical surface, one should always have corrections to vertexes which will inevitably vanish as $l \to \infty$. More informative is the case when the deviation moved the system off the surface. Then in the Hamiltonian $\Delta H(l)$ additional terms related to the external field or $\Delta T = T - T_c$ should also be included. Such terms will destabilize the situation and the system leaves the fixed point $H = H^*$, if one would not take measures eliminate them. This intuitive description can be formalized. Let us define the eigenvectors $\mathcal{O}_i$ of the linear RG operator Eq. (3.27)

$$\hat{L}\mathcal{O}_i = \lambda_i \mathcal{O}_i, \qquad (3.29)$$

where λ_i are eigenvalues. Let us also assume that the vectors $\mathcal{O}_i$ form a full set. In such a case the Hamiltonian $H(l)$ can be represented as

$$H(l) = H^* + \Delta H(l) = H^* + \sum_i \mu_i(l)\mathcal{O}_i, \qquad (3.30)$$

where the factors $\mu_i(l)$, usually called as "scaling fields", satisfy the equations

$$\frac{\partial \mu_i(l)}{\partial l} = \lambda_i \mu_i(l). \qquad (3.31)$$

With the help of Eq. (3.31) the expression for the Hamiltonian $H(l)$ can be rewritten in a more convenient form for further analysis

$$H(l) = H^* + \sum_i \mu_i e^{\lambda_i l}\mathcal{O}_i. \qquad (3.32)$$

From this equation one can see that if some of the eigenvalues have $\text{Re}\lambda_i > 0$ the system could reach the fixed point only when $\mu_i = 0$. Nonetheless, one should not consider all of the eigenvectors $\mathcal{O}_i$ with $\text{Re}\lambda_i > 0$ to be that important and require $\mu_i = 0$ in order to reach critical point. There is one eigenvector which plays a special role, — this is the eigenvector $\mathcal{O}_0 = V$ with the eigenvalue $\lambda_0 = d$. Under the RG transformation the volume contracts and is related to the volume of the system $V(0)$ as follows $V = \exp(-dl)V(0)$. The scaling field corresponding to this eigenvector increases with l as $\mu_0(l) = \mu_0 \exp(dl)$. As a result the product $\mu_0(l)V = \mu_0 V(0)$ remains unchanged, and this constant does not have any relation to the critical behavior. The physical sense of it can be revealed if one recalls that the free energy F, defined as a logarithm of the partition function is an invariant of the RG transformation. Therefore, the constant in the Hamiltonian proportional to the volume gives contribution to the regular part of the free energy. This fact enabled Wilson and other authors to treat this term independently from other terms in the Hamiltonian. However, such a treatment should be made quite carefully, as the terms with the second variational derivatives create contributions proportional to the volume. The special role of the eigenvector $\mathcal{O}_0$ is also emphasized by its independence of the particular form of the Hamiltonian H^*. Whatever the Hamiltonian is, $\mathcal{O}_0$ is always equal to V, and its eigenvalue is always equal to d.

Thus, if one eliminates the eigenvector $\mathcal{O}_0$ from consideration, then Eq. (3.32) allows us to convey the general analysis of the behavior of the system in the vicinity of the critical point. Before starting this analysis let us first make some comments concerning the eigenvalues λ_i. It is quite plausible that all λ_i should be real for the eigenvectors having physical sense. This statement up to now has not been rigorously proved. One can find discussion of this problem and the reasoning done in the framework of the perturbation theory to the second order of ϵ in the paper by Wallace and Zia (1974). The consideration, not restricted by the perturbation theory, is given in the Appendix (see also Lisyansky *et al* 1992), where the rigorous proof is made in the framework of the so-called local approximation. In the following we assume this statement as proven, or at least one may consider that we are interested only in the case of real λ_i. Then, if all $\lambda_i < 0$, the Hamiltonian H^* does not describe critical behavior. In this case one is not compelled to fix any of the physical parameters (temperature, pressure, density of fields and the like) to reach the fixed point. In other words, a fixed point of such a kind describes some inherent symmetry property of the chosen Hamiltonian with respect to the RG transformation. Another case is realized when among the negative eigenvalues there is a number of positive λ_i. Then, the fixed point can be reached only if some external parameters are fixed so as to nullify the consequent values of μ_i.

Let us consider the following example. Suppose that there is only one eigenvector $\mathcal{O}_1$ conserving the symmetry of the Hamiltonian with $\lambda_1 > 0$.[8] Let us recall that the initial Hamiltonian is defined by $H = F/T \equiv \beta F$. Then the fixed-point Hamiltonian can be also represented as a series $H^* = \sum_i \mu_i \mathcal{O}_i$ along with the Hamiltonian $H = \sum_i \beta\mu_i^0 \mathcal{O}_i$. This series leads to the following representation for ΔH

$$\Delta H = \sum_i \mu_i \mathcal{O}_i = \sum_i (\beta\mu_i^0 - \mu_i^*)\mathcal{O}_i.$$

If we also assume that the trial Hamiltonian is only a little different from the fixed one, then one is compelled to put $\mu_1 = 0$. The latter can be done by fixing temperature $\beta = \beta_c$, where β_c is defined with the help of

$$\tau \propto \mu_1 = \beta\mu_1^0 - \mu_1^* = \mu_1^0(\beta - \beta_c),$$

i.e. $\beta_c = \mu_1^*/\mu_1^0$.

Thus, in this example the condition for criticality is $\beta = \beta_c$ and the considered situation corresponds to conventional second-order phase transition. The situation with polycritical behavior corresponds to more than one essential eigenvector. As an instance, the tricritical point should have two eigenvectors, conserving symmetry of the Hamiltonian, with $\lambda_{1,2} > 0$. The required physical parameter to nullify μ_1 and μ_2 additional to the temperature could be the density of matter, pressure, magnetic field, and so on, depending on the particular physical situation under the consideration. The eigenvectors with positive eigenvalues are called relevant, due to their crucial role in the determination of the critical behavior. The influence of the eigenvectors with $\lambda_i < 0$, is vanishing in the limit $l \to \infty$ and they are called irrelevant. Besides the relevant and irrelevant eigenvectors there can also be eigenvectors with $\lambda_i = 0$, which naturally should be called marginal. If $\mathcal{O}_i$ is a marginal eigenvector, then in the linear approximation the Hamiltonian $H^* + \mu_i \mathcal{O}_i$ will also be a fixed-point Hamiltonian for any value of μ_i. Such a situation is realized for the Bakster's eight-vertex model (1982).

In the general case, nonlinear terms in Eq. (3.26) could destroy this property. And only the study of the nonlinear contributions could show whether the parameter μ_i has an arbitrary value at the fixed point or not. If it happens that μ_i has an arbitrary value then the universality of the critical behavior will be lost and critical parameters of the system (critical exponents, universal ratio of amplitudes) are defined by the particular form of the initial Hamiltonian.

[8] The conservation of the symmetry of the initial Hamiltonian does not eliminate the spontaneous break of the symmetry in the system.

In summary of this discussion we should conclude that there are four types of eigenvectors of the linear RG operator:

1. Special eigenvector (volume) with $\lambda_0 = d$.
2. Relevant eigenvectors with $\lambda_i > 0$.
3. Marginal eigenvectors with $\lambda_i = 0$.
4. Irrelevant eigenvectors with $\lambda_i < 0$.

The condition of the critical behavior is $\mu_i = 0$ for all relevant eigenvectors.

3.3.3 Redundant Eigenvectors

Unfortunately, the RG transformation Eq. (3.20) also admits the existence of eigenvectors which do not have physical interpretations, some of which have been mentioned earlier. These eigenvectors are called, after Wegner (1976), "redundant". The discussion given in Sect. 3.3.1 may attract the reader's attention to some weak points and drawbacks of the RG equation derivation. The problem of redundant eigenvectors accounts for the drawbacks which were recognized long ago by Bell and Wilson (1975). In short, one could say, that any kind of renormalization procedure consisting only of the two steps, coarsening and dilation in $\boldsymbol{q}$-space, has an additional liberty which must be restrained by some additional physical considerations. The theory of critical behavior in a compact form (free from any reflections outside the framework of its formulation) is considered in the Chapter 5. Here, we only introduce the reader into the circle of the traditional terminology. Bellow, as an instance, a three-parametric class of redundant eigenvectors is built. Eq. (3.29), defining eigenvectors, will be for convenience represented in a somewhat different form. Let us return to the value $\rho = \exp(-H)$ and to the consequent equation defining its evolution with the change of l

$$\frac{\partial \rho}{\partial l} = \left(\frac{\partial \rho}{\partial l}\right)_c + \left(\frac{\partial \rho}{\partial l}\right)_d = \hat{\Pi}\rho, \tag{3.33}$$

where

$$\hat{\Pi} = dV\frac{\partial}{\partial V} + \int_q \left\{ \left(\frac{d}{2}\phi_q + \boldsymbol{q}\frac{\partial \phi_q}{\partial \boldsymbol{q}}\right)\frac{\delta}{\delta \phi_q} + h(\boldsymbol{q})\frac{\delta}{\delta \phi_q}\left(\phi_q + \frac{\delta}{\delta \phi_{-q}}\right)\right\}. \tag{3.34}$$

The fixed point and the linear RG operator are now defined by

$$\hat{\Pi}e^{-H^*} = 0 \text{ , and } \hat{L} = e^{H^*}\hat{\Pi}e^{-H^*}. \tag{3.35}$$

From these equations one can derive the following equation for the eigenvectors

$$e^{H^*}\hat{\Pi}e^{-H^*}\mathcal{O}_i = \lambda_i\mathcal{O}_i. \tag{3.36}$$

Let us make some transformation of variables in the fixed-point Hamiltonian H^* ($H^* = H^*_{\xi=0}$). In the differential representation this transformation is

$$\frac{\partial e^{-H^*_\xi}}{\partial \xi} = \hat{U}e^{-H^*_\xi}, \tag{3.37}$$

where $\hat{U}$ is the transformation generator. Some of the eigenvectors $\mathcal{O}_\lambda$ can be represented in the form $\mathcal{O}_\lambda = e^{H^*}\hat{U}e^{-H^*}$. For these eigenvectors Eq. (3.36) (accounting for Eq. (3.35) defining the fixed point Hamiltonian) can be rewritten as

$$e^{H^*}[\hat{\Pi}\hat{U}]_-e^{-H^*} = \lambda e^{H^*}\hat{U}e^{-H^*}, \tag{3.38}$$

where $[\cdots]_-$ means the commutator.

Let us restrict the consideration only by the vectors $\mathcal{O}_\lambda$ having the same structure as the Hamiltonian Eq. (3.24)

$$\mathcal{O}_\lambda = V\mathcal{O}_{\lambda,0} + \sum_k \left\{ \sum_{\alpha_1} \int_{q_1} \cdots \sum_{\alpha_k} \int_{q_k} \mathcal{O}^{\alpha_1 \cdots \alpha_k}_{\lambda,k}(\boldsymbol{q}_1 \cdots \boldsymbol{q}_k) \right.$$
$$\left. \times\, \delta\left(\sum_{i=1}^k \boldsymbol{q}_i\right) \prod_{i=1}^k \phi^{\alpha_i}_{q_i} \right\}, \tag{3.39}$$

where $\mathcal{O}^{\alpha_1 \cdots \alpha_k}_{\lambda,k}(\boldsymbol{q}_1 \cdots \boldsymbol{q}_k)$ are "c-number" functions. The vectors represented by Eq. (3.39) have an additional property which can be obtained as a result of the direct action of the operator $\hat{\Pi}$ defined by Eq. (3.34)

$$(d-\lambda)\mathcal{O}_{\lambda,0} + \sum_\alpha \int_q h(\boldsymbol{q})\mathcal{O}^{\alpha\alpha}_{\lambda,2}(\boldsymbol{q}) = 0. \tag{3.40}$$

The procedure of calculations of the vectors $\mathcal{O}_\lambda$ can be essentially simplified with the help of Eq. (3.40). This equation relates the term proportional to the volume in Eq. (3.39) with the $\mathcal{O}_{\lambda,2}$ amplitudes. The complications in calculations originate from the noncommutability of the two operations: the partial derivative with respect to the volume and the second variational derivatives. The latter can be verified by the direct application of the operator Eq. (3.34) to the vector Eq. (3.39). The second variational derivative in the operator $\hat{\Pi}$ produces terms proportional to volume when it is applied to the square in $\boldsymbol{\phi}_q$ part of the vector having the form given by

Eq. (3.39). In order not to complicate calculations, it is possible with the help of Eq. (3.40) to divide the terms in Eq. (3.38) into those proportional to volume and consider the remaining terms as if $V = 0$. Such a formal approach substantially simplifies all calculations. Applying this method to Eq. (3.38) we arrive at

$$[\hat{\Pi}\hat{U}]_- \Big|_{V=0} + \hat{B}\hat{\Pi}\Big|_{V=0} = \lambda\,\hat{U}\Big|_{V=0}, \tag{3.41}$$

where $\hat{B}$ is a quite arbitrary operator. Let us narrow the class of the solutions of Eq. (3.41) by taking $\hat{B} = 0$ and restricting consideration to only those transformations which conserve the partition function. The latter restriction is not enforced by the physical considerations, because Eq. (3.37) is completely formal and has no relation to measurable values. Hence, this restriction can be considered as an artificial requirement narrowing the class of possible solutions. For this class the operator $\hat{U}$ can be represented in the form

$$\hat{U} = \int_q \frac{\delta \hat{\boldsymbol{U}}_q}{\delta \boldsymbol{\phi}_q},$$

where $\hat{\boldsymbol{U}}_q$ is some arbitrary n-component vector operator. The eigenvectors $\mathcal{O}_\lambda$ with the help of this operators can be transformed to

$$\mathcal{O}_\lambda = -\int_q \left(\boldsymbol{\Psi}_q \cdot \frac{\delta H^*}{\delta \boldsymbol{\phi}_q} - \frac{\delta \boldsymbol{\Psi}_q}{\delta \boldsymbol{\phi}_q} \right), \tag{3.42}$$

where

$$\boldsymbol{\Psi}_q = e^{H^*} \hat{\boldsymbol{U}}_q e^{-H^*}. \tag{3.43}$$

The eigenvectors defined by Eqs. (3.42) and (3.43) were denoted by Wegner (1976) as redundant. As is seen from the previous consideration, the class of redundant eigenvectors is essentially wider. Nonetheless, even the eigenvectors represented in such a form generate a very wide class. Here, as an example we will define some of the redundant vectors (which can be simply interpreted) in an explicit form. Let us represent the operator $\hat{\boldsymbol{U}}_q$ in the simplest case with the help of three arbitrary functions $U_{1,q}$, $U_{2,q}$, and $\boldsymbol{U}_{3,q}$ as

$$\hat{\boldsymbol{U}}_q = U_{1,q}\boldsymbol{\phi}_q + U_{2,q}\frac{\delta}{\delta \boldsymbol{\phi}_{-q}} + \boldsymbol{U}_{3,q}. \tag{3.44}$$

Using Eq. (3.41) we obtain the following equations for these functions

$$\begin{aligned}\lambda U_{1,q} + \boldsymbol{q}\frac{\partial U_{1,q}}{\partial \boldsymbol{q}} &= 0, \\ (\lambda + 2h(\boldsymbol{q}))U_{2,q} + \boldsymbol{q}\frac{\partial U_{2,q}}{\partial \boldsymbol{q}} &= 2h(\boldsymbol{q}), \\ (\lambda + d/2 + h(\boldsymbol{q}))\boldsymbol{U}_{3,q} + \boldsymbol{q}\frac{\partial \boldsymbol{U}_{3,q}}{\partial \boldsymbol{q}} &= 0.\end{aligned} \tag{3.45}$$

This system of equations has the following solutions

$$\begin{aligned}U_{1,q} &= \tfrac{c_1}{q^\lambda}, \\ U_{2,q} &= \tfrac{c_1}{q^\lambda} + \tfrac{c_2 \exp(-2q^2)}{q^{\lambda+2h(0)}}, \\ \boldsymbol{U}_{3,q} &= \frac{\boldsymbol{c}_3 \exp(-q^2)}{q^{(d+2\lambda)/2+h(0)}},\end{aligned} \tag{3.46}$$

where c_1, c_2, and $\boldsymbol{c}_3$ are arbitrary constants. When $d > \lambda > d - 2h(0)$, the constants c_2 and $\boldsymbol{c}_3$ are equal to zero. For the case when $d - 2h(0) > \lambda > d/2 - h(0)$ only the vector constant $\boldsymbol{c}_3$ is equal to zero. Finally, when $d/2 - h(0) > \lambda$ all three constants can be different from zero. Apart from the solution defined by Eq. (3.46) there are also solutions having a δ-like singularity. If one inserts, into Eq. (3.45) $U_i = c_i\delta(\boldsymbol{q})$, then the following relations should be satisfied

$$\begin{aligned}(\lambda - d)c_1 &= 0, \\ (\lambda + 2h(0) - d)c_2 &= 2h(0)c_1, \\ (\lambda - d/2 + h(0))\boldsymbol{c}_3 &= 0.\end{aligned} \tag{3.47}$$

These equations give the following relations: when $\lambda = d$ $c_1 \neq 0$, $c_2 = c_1$, and $\boldsymbol{c}_3 = 0$; when $\lambda = d - 2h(0)$ $c_1 = \boldsymbol{c}_3 = 0$, and $c_2 \neq 0$; when $\lambda = d/2 - h(0)$ $\boldsymbol{c}_3 \neq 0$ $(\lambda \neq 0)$. Let us introduce notations for eigenvectors $\mathcal{O}_{\lambda_i}$ $(i = 1, 2, 3)$, where $\lambda_1 = d$, $\lambda_2 = d - 2h(0)$, and $\lambda_3 = d/2 - h(0)$. The functions $\boldsymbol{\Psi}_{i,q}$ can be obtained from Eq. (3.43) as

$$\begin{aligned} \boldsymbol{\Psi}_{1,q} &= c_1\delta(\boldsymbol{q})\left(\phi_q - \frac{\delta H^*}{\delta\phi_q}\right), \\ \boldsymbol{\Psi}_{2,q} &= c_2\delta(\boldsymbol{q})\frac{\delta H^*}{\delta\phi_q}, \\ \boldsymbol{\Psi}_{3,q} &= c_3\delta(\boldsymbol{q}). \end{aligned}$$

The eigenvectors $\mathcal{O}_{\lambda_i}$ have the following expressions

$$\begin{aligned} \mathcal{O}_{\lambda_1} &= \left.\frac{\delta^2 H^*}{\delta\phi_0\delta\phi_0}\right|_{V=0} - \frac{\delta H^*}{\delta\phi_0}\frac{\delta H^*}{\delta\phi_0} + \phi_0\frac{\delta H^*}{\delta\phi_0} + V\mathcal{O}_{\lambda_1,0}, \\ \mathcal{O}_{\lambda_2} &= \left.\frac{\delta^2 H^*}{\delta\phi_0\delta\phi_0}\right|_{V=0} - \frac{\delta H^*}{\delta\phi_0}\frac{\delta H^*}{\delta\phi_0} + V\mathcal{O}_{\lambda_2,0}, \\ \mathcal{O}_{\lambda_3} &= \frac{\delta H^*}{\delta\phi_0}, \end{aligned} \tag{3.48}$$

where the amplitudes $\mathcal{O}_{\lambda_1,0}$, $\mathcal{O}_{\lambda_2,0}$ should be defined with the help of Eq. (3.40). The eigenvectors in Eq. (3.48) have been obtained with the assumption that the fixed-point Hamiltonian is an even functional of ϕ_q (which is usually the case). We should also add that the definitions in Eq. (3.48) are made up to some arbitrary constants which could not be defined. It is easy to show that for the eigenvector $\mathcal{O}_{\lambda_1}$ such a constant is practically always equal to zero. Really, from Eq. (3.40) one can obtain an additional condition when $\lambda_1 = d$,

$$\begin{aligned} c_1 &\left\{ \frac{3}{2}\sum_{\alpha,\beta}\int_q h(\boldsymbol{q})\, g_4^{*\alpha\alpha\beta\beta}(0,0,\boldsymbol{q},-\boldsymbol{q}) \right. \\ &\left. + \sum_{\alpha,\beta} h(0)\left[\frac{g_2^{*\alpha\alpha}(0)}{n} - g_2^{*\alpha\beta}(0)g_2^{*\beta\alpha}(0)\right]\right\} = 0, \end{aligned} \tag{3.49}$$

where g_2^* and g_4^* are the second and the fourth vertexes of the fixed Hamiltonian. If $c_1 \neq 0$, then Eq. (3.49) relates the constants of the fixed Hamiltonian. This condition is very restrictive and it is difficult to imagine a Hamiltonian which could satisfy it. At least, one can verify in the ϵ- expansion that this condition cannot be satisfied for the conventional Landau model. Therefore, c_1 should be equal to zero and the eigenvector $\mathcal{O}_{\lambda_1}$ is reduced to V, i.e. to the special eigenvector. The vectors $\mathcal{O}_{\lambda_2}$ and $\mathcal{O}_{\lambda_3}$ do not contain such restrictive conditions. They have the eigenvalues $\lambda_i > 0$ but, nonetheless, they do not contribute to the critical behavior. Therefore, they should be considered as redundant. Some of the properties of the vector $\mathcal{O}_{\lambda_3}$ were studied by Wegner in detail (1976).

It is necessary to mention that when the fixed Hamiltonian is even with respect to ϕ_q then the eigenvectors of the linear RG can be classified by parity. In this case the first two terms in Eq. (3.44) generate even vectors and $U_{3,q}$ generates odd vectors. Let us now consider a very important class of marginal redundant vectors, i.e. vectors with $\lambda = 0$. They generate the transformation of H^* into H^*_ξ (see Eq. (3.37)), where H^*_ξ is also a fixed point Hamiltonian. From Eq. (3.37) it follows that $\exp(-H^*_\xi) = \exp(\xi \hat{U}) \exp(-H^*)$, and, as a result of the commutability of $\hat{\Pi}$ and $\hat{U}$, we have $\hat{\Pi} \exp(-H^*_\xi) = 0$. Thus any of the operators of such a kind ($[\hat{\Pi}\hat{U}]_- = 0$), generates from the given H^* a one parametric set of the fixed-point Hamiltonians H^*_ξ, i.e. a line of the fixed-point Hamiltonians. Using Eqs. (3.45) and (3.46), by taking $c_1 \neq 0$, $c_2 = 0$ and *vice versa*, one finds the following two vectors ($\mathcal{O}_{01,2} = -\partial H^*_\xi / \partial \xi$)

$$\begin{aligned} \hat{U}_1 &= \int_q \frac{\delta}{\delta \phi_q} \left(\phi_q + \frac{\delta}{\delta \phi_{-q}} \right), \\ \mathcal{O}_{01} &= \int_q \left[\frac{\delta^2 H^*}{\delta \phi_q \delta \phi_{-q}} - \frac{\delta H^*}{\delta \phi_q} \cdot \frac{\delta H^*}{\delta \phi_{-q}} + \phi_q \cdot \frac{\delta H^*}{\delta \phi_q} \right], \end{aligned} \tag{3.50}$$

and

$$\begin{aligned} \hat{U}_2 &= \int_q e^{-2q^2} q^{-2h(0)} \frac{\delta^2}{\delta \phi_q \delta \phi_{-q}}, \\ \mathcal{O}_{02} &= \int_q e^{-2q^2} q^{-2h(0)} \left[\frac{\delta^2 H^*}{\delta \phi_q \delta \phi_{-q}} - \frac{\delta H^*}{\delta \phi_q} \cdot \frac{\delta H^*}{\delta \phi_{-q}} \right]. \end{aligned} \tag{3.51}$$

The vector $\mathcal{O}_{01}$ was studied in detail by Riedel *et al* (1985). To eliminate this vector they developed an original procedure having the following foundation. In order to have a fixed point in Eq. (3.20) the function $h(0,l)$ should be appropriately chosen. Bell and Wilson (1975) related this problem to the invariance of the Hamiltonian with respect to the transformation $\phi_q \to e^\xi \phi_q$. The resulting transformation can be written as

$$\exp(-H_\xi\{e^\xi \phi_q\}) = C(\xi) \int_{[\phi'_q]} \exp\left\{ -\frac{|\phi_q e^\xi - \phi'_q|^2}{2(e^{2\xi} - 1)} - H\{\phi'_q\} \right\}, \tag{3.52}$$

where

$$C(\xi) = \exp\left\{ -\frac{n}{2} V \ln[2\pi(e^{2\xi} - 1)] \right\}.$$

This transformation generates the eigenvector $\mathcal{O}_{01}$. Let us consider the vector $\mathcal{O}_{01}(l) = -\partial H_\xi(l)/\partial\xi$ ($\mathcal{O}_{01}(\infty) \equiv \mathcal{O}_{01}$). This vector can be represented as

$$\mathcal{O}_{01}(l) = \sum_i \chi_i(l)\mathcal{O}_i$$

with the help of the eigenvectors. Let us also impose an additional condition. The deviation from the fixed point should be always perpendicular to the line constituted by the transformation. This means the absence of the projection on the direction of the vector $\mathcal{O}_{01}$. In order to fulfill this condition one needs the orthogonality of the vectors $\mathcal{O}_{01}(l)$ and $\partial H(l)/\partial l$. Taking into account Eq. (3.30) we obtain the orthogonality condition in the form[9]

$$\sum_i \chi_i(l)\frac{\partial\mu_i(l)}{\partial l} = 0. \tag{3.53}$$

Thus, by defining the function $h(0,l)$ from Eq. (3.53), Riedel *et al* (1985) eliminated the redundant eigenvector $\mathcal{O}_{01}$ from the series for the Hamiltonian $H(l)$. As is seen, there is an infinite number of redundant eigenvectors. Hence, the question how to deal with other redundant eigenvectors cannot be disregarded. The additional requirement of analyticity in the limit $q \to 0$ may partially help. At least this requirement excludes eigenvectors like the vector $\mathcal{O}_{02}$. Nonetheless, it is impossible to imagine the elimination of all redundant eigenvectors, which do not have such a trivial representation for $\hat{U}_q$ as given by Eq. (3.44). Unfortunately in this RG scheme one does not have any more free functions which could be used for the elimination of other redundant vectors. Therefore, in such a calculational procedure there is always some ambiguity left. Nonetheless, the results of computer simulations by Riedel *et al* (1985) showed that the accuracy of the obtained critical exponents is quite high.

3.3.4 Scaling Properties and Critical Exponents

The definition of the critical exponents in the RG approach can be reduced to the deliberation of the scaling properties of measurable quantities in the vicinity of the fixed point. The density of the free energy $\mathcal{F} = F/V(0)$ has the simplest scaling properties. As the partition function Z and, therefore, the free energy $F = -T\ln Z$ are invariants of the RG transformation, we have

$$F = \mathcal{F}(0)V(0) = \mathcal{F}(l)V(l),$$

---

[9] Strictly speaking this procedure requires definition of the Hilbert space and the scalar product, which so far has not been done due to the mathematical difficulties.

from which

$$\mathcal{F} = e^{dl}\mathcal{F}(l).$$

On the other hand, in the vicinity of the fixed point, the Hamiltonian has the form $H = H^* + \Delta H\{\mu_i e^{\lambda_i l}\}$. Hence, for the free energy density one can obtain the following scaling relation

$$\mathcal{F}\{\mu_i\} = e^{-dl}\mathcal{F}\{\mu_i e^{\lambda_i l}\}.$$

If we explicitly extract the regular part related to the term $\mu_0(l)V(l)$ in the Hamiltonian (see Sect. 3.3.2), we arrive at

$$\mathcal{F} = \mu_0 + e^{-dl}\mathcal{F}_s\{\mu_i e^{\lambda_i l}\}. \tag{3.54}$$

Spatially Homogeneous Perturbations

Let us at first consider such deviations from the fixed point which conserve the spatial homogeneity of the system. We will add to the Hamiltonian H^* two terms containing the relevant eigenvectors $\mathcal{O}_1$, $\mathcal{O}_h$ and a term with some irrelevant eigenvector $\mathcal{O}_i$. As will be seen from the following, the generalization for the case of any number of irrelevant eigenvectors is trivial. The vector $\mathcal{O}_1$ is the symmetry conserving eigenvector related to the change of temperature and therefore, the scaling field corresponding to it can be taken as $\mu_1 = \tau$. The vector $\mathcal{O}_h$ is an odd n-component isotopic eigenvector related to the external field h. Hence, without any restriction in generality, we can assume the corresponding scaling field to be equal to h. The equalities $\mu_1 = \tau$ and $\mu_h = h$ can be written without any proportionality factors, as there are no conditions on the normalization of the eigenvectors. Thus, the factor in these equalities will always be referred to the corresponding eigenvectors. Therefore, the Hamiltonian can be written as

$$H = H^* + \tau\mathcal{O}_1 + h\mathcal{O}_h + \mu_i\mathcal{O}_i. \tag{3.55}$$

Using Eq. (3.54) one can rewrite the singular part of the density of the free energy as

$$\mathcal{F}_s(\tau, h, \mu_i) = e^{-dl}\mathcal{F}_s(\tau e^{\lambda_1 l}, h e^{\lambda_h l}, \mu_i e^{\lambda_i l}). \tag{3.56}$$

Let us take $h = 0$ and choose l to satisfy the equality $|\tau|e^{\lambda_1 l} = 1$. We will also suppose that the value $|\tau|$ is quite small, in order that the deviation from the fixed point be described only by the linear part of the RG equation. As a result, from Eq. (3.56) one obtains

$$\mathcal{F}_s = |\tau|^{2-\alpha}\mathcal{F}_s\left(\pm 1, 0, \frac{\mu_i}{|\tau|^{\Delta_i}}\right), \tag{3.57}$$

where

$$\frac{d}{\lambda_1} = 2 - \alpha \; ; \; \Delta_i = \frac{\lambda_i}{\lambda_1} \; . \tag{3.58}$$

If there is a finite value for $\mathcal{F}_s(\pm 1, 0, 0)$, then when $\tau \to 0$, taking also into account that $\Delta_i < 0$,[10] one can obtain the main contribution into the specific heat ($C = -T\partial^2(T\mathcal{F})/\partial T^2$) in the form

$$C_s = |\tau|^{-\alpha} \mathcal{F}_s(\pm 1, 0, 0).$$

From this equation one can see that introduced in Eqs. (3.57) and (3.58) value α is the critical exponent for the specific heat.

In the general case, when $h \neq 0$, the relation

$$\mathcal{F}_s = |\tau|^{2-\alpha} \mathcal{F}_s \left(\pm 1, \frac{h}{|\tau|^{\Delta}}, \frac{\mu_i}{|\tau|^{\Delta_i}} \right) , \tag{3.59}$$

where $\Delta = \lambda_h / \lambda_1$ ($\Delta > 0$), follows from the scaling relation Eq. (3.56).

In those cases, when the value $\mathcal{F}_s$ can be expanded in the power series over μ_i, Eq. (3.59) leads to the following addition to the main asymptote when $\tau \to 0$

$$\mathcal{F}_s = |\tau|^{2-\alpha} f_{\pm} \left(\frac{h}{|\tau|^{\Delta}} \right) + \mu_i |\tau|^{2-\alpha-\Delta_i} f_{i\pm} \left(\frac{h}{|\tau|^{\Delta}} \right) , \tag{3.60}$$

where

$$f_{\pm}(x) = \mathcal{F}_s(\pm 1, x, 0) \; , \; f_{i\pm}(x) = \left. \frac{\partial \mathcal{F}_s(\pm 1, x, \mu_i)}{\partial \mu_i} \right|_{\mu_i = 0} . \tag{3.61}$$

If we restrict the consideration only by the first term in Eq. (3.60), the equation of state for the magnetization in the system can be obtained as

$$m_s = \frac{1}{V} \langle \mathcal{O}_h \rangle = \frac{\partial \mathcal{F}_s}{\partial h} = |\tau|^{\beta} f'_{\pm} \left(\frac{h}{|\tau|^{\Delta}} \right) , \tag{3.62}$$

where $\beta = 2 - \alpha - \Delta$ and $f'_{\pm} = \partial f_{\pm} / \partial x$.

In the same manner one can obtain expressions for any other mean value of the eigenvectors of the linear RG as

$$\frac{1}{V} \langle \mathcal{O}_i \rangle = \frac{\partial \mathcal{F}_s}{\partial \mu_i} = |\tau|^{\beta_i} f_{i\pm} \left(\frac{h}{|\tau|^{\Delta}} \right) , \tag{3.63}$$

[10] This is the consequence resulting from the irrelevant nature of the vector $\mathcal{O}_i$ (see Sect. 3.3.2).

where $\beta_i = 2 - \alpha - \Delta_i$.

The singular part of the homogeneous susceptibility χ_s can be easily obtained from Eq. (3.62) by the simple differentiation with respect to h

$$\chi_s = \frac{\partial m_s}{\partial h} = |\tau|^\gamma f''_\pm \left(\frac{h}{|\tau|^\Delta} \right), \tag{3.64}$$

where $\gamma = 2\Delta - 2 + \alpha$.

This procedure also leads to the following expression for the generalized susceptibilities related to the irrelevant eigenvectors $\mathcal{O}_i$ and $\mathcal{O}_j$

$$\chi_{ij} = \frac{\partial^2 \mathcal{F}_s}{\partial \mu_i \partial \mu_j} = |\tau|^{\gamma_{ij}} f_{ij\pm} \left(\frac{h}{|\tau|^\Delta} \right), \tag{3.65}$$

where f_{ij} is the second derivative with respect to the consequent scaling fields taken at $\mu_i = \mu_j = 0$, and $\gamma_{ij} = \Delta_i + \Delta_j - 2 + \alpha$.

In the expansion Eq. (3.60) we have assumed that the derivatives of Eq. (3.61) exist. Strictly speaking this is not always the case. When these derivatives are absent, all scaling properties may be substantially different. Here is an example, illustrating such a possibility. When $d > 4$ the ϕ_q dependent part of the fixed Hamiltonian at $q \ll 1$ has the form (see Sect. 3.3.5) $H^* = \int_q q^2 {\phi_q}^2/2$. The linear RG gives for this Hamiltonian the following two, symmetry conserving, eigenvectors $\mathcal{O}_1$ and $\mathcal{O}_2$

$$\mathcal{O}_1 = \int_q |\phi_q|^2 + \text{const}, \tag{3.66}$$

$$\mathcal{O}_2 = \int_{q_1 \cdots q_4} \delta \left(\sum_{i=1}^4 q_i \right) (\phi_{q_1} \cdot \phi_{q_2})(\phi_{q_3} \cdot \phi_{q_4}) - K \int_q |\phi_q|^2 + \text{const}, \tag{3.67}$$

The vector $\mathcal{O}_1$ is a relevant eigenvector with $\lambda_1 = 2$. The vector $\mathcal{O}_2$ is irrelevant, because it has $\lambda_2 = 4 - d < 0$ for $d > 4$. When $\tau > 0$, one can easily calculate the free energy for the Hamiltonian $H_0 = H^* + \tau\mathcal{O}_1$, because H_0 is a Gaussian positively defined form. Below the critical temperature ($\tau < 0$) part of the modes have negative energy and therefore, they will give divergent contribution to the partition function. To eliminate the divergence one has to add to the Hamiltonian the term $\mu_2\mathcal{O}_2$ with $\mu_2 > 0$. In this case the expansion of Eq. (3.59) is not valid, not only because of the indefiniteness of $f_i(x)$ but also due to the absence of $f(x)$ when $\mu_2 = 0$. Hence, one is compelled to use Eq. (3.59) with $\Delta = (d+2)/2$, $\Delta_2 = -(d-4)/2$ and $\alpha = 2 - d/2$, i.e.

$$\mathcal{F}_s = |\tau|^{d/2} \mathcal{F}_s \left(\pm 1 \, , \, \frac{h}{|\tau|^{(d+4)/2}} \, , \, \mu_2 |\tau|^{(d-4)/2} \right).$$

From this relation, after taking the derivative with respect to h, we arrive at the following equation of state

$$m_s = |\tau|^{(d-2)/2} f'_{\pm} \left(\frac{h}{|\tau|^{(d+4)/2}} \, , \, \mu_2 |\tau|^{(d-4)/2} \right) . \tag{3.68}$$

Unfortunately, this equation is not sufficient to define the exponent β, as it was the case of Eq. (3.62). Now, when $\mu_2 \to 0$ the magnetization, as well as the partition function are divergent if $\tau < 0$. Thus, in order to find the exponent β, one has to know behavior of m_s when $\mu_2 \to 0$. Luckily, such behavior can be easily found when $d > 4$. In this case the problem has an asymptotically exact solution (at $V \to \infty$.) This solution is realized as the Gaussian form of the general type in which coefficients are defined with the help of the thermodynamic variational principle. Now, the factor K in Eq. (3.67) exactly compensates the renormalization of critical temperature originating from the ϕ^4-term. As a result, when $h = 0$ one has $m_s \propto \mu^{-1/2}$. Therefore, with the help of Eq. (3.68) we obtain the relations

$$m_s \propto |\tau|^{(d-2)/4} \mu_2^{-1/2} |\tau|^{-(d-4)/4} \propto |\tau|^{1/2},$$

i.e. $\beta = 1/2$. The dependence of μ_s on the magnetic field is also defined nonconventionally at $\tau = 0$. Really, when $d < 0$ by taking $h/|\tau|^{\Delta} = 1$ one can find

$$m_s = |h|^{1/\delta} f_{\pm}(1), \tag{3.69}$$

where

$$\delta = \frac{\Delta}{2 - \alpha - \Delta} \, . \tag{3.70}$$

From this equation one could expect that δ is equal to $(d+2)/(d-2)$ when $d > 4$. This is not correct and the reason is the same: $\mathcal{F}_s$ does not exist at $\mu_2 = 0$. Thus, to find the correct value for δ one should use the exact solution, or, using relations like Eq. (3.56) at $\tau = 0$, and defining m_s in the form

$$m_s = e^{-\frac{d-2}{2} l} f_h \left(h e^{\frac{d+2}{2} l} \, , \, \mu_2 e^{(4-d) l} \right) , \tag{3.71}$$

where f_h is the derivative of $\mathcal{F}_s$ with respect to h. Let us take $he^{(d+2)l/2} = 1$, then Eq. (3.71) can be rewritten as

$$m_s = h^{\frac{d-2}{d+2}} f_h \left(1 \, , \, \mu_2 h^{2 \frac{4-d}{d+2}} \right) . \tag{3.72}$$

Now, taking into account that at $\tau = 0$ the magnetization $m_s \propto \mu_2^{-1/3}$ one can find

$$m_s \propto h^{(d-2)/(d+2)} \mu_2^{-1/3} h^{2(4-d)/3(d+2)} \propto h^{1/3},$$

i.e. $\delta = 3$, as follows from the Landau theory.

Thus, not complying with their name — "irrelevant", these eigenvectors sometimes can play a very essential role. They may lead to the change of the conventional scaling laws and to the loss of the universality.

Local Perturbations

Until now, we have considered only the translationally-invariant perturbations of the fixed Hamiltonian. To define the behavior of correlation functions in the critical region, one needs to study the influence of local eigenvectors $\mathcal{O}_i(\boldsymbol{x})$ or $\mathcal{O}_i(\boldsymbol{q}) = \int d\boldsymbol{x}\mathcal{O}_i(\boldsymbol{x})\exp(-i\boldsymbol{q}\boldsymbol{x})$, which satisfy the equation of the linear RG

$$\frac{\partial \mathcal{O}_i(\boldsymbol{q}, l)}{\partial l} = \hat{L}\mathcal{O}_i(\boldsymbol{q}, l) = \lambda_i \mathcal{O}_i(\boldsymbol{q}, l).$$

The formal solution of this equation is

$$\mathcal{O}_i(\boldsymbol{q}, l) = e^{l\hat{L}}\mathcal{O}_i(\boldsymbol{q}) = e^{\lambda_i l}\mathcal{O}_i(\boldsymbol{q}e^l) .$$

By writing the Hamiltonian in the form

$$H(l) = H^* + \tau e^{\lambda_1 l}\mathcal{O}_1 + \mu_i \mathcal{O}_i(\boldsymbol{q}e^l) + \mu_k \mathcal{O}_k(-\boldsymbol{q}e^l)$$

and using the traditional definition for the correlation function G_{ik} as the second derivative from the free energy density

$$G_{ik}(\boldsymbol{q}, \tau) = \left.\frac{\partial^2 \mathcal{F}_s}{\partial \mu_i \partial \mu_k}\right|_{\mu_i = \mu_i = 0} ,$$

one can obtain the following scaling relation

$$G_{ik}(\boldsymbol{q}, \tau) = e^{(\lambda_i + \lambda_k - d)l} G_{ik}(\boldsymbol{q}e^l, \tau e^{\lambda_1 l}) . \tag{3.73}$$

By taking as before $|\tau| e^{\lambda_1 l} = 1$ we arrive at

$$G_{ik}(\boldsymbol{q}, \tau) = |\tau|^{-\gamma_{ik}} G_{ik}(\boldsymbol{q}|\tau|^{-1/\lambda_1}, \pm 1) = |\tau|^{-\gamma_{ik}} g^{\pm}_{ik}(\boldsymbol{q}\xi), \tag{3.74}$$

where the correlation length has been introduced by

$$\xi = |\tau|^{-\nu} , \ \nu = \frac{1}{\lambda_1} . \tag{3.75}$$

If we now choose in $qe^l = 1$ Eq. (3.73) and also put $\mu_i = \mu_k = \mu_h$, then the expression for the correlation function of the order parameter at the critical point can be obtained as

$$G(\boldsymbol{q}, 0) = q^{-2+\eta}\, G(1, 0) ,$$

where

$$\eta = 2 + d - 2\lambda_h \ . \tag{3.76}$$

From Eqs. (3.58), (3.64), (3.70), (3.75), and (3.76) one can find the scaling laws (see also Sect. 1.7) in the following form

$$\begin{array}{rclrcl} \lambda_1 & = & \frac{1}{\nu}, & \Delta & = & \left(\frac{d+2-\eta}{2}\right)\nu, \\ d & = & \frac{2-\alpha}{\nu}, & \beta & = & \frac{(2-\alpha)(d-2+\eta)}{2d}, \\ \gamma & = & (2-\eta)\nu, & \delta & = & \frac{d+2-\eta}{d-2+\eta}. \end{array} \tag{3.77}$$

From the derivation given in this section one can see that the exponents α, β, and ν have the same values above and below T_c.

In concluding this section, it is necessary to make the following comments:

1. It follows from Eq. (3.74) that the two-point correlation function G_{ik} can be either divergent or not. The divergence appears when $\boldsymbol{q} \to 0$ at the critical point only when $\lambda_i + \lambda_k > d$.

2. The traditionally considered correlation function of the field ϕ_q behaves, in the vicinity of the critical point, like the correlation function of the eigenvector $\mathcal{O}_h(\boldsymbol{q})$.

The latter comment follows from the expansion of the field ϕ_q in the series

$$\phi_q = \sum_i C_i \mathcal{O}_i(\boldsymbol{q}) \ .$$

In this expansion the main asymptotic contribution (at $l \to \infty$) appears from the eigenvector with the maximum λ_i, i.e. from the eigenvector $\mathcal{O}_h$.

3.3.5 Gaussian Fixed Point

The fixed point RG equation has one very simple exact solution in the form of the Gaussian functional

$$H^* = V g_0^* + \int_q \frac{g_2^*(\boldsymbol{q})}{2} |\phi_q|^2,$$

where the vertexes g_0^* and g_2^* can be easily found from Eq. (3.25), which gives the following relations

$$g_0^* + n\int_q h(\boldsymbol{q})[g_2^*(\boldsymbol{q})+1] = 0\,, \tag{3.78}$$

$$2h(\boldsymbol{q})[1-g_2^*(\boldsymbol{q})]g_2^*(\boldsymbol{q}) - \boldsymbol{q}\frac{\partial g_2^*(\boldsymbol{q})}{\partial \boldsymbol{q}} = 0\,. \tag{3.79}$$

One can define the solution for the value g_0^* using Eq. (3.78), if the function $g_2^*(\boldsymbol{q})$ is known. In turn, this function can be obtained by the direct integration of Eq. (3.79), which, accounting for the definition of $h(\boldsymbol{q})$ (see Eq. (3.14)), gives

$$g_2^*(\boldsymbol{q}) = \frac{Aq^{2h(0)}}{Aq^{2h(0)} + \exp(-q^2)}, \tag{3.80}$$

where A is an arbitrary positive constant of integration. For the short range interaction potentials of the initial physical Hamiltonian (see Chapter 2), the vertex $g_2^*(\boldsymbol{q})$ should be an analytical function of $\boldsymbol{q}$. Therefore, $h(0)$ can only be taken as an integer number. The conventional critical behavior, corresponding to the Landau model, has $h(0) = 1$.

Eigenvectors and Eigenvalues

Let us find the eigenvectors and eigenvalues for the Gaussian fixed Hamiltonian. The equation for the definition of the eigenvalues can be represented as follows

$$\hat{L}\mathcal{O} = \hat{L}_1\mathcal{O} + \hat{L}_2\mathcal{O} = \lambda\mathcal{O}, \tag{3.81}$$

where the operators are defined by

$$\hat{L}_1 = dV\frac{\partial}{\partial V} + \int_q \left\{\frac{d+2h(\boldsymbol{q})[1-2g_2^*(\boldsymbol{q})]}{2}\phi_q + \left(\boldsymbol{q}\frac{\partial}{\partial \boldsymbol{q}}\right)\phi_q\right\}\cdot\frac{\delta}{\delta\phi_q} \tag{3.82}$$

and

$$\hat{L}_2 = \int_q h(\boldsymbol{q})\frac{\delta^2}{\delta\phi_q\delta\phi_{-q}}\,. \tag{3.83}$$

The structure of the operators $\hat{L}_1$ and $\hat{L}_2$ allows for the elimination of the operator $\hat{L}_2$ from the Eq. (3.81) with the help of a canonical transformation. The generator for this transformation

$$\hat{K} = \int_q K(\boldsymbol{q})\frac{\delta^2}{\delta\phi_q\delta\phi_{-q}},$$

transforms Eq. (3.81) as follows

$$\hat{\overline{L}}\,\overline{\mathcal{O}} = \left\{\hat{L}_1(-1) + \hat{L}_2\right\}\overline{\mathcal{O}} = \lambda\overline{\mathcal{O}}, \tag{3.84}$$

where

$$\begin{aligned} \overline{\mathcal{O}} &= e^{-\hat{K}}\mathcal{O}, \\ \hat{\overline{L}} &= e^{-\hat{K}}\hat{L}e^{\hat{K}}, \\ \hat{L}_1(\lambda) &= e^{\lambda\hat{K}}\hat{L}_1 e^{-\lambda\hat{K}}. \end{aligned} \tag{3.85}$$

By differentiation of Eq. (3.85) with respect to λ, one arrives at the equation

$$\frac{\partial\hat{L}_1(\lambda)}{\partial\lambda} = [\hat{K}\hat{L}_1(\lambda)]_- = \hat{C}, \tag{3.86}$$

where the operator $\hat{C}$ is defined by

$$\hat{C} = \int_q \left\{2h(\boldsymbol{q})[1 - 2g_2^*(\boldsymbol{q})]K(\boldsymbol{q}) + \left(\boldsymbol{q}\frac{\partial}{\partial\boldsymbol{q}}\right)K(\boldsymbol{q})\right\}\frac{\delta^2}{\delta\phi_q\delta\phi_q}. \tag{3.87}$$

The solution of Eq. (3.86) gives the following expression for the operator $\hat{L}_1(-1)$

$$\hat{L}_1(-1) = \hat{L}_1 - \hat{C}\,.$$

So far, the operator $\hat{K}$ as well as the operator $\hat{C}$ have not been defined. We will define them from the requirement for the elimination of the operator $\hat{L}_2$ in the transformed operator $\hat{\overline{L}}$. To find this requirement, it is necessary to set the equality $\hat{L}_2 = \hat{C}$, which gives the following equation for the function $K(\boldsymbol{q})$

$$2h(\boldsymbol{q})[1 - 2g_2^*(\boldsymbol{q})]K(\boldsymbol{q}) + \left(\boldsymbol{q}\frac{\partial}{\partial\boldsymbol{q}}\right)K(\boldsymbol{q}) = h(\boldsymbol{q}). \tag{3.88}$$

Inserting the function $h(\boldsymbol{q})$ expressed with the help of Eq. (3.79) as a functional of $g_2^*(\boldsymbol{q})$, into Eq. (3.88) one can get the solution in the form

$$K(\boldsymbol{q}) = \frac{1}{2(1 - g_2^*(\boldsymbol{q}))}\,.$$

Strictly speaking, this procedure is not completely correct, as the commutator of the operator $\hat{K}$ with the term $V\partial/\partial V$ is not zero, thus violating Eq. (3.86). Nonetheless, one can use the same method which was applied in Sect. 3.3.3. This method is only applicable to the eigenvectors representable in the form Eq. (3.39). It allows us to divide the terms depending on the volume in the initial Eq. (3.81) from the other contributions, which formally reduces the problem to the case with $V = 0$. Only in this sense Eq. (3.86) and the following procedure of elimination of the operator $\hat{L}_2$ can be considered as well founded ones.

A few words should be said as to why it is convenient to eliminate the operator $\hat{L}_2$ from the structure of the full operator $\hat{L}$. The action of the operator $\hat{L}_1$ on a functional of Landau field operators conserves the number of these operators in every term in the power series of ϕ_q. The operator $\hat{L}_2$ changes the number of these operators. Therefore, by the elimination of $\hat{L}_2$ we have arrived at the eigenvectors $\overline{\mathcal{O}}$ which are characterized by the fixed number of field operators m. Restricting the following consideration to the case of the space homogeneous eigenvectors having the symmetry of the initial Hamiltonian, we can represent the eigenvectors in the form

$$\overline{\mathcal{O}}_{mr\kappa_r} = \frac{\delta_{2k,m}}{m!} \int_{q_1} \cdots \int_{q_k} P^m_{r\kappa_r}(\boldsymbol{q}_1 \boldsymbol{q}_{1'}, \cdots, \boldsymbol{q}_k \boldsymbol{q}_{k'}) \times \delta\left(\sum_{i=1}^{k} (\boldsymbol{q}_i + \boldsymbol{q}_{i'})\right) \prod_{i=1}^{k} f(\boldsymbol{q}_i) f(\boldsymbol{q}_{i'}) (\phi_{q_i} \phi_{q_{i'}}), \tag{3.89}$$

where $f(\boldsymbol{q})$ is a function to be defined, $P^m_{r\kappa_r}$ is a homogeneous polynomial of the $2r$-order symmetrical with respect to the arbitrary permutations of the pair of momenta $(\boldsymbol{q}_i \boldsymbol{q}_{i'})$ and $(\boldsymbol{q}_k \boldsymbol{q}_{k'})$, and κ_r is an index characterizing the polynomial type at the fixed homogeneity order and fixed m. Riedel *et al* (1985) suggested the "orthogonality" condition, which can help in the construction of these polynomials

$$P^m_{r\kappa_r}(\boldsymbol{\nabla}_{q_1} \cdots \boldsymbol{\nabla}_{q_m}) P^m_{r\kappa'_r}(\boldsymbol{q}_1 \cdots \boldsymbol{q}_m) = 0, \tag{3.90}$$

when $\kappa_r \neq \kappa'_r$. With the help of the representation Eq. (3.89) from Eq. (3.84) one can get the following equation for the definition of the function $f(\boldsymbol{q})$

$$\{h(\boldsymbol{q})[1 - 2g^*_2(\boldsymbol{q})] - h(0)\} f(\boldsymbol{q}) - \boldsymbol{q} \cdot \frac{df(\boldsymbol{q})}{d\boldsymbol{q}} = 0 \tag{3.91}$$

and the eigenvalues λ_{m2r}

$$\lambda_{m2r} = d - m\left(\frac{d}{2} - h(0)\right) - 2r\,. \tag{3.92}$$

To define the function $f(\boldsymbol{q})$ from Eq. (3.91) one has to use boundary condition $f(0) = \text{const} \neq 0$. As a result

$$f(\boldsymbol{q}) = \frac{\sqrt{g^*_2(\boldsymbol{q})[1 - g^*_2(\boldsymbol{q})]}}{|\boldsymbol{q}|^{h(0)}}.$$

Thus, the definition of the functions $f(\boldsymbol{q})$ completely solves the problem of the definition of the translationally-invariant, even with respect to ϕ, eigenvectors for the $\mathcal{O}(n)$ symmetrical case. Only these eigenvectors do not break the symmetry of the fixed Hamiltonian and give analytical behavior of all higher vertexes with respect to $\boldsymbol{q}$ ($h(0) = 1$). The final expression for the eigenvectors $\mathcal{O}_{mr\kappa_r}$ can be easily restored by taking into account equations defining all parameters of the canonical transformation (see Eq. (3.85))

$$\mathcal{O}_{mr\kappa_r} = e^{\hat{K}}\overline{\mathcal{O}}_{mr\kappa_r}.$$

It is possible to show that the eigenvectors with $r = 0$ are not redundant. The subsequent consideration was made by Wegner (1976). He also gave an example of a redundant eigenvector when $r \neq 0$. This is the vector $\mathcal{O}_{22}$. When $m = 2$ we do not need to introduce the index κ_r as the homogeneity of space gives $P_{2r} = (\boldsymbol{q}^2)^r$. There is also a requirement of analyticity with respect to $\boldsymbol{q}$ for the nonlocal vertexes in the Hamiltonian. The latter leads to the appearance of only even powers of $\boldsymbol{q}$. From Eq. (3.92) one can see that $\lambda_{m0} > 0$ at any m if $2h(0) > d$. This case corresponds to the infinite number of relevant eigenvectors and, hence, the critical state cannot ever be achieved. Therefore, for critical behavior we should have $d > 2h(0)$. One can find the number of relevant eigenvectors affecting the critical behavior recalling that the case with $m = 0$ corresponds to the special eigenvector. This can be done with the help of Eq. (3.92). The eigenvector $\mathcal{O}_{m0}$ with given $m = 2\sigma$ (at $r = 0$) will be relevant if $d < d_\sigma = 2\sigma/(\sigma - 1)$. The following eigenvector $\mathcal{O}_{m+1,0}$ is irrelevant when $d \geq d_{\sigma+1}$. This means, that when $d_{\sigma+1} \leq d < d_\sigma$ the number of relevant eigenvectors is equal to σ. Therefore, $\sigma = 1$ when $d \geq 4$. The latter means that the fixed-point Hamiltonian at $h(0) = 1$ describes the conventional critical behavior. When $3 \leq d < 4$ we have $\sigma = 2$ and the Gaussian functional describes the tricritical behavior.

3.3.6 The Scaling-Field Method

As is seen from the discussion in the previous section, the conventional critical behavior cannot be described by the Gaussian fixed-point Hamiltonian when $d < 4$. For this case we should have a fixed point with the only one symmetry-conserving relevant eigenvector. Unfortunately, there are no simple methods for solving nonlinear equations in functional derivatives. Currently, the most advanced approximate method for the solution of Wilson's equation is the scaling-field method (SFM). This method had been developed by Golner and Riedel in 1975. They used the expansion over the scaling fields (see Eq. (3.30)) proposed by Wegner (1972) in order to obtain an infinite set of equations in ordinary derivatives. Another method, which had been developed by Wilson and Kogut in 1974, reduces Eq. (3.20) to an infinite set of coupled equations in partial derivatives for the vertexes $g_m(\boldsymbol{q}_1, \cdots, \boldsymbol{q}_m, l)$. In this approach the Gaussian fixed point was discussed and the ϵ-expansion developed for the nontrivial fixed point in the lowest orders. Shukla and Green (1974, 1975) calculated the critical fixed point to the second order of ϵ.[11] The exponent η was also calculated in the framework of this approach by Rudnick (1975) and by Golner and Riedel (1976).

The scaling field method was being further improved by Riedel *et al* (1985). This method, like the Wilson's approach, does not have simple calculational schemes, nonetheless the developed approximations proved to be very effective in calculations of such global properties for the three-dimensional case, like phase diagrams, scaling functions, crossover phenomena and so on. The equations of the SFM can be derived as follows. Let us insert into Eq. (3.27) the Hamiltonian in the form of the expansion Eq. (3.30), then we arrive at

$$\begin{aligned}\sum_i \mathcal{O}_i \frac{d\mu_i}{dl} &= \sum_i \lambda_i^0 \mu_i \mathcal{O}_i - \sum_{ik} \mu_i \mu_k \int_q h(\boldsymbol{q}, l) \frac{\delta \mathcal{O}_i}{\delta \phi_q} \cdot \frac{\delta \mathcal{O}_k}{\delta \phi_{-q}} \\ &- Q(l) \left\{ \sum_i \mu_i \int_q \left[\frac{\delta}{\delta \phi_{-q}} - 2 \frac{\delta H^*}{\delta \phi_{-q}} + \phi_q \right] \cdot \frac{\delta \mathcal{O}_i}{\delta \phi_q} \right. \\ &+ \left. \int_q \left[\left(\frac{\delta}{\delta \phi_{-q}} - \frac{\delta H^*}{\delta \phi_{-q}} + \phi_q \right) \cdot \frac{\delta H^*}{\delta \phi_q} - nV \right] \right\}, \end{aligned} \qquad (3.93)$$

where $Q(l) = h^*(0) - h(0, l)$, as usual for λ_i^0, H^* and $h^*(0)$ one has to take the Gaussian values.

Let us express all the functionals of the Landau fields ϕ_q in Eq. (3.93) as linear rows of the eigenvectors $\mathcal{O}_i$. This gives for the bilinear terms in the first row of Eq. (3.93)

$$\int_q h(\boldsymbol{q}, l) \frac{\delta \mathcal{O}_i}{\delta \phi_q} \cdot \frac{\delta \mathcal{O}_k}{\delta \phi_{-q}} = - \sum_j a_{jik} \mathcal{O}_j, \qquad (3.94)$$

just as well as the following relations applicable to the second row of Eq. (3.93)

$$Q(l) \int_q \left[\frac{\delta}{\delta \phi_{-q}} - 2 \frac{\delta H^*}{\delta \phi_{-q}} + \phi_q \right] \cdot \frac{\delta \mathcal{O}_i}{\delta \phi_q} = - \sum_j a_{ji} \mathcal{O}_j, \qquad (3.95)$$

[11] The critical exponents were found in the first order excepting η, for which the first nonvanishing order is ϵ^2.

and similar expressions for the last row

$$Q(l) \int_q \left[\left(\frac{\delta}{\delta\phi_{-q}} - \frac{\delta H^*}{\delta\phi_{-q}} + \phi_q \right) \cdot \frac{\delta H^*}{\delta\phi_q} - nV \right] = -\sum_i a_i \mathcal{O}_i. \qquad (3.96)$$

Using these equations one can transform Eq. (3.93) into an infinite set of ordinary differential equations

$$\frac{d\mu_i}{dl} = \lambda_i^0 \mu_i + \sum_{jk} a_{ijk}\mu_j\mu_k + \sum_j a_{ij}\mu_j + a_i. \qquad (3.97)$$

The set of Eq. (3.97) is not a full set, even though the number of equations is exactly the number of unknown functions μ_i. All coefficients in this set a_{ijk}, a_{ij} and a_i are dependent on the unknown function $Q(l)$. This dependencies can be represented in the form

$$a_{ijk} = a'_{ijk} + Qa''_{ijk}\ ,\ a_{ij} = Qa''_{ij}\ ,\ a_i = Qa''_i. \qquad (3.98)$$

Thus, it is necessary to complete the set in Eq. (3.97) with an additional equation, which would define the function $Q(l)$. This equation can be found from the condition for the elimination of the projection on the direction of the redundant operator $\mathcal{O}_{01}$. In order to obtain this additional equation one has to use the definition for the operator $\mathcal{O}_{01}$ (see Eq. (3.50)), the orthogonality condition Eq. (3.53), take into account Eq. (3.98), and express the values $\chi_i(l)$ in terms of the scaling fields μ_i

$$\chi_i(l) = a''_i + \sum_j a''_{ij}\mu_j + \sum_{jk} a''_{ijk}\mu_j\mu_k. \qquad (3.99)$$

The following analysis completely corresponds to the calculational scheme considered in Sect. 3.3.2. The fixed points are defined from the conditions $d\mu_i^*/dl = 0$ when $Q = Q^*$ — some constant fixed by the condition Eq. (3.53). Evidently, for the Gaussian model we have all $\mu_i^* = 0$ and $Q^* = 0$. The eigenvalues generated by the nontrivial fixed point ($\mu_i^* \neq 0$) should be defined from a set of linear equations for the SFM. From the system Eq. (3.97), by introducing small deviations $\delta\mu_i(l) = (\mu_i - \mu_i^*)\exp(\lambda l)$, one can obtain the following linear set of equations

$$\lambda\delta\mu_i(0) = \sum_j Y_{ij}\delta\mu_j(l), \qquad (3.100)$$

where

$$Y_{ij} = \lambda_i^0\delta_{ij} + 2\sum_k a_{ijk}(Q^*)\mu_k^* + a_{ij}^*(Q^*). \qquad (3.101)$$

The standard requirement $\det(\lambda - Y) = 0$ gives an algebraic equation defining the eigenvalues λ. It is also interesting to reveal the physical sense of the value Q^*. This value is directly related to the exponent η. In order to find this relation, let us consider the correlation function $G(\boldsymbol{q}, l)$ which is defined by

$$G(\boldsymbol{q}, l) = \frac{1}{ZV} \int D\phi \exp(-H_l\{\phi\}) \phi_q \phi_{-q} = \frac{\langle \phi_q \phi_{-q} \rangle_l}{V} . \tag{3.102}$$

Taking into account Eq. (3.33) and Eq. (3.34), after differentiation of Eq. (3.102) with respect to l we arrive at

$$\frac{\partial G}{\partial l} = \frac{1}{ZV} \int D\phi \left\{ \hat{\Pi} \exp(-H_l) \right\} \phi_q \phi_{-q} + d \frac{\partial H_l}{\partial V} \langle \phi_q \phi_{-q} \rangle_l - dG(\boldsymbol{q}, l).$$

By carrying out the integration in the functional integral[12] one can obtain the following simple equation

$$\frac{\partial G}{\partial l} = -2h(\boldsymbol{q})[G(\boldsymbol{q}, l) - n] - \boldsymbol{q} \cdot \frac{\partial G}{\partial \boldsymbol{q}} .$$

The solution of this equation can be represented in the form

$$G(\boldsymbol{q}e^l, l) = n + [G(\boldsymbol{q}, \tau) - n] \exp\left\{ -2 \int_0^l dl' h(\boldsymbol{q}, l') \right\} . \tag{3.103}$$

In the derivation of Eq. (3.103) the initial condition resulting from the definition, Eq. (3.102) has been taken into account. This condition says that the correlator $G(\boldsymbol{q}, l)$ at $l = 0$ is a conventional single particle propagator $G(\boldsymbol{q}, \tau)$. Let us turn $l \to \infty$ considering that the fixed point can only be reached when $\tau = 0$. As we are going to obtain the asymptotic behavior let us also impose the constrain $q \exp l = 1$. Then from Eq. (3.103) and Eq. (3.76) also taking into account that $Q(l) = h^*(0) - h(0, l)$ one can obtain the following relation

$$2Q^* = \eta . \tag{3.104}$$

This concludes the general consideration of the calculational procedure in the framework of the SFM. However, any concrete calculation requires explicit evaluations of the factors a_{ijk} and a_{ij} from Eq. (3.94) and Eq. (3.95). The evaluation procedure does not contain any principal difficulties, but it is very cumbersome. The calculations for some of the coefficients a_{ijk} can be found in the Wegner's review (1976). The general method based on the application of an unconventional graph technique was developed by Riedel *et al* in 1985.

---

[12]This integration can be made by parts in such a way that the variational derivatives will act on the field variables ϕ_q.

3.3.7 ϵ-Expansion in Scaling Fields

Let us consider the scheme of analytical expansions in the power series of $\epsilon = d_\sigma - d$ in the SFM for the isotropic n-component model.[13] As deviations from the Gaussian fixed point are assumed to be small, one can expect that μ_i^* are of the order of ϵ or less, while $Q^* \propto \epsilon^2$. The structure of the set Eq. (3.97) shows that μ_i^* could be simply defined for those λ_i^0 which are of the order of unity. The equations containing small $\lambda_i^0 \propto \epsilon$ require treating of the nonlinear terms in Eq. (3.97) on equal footing with the linear ones. As a result, μ_i defined from these equations will be of the order of ϵ, while the scaling amplitudes for $\lambda_i^0 \approx 1$ are of the order of ϵ^2. This means that first we should consider the equations with small λ_i^0. As is seen from Eq. (3.92), we have two such equations. They define the evolution of $\mu_{2,2}(l)$ $(\lambda_{2,2}^0 = 0)$

$$\frac{d\mu_{2,2}}{dl} = \sum_{ik} a_{2,2;j;k}\, \mu_j \mu_k + Q\left\{\sum_j a''_{2,2;j}\, \mu_j + a''_{2,2}\right\} \tag{3.105}$$

and the evolution of $\mu_{2\sigma,0}(l)$ $(\lambda_{2\sigma,0}^0 = (\sigma - 1)\epsilon)$

$$\frac{d\mu_{2\sigma,0}}{dl} = \lambda_{2\sigma,0}^0\, \mu_{2\sigma,0} + \sum_{jk} a_{2\sigma,0;j;k}\, \mu_j \mu_k$$

$$+Q\left\{\sum_j a''_{2\sigma,0;j}\, \mu_j + a''_{2\sigma,0}\right\}. \tag{3.106}$$

It is necessary to remind that $2\sigma = m$ (see Sect. 3.3.5), the indexes j, k denotes the sets $\{k, 2r, \kappa_r\}$, and when $r \leq 1$ the index κ_r can be omitted.

For the following calculations we also need to find the coefficients a_i from Eq. (3.96). This can be done by inserting the Gaussian Hamiltonian H^* defined in Sect. 3.3.5 into Eq. (3.96). As a result the left part of this equation takes the form

$$Q\left\{nV\int_q [1 - g_2^*(\boldsymbol{q})] - \int_q g_2^*(\boldsymbol{q})[1 - g_2^*(\boldsymbol{q})]|\phi_q|^2\right\}$$

$$= Q\left\{n\int_q [1 - g_2^*(\boldsymbol{q})]\mathcal{O}_0 - 2A\mathcal{O}_{2,2}\right\}. \tag{3.107}$$

This equation shows that only two coefficients are different from zero, i.e. a_0 and $a_{2,2}$. Now leaving only the lowest order terms in Eq. (3.106) we arrive at the following fixed-point equation

$$(\sigma - 1)\epsilon\mu_{2\sigma,0}^* + a_{2\sigma,0;2\sigma,0;2\sigma,0}(\mu_{2\sigma,0}^*)^2 = 0. \tag{3.108}$$

[13] Here $d_\sigma = 2\sigma/(\sigma - 1)$, see Sect. 3.3.5.

From this equation one can see that for the function $\mu^*_{2\sigma,0}(d)$ the point $d = d_\sigma$ is a branch point. At this point two lines cross: the trivial solution $\mu^*_{2\sigma,0}(d) = 0$ and the nontrivial one

$$\mu^*_{2\sigma,0}(d) = \frac{-(\sigma-1)\epsilon}{a_{2\sigma,0;2\sigma,0;2\sigma,0}}.$$

For the trivial solution when $\epsilon > 0$ we also have an additional relevant eigenvector. As we will see from the following analysis, the nontrivial solution conserves the number of relevant eigenvectors independently of the sign of ϵ.

Before the formal analysis of Eq. (3.105), let us consider the procedure for the elimination of the redundant eigenvector $\mathcal{O}_{01}$. With the help of the explicit representation of functions $\chi_i(l)$ Eq. (3.99), one can rewrite Eq. (3.53) in the first nonvanishing approximation as

$$\chi_{2,2}(l)\frac{d\mu_{2,2}}{dl} \approx a''_{2,2}\frac{d\mu_{2,2}}{dl} = 0.$$

From Eq. (3.107) it follows that $a''_{2,2} = -2A$, and, as a consequence, one can get the condition of the $\mathcal{O}_{2,2}$ elimination as $\mu_{2,2} = 0$. The latter could also be anticipated from the fact that for the Gaussian fixed point the eigenvector $\mathcal{O}_{2,2}$ is redundant. All these facts help to determine the function $Q(l)$ from Eq. (3.105) as

$$Q(l) = \frac{a_{2,2;2\sigma,0;2\sigma,0}\,[\mu_{2\sigma}(l)]^2}{2A}$$

or at the fixed point as

$$Q^* = \frac{\eta}{2} = \frac{(\sigma-1)^2\epsilon^2 a_{2,2;2\sigma,0;2\sigma,0}}{2Aa^2_{2\sigma,0;2\sigma,0;2\sigma,0}}. \tag{3.109}$$

Thus the new fixed-point Hamiltonian can be written in the first nonvanishing order as

$$H^* = H^*_G - \frac{(\sigma-1)\epsilon}{a_{2\sigma,0;2\sigma,0;2\sigma,0}}\mathcal{O}_{2\sigma,0}\,. \tag{3.110}$$

In this equation we have taken into account that $\mu^*_i \propto \epsilon^2$ when $i \neq \{2\sigma, 0\}$. The procedure for the diagonalization of the matrix Y_{ij} can also be simply realized as the matrix has a block diagonal form. All the matrix elements which should be diagonalized belong to the degenerate subspace in which $\lambda^0_i(d_\sigma) = \lambda^0_j(d_\sigma)$ and they have the structure

$$Y_{ij} = Y(i, r, \kappa_i | i, r, \kappa_j).$$

Considering only the eigenvalues with $r = 0$, one can obtain the following relations in the first nonvanishing order, using Eq. (3.101)

$$\lambda_{m,0} = \lambda^0_{m,0}(d_\sigma - \epsilon) - \frac{2a_{m,0;m,0;2\sigma,0}(\sigma - 1)\epsilon}{a_{2\sigma,0;2\sigma,0;2\sigma,0}}. \tag{3.111}$$

This equation shows that instead of the eigenvalue $\lambda^0_{2\sigma,0}$ related to the relevant, in the Gaussian fixed point, eigenvector $\mathcal{O}_{2\sigma,0}$, we now have the negative value equal to $-\lambda^0_{2\sigma,0}$. Therefore, the eigenvector $\mathcal{O}_{2\sigma,0}$ is an irrelevant vector when $\epsilon > 0$.

The formulae Eqs. (3.109) and (3.111) will give the solution for the new fixed point if one could know the coefficients a_{ijk}. Here we only show the method of their definition. The most important point of this method is the so-called linked-contraction theorem formulated by Riedel *et al* (1985). This theorem can be written as follows

$$\left\{ e^{-2\hat{K}_{12}} \int_q h(\boldsymbol{q}, l) \frac{\delta\overline{\mathcal{O}}_i[\boldsymbol{\phi}^{(1)}]}{\delta\boldsymbol{\phi}^{(1)}_q} \cdot \frac{\delta\overline{\mathcal{O}}_k[\boldsymbol{\phi}^{(2)}]}{\delta\boldsymbol{\phi}^{(2)}_{-q}} \right\} = -\sum_j a_{jik}\overline{\mathcal{O}}_j, \tag{3.112}$$

where in the left-hand side after the action of the operator $e^{-2\hat{K}_{12}}$ one has to take $\boldsymbol{\phi}^{(1)} = \boldsymbol{\phi}^{(2)} = \boldsymbol{\phi}$. The vectors $\overline{\mathcal{O}}_i$, as well as the operator $\hat{K}$, are defined in Sect. 3.3.5. In this modification they have the form

$$\hat{K}_{lp} = \int_q K(\boldsymbol{q}) \frac{\delta^2}{\delta\boldsymbol{\phi}^{(l)}_q \delta\boldsymbol{\phi}^{(p)}_{-q}} \, .$$

In order to prove this theorem let us consider the functional $\Psi(\lambda) = \exp[\lambda\hat{K}]AB$, where A and B are arbitrary functionals of $\boldsymbol{\phi}$. After differentiation of this functional with respect to λ we arrive at

$$\left.\frac{d\Psi}{d\lambda}\right|_{\lambda=0} = \hat{K}AB \equiv \Big(\hat{K}_{11}A[\boldsymbol{\phi}^{(1)}]B[\boldsymbol{\phi}^{(2)}]$$

$$\left. +2\hat{K}_{12}A[\boldsymbol{\phi}^{(1)}]B[\boldsymbol{\phi}^{(2)}] + \hat{K}_{22}A[\boldsymbol{\phi}^{(1)}]B[\boldsymbol{\phi}^{(2)}]\Big)\right|_{\boldsymbol{\phi}^{(1)}=\boldsymbol{\phi}^{(2)}=\boldsymbol{\phi}} .$$

From this relation follows the representation

$$\Psi(1) = e^{-\hat{K}}AB = \exp\left\{-(\hat{K}_{11} + 2\hat{K}_{12} + \hat{K}_{22})\right\} A[\boldsymbol{\phi}^{(1)}]B[\boldsymbol{\phi}^{(2)}], \tag{3.113}$$

where again we have to set $\boldsymbol{\phi}^{(1)} = \boldsymbol{\phi}^{(2)} = \boldsymbol{\phi}$ after the action of all operators.

Now if we substitute in Eq. (3.94) the vectors $\overline{\mathcal{O}}_i$ for the eigenvectors $\mathcal{O}_i$ using the definitions from Eq. (3.85) and apply Eq. (3.113), we arrive at the linked-contraction theorem Eq. (3.112). This theorem allows us to avoid the summation over the index denoting the number of the field operators ϕ in the vectors $\overline{\mathcal{O}}_i$ and to rewrite Eq. (3.112) in the form of independent equations

$$\frac{1}{k!}\left\{\int_q \frac{1}{g_2^*(\boldsymbol{q})}\frac{\delta^2}{\delta\phi_q^{(1)}\delta\phi_q^{(2)}}\right\}^k \int_q h(\boldsymbol{q},l)\frac{\delta^2}{\delta\phi_q^{(1)}\delta\phi_q^{(2)}}\overline{\mathcal{O}}_j^{(1)}\overline{\mathcal{O}}_l^{(2)}$$

$$= -\sum_{r_i,\kappa_i} a_{i,r_i,\kappa_i;j;l}\,\overline{\mathcal{O}}_{i,r_i,\kappa_i}, \tag{3.114}$$

where the following conditions are imposed on the numbers $k,\ i,\ j,\ l$: $i = j + l - 2(k+1)$; $k \le j - 1$; $k \le l - 1$. The final procedure of the calculations consists in the carrying out of the variational derivatives and a number of integrals over $\boldsymbol{q}$. We will not follow the details of these trivial but very cumbersome calculations here, but only give the final answers. The critical point has $\sigma = 2$ and $d_\sigma = 4$

$$\eta = \frac{(n+2)\epsilon^2}{2(n+8)^2} + o(\epsilon^3) \tag{3.115}$$

and eigenvalues

$$\lambda_{m0} = \lambda_{m0}^0(d = 4-\epsilon) - \frac{m(n+3m-4)\epsilon}{2(n+8)} + o(\epsilon^2). \tag{3.116}$$

The tricritical point has $\sigma = 3$ and $d_\sigma = 3$

$$\eta = \frac{(n+2)(n+4)\epsilon^2}{12(3n+22)^2} + o(\epsilon^3) \tag{3.117}$$

and eigenvalues

$$\lambda_{m0} = \lambda_{m0}^0(d = 3-\epsilon) - \frac{m(m-2)(3n+5m-8)\epsilon}{6(3n+22)} + o(\epsilon^2). \tag{3.118}$$

The eigenvalues $\lambda_{mr\kappa_r}$ when $r = 2$ and $r = 4$ were calculated by Riedel *et al* in 1985. References to earlier calculations with $r = 0$ can be found in the review by Wegner (1976).

In conclusion, it is necessary to mention that the analysis of the Wilson's RG equation to some extent resembles quantum mechanical (QM) problems. In the RG scheme we have, instead of the space variables, an infinite number of Landau field variables. The most important difference consists in the nonlinear nature of the RG equation which accounts for the existence of fixed points. The problem of the definition of the critical exponents can be compared with the problem of the definition of a particle spectrum in a given potential field. In the case of the RG linear equation, the role of the potential in QM can be ascribed to the fixed-point Hamiltonian. An essential difference between the linear RG and QM is in the absence of the definition of the scalar product and the norm in the RG. The absence of the normalization in the RG is usually replaced by the analyticity requirements. As the QM allows for the application of numerical analysis so does the RG. Especially effective numerical analysis in the RG was developed by Newman and Riedel in 1984. Their approach is based on the SFM representation of the Wilson's equation. They proposed a procedure for generation in a nonperturbative and unbiased fashion, sequences of truncations to the infinite hierarchy of the SFM equations. To provide a self-consistent criterion they invented a "principle of balance", which was found empirically. The concept of balance can be crudely formulated as: in truncated expansions different competing eigenvectors should be represented in a balanced way. This approach was applied to yield leading critical exponents ν, η, and several of the correction-to-scaling exponents Δ_i (see Sect. 3.3.4) to high precision.

Chapter 4

Field Theoretical RG

4.1 Introduction

The field theoretical RG, developed to eliminate divergences in relativistic quantum field theories, has been a very effective calculational tool in critical thermodynamics. Usually, a thermodynamic system can be characterized by least two characteristic lengths: the correlation length ξ and the lattice cell size $a \propto \Lambda^{-1}$. However, in the critical region one has the inequality $\xi \gg a$, which leaves only one characteristic length in the problem. At the critical point, the interesting region of wave vectors in the correlation functions satisfies the condition $q \ll \Lambda$. This means that one may just as well consider the limit $\Lambda \to \infty$. Nonetheless, this limit should be treated carefully, as the temperature renormalization and some factors in the correlation functions diverge with increasing Λ. The conventional approach to this problem is to consider diverging coefficients as finite experimental quantities in all measurable values. Such an approach is essentially the same as the procedure for the elimination of divergent terms in quantum field theory. As in the field theory, this procedure, applied to the phase transition theory, introduces some freedom. This freedom can be restricted with the help of adding counter-term to the Hamiltonian. These counter-terms should have the same structure as particular terms of the zero-order Hamiltonian (renormalization of T_c and field operators) and with the interaction Hamiltonian (renormalization of interaction constant). The renormalization transformations of correlation functions and interaction constants satisfy all the necessary group properties and, therefore, they constitute a renormalization group. This group is a genuine Lie group as it has an inverse element, which is not present in Wilson's approach. The group properties lead to some constraints on the possible structures of correlation functions and their dependencies on interaction constants. The conventional perturbation theory calculations do not posses the RG covariance. The requirement of the RG covariance imposed on such calculations helps to find leading asymptotes. The latter can be done by comparisons of perturbation series with formal expressions for correlators satisfying the Lie group equations.

Laying aside some specificity of the initial definitions, one may conclude that applications of the field theoretical RG to the theory of phase transitions are identical with the consequent applications in the field theory. In reality, however, this is not the case. The most important difference can be seen as follows. In the field theory we ascribe the physical meaning to renormalized correlators, while in the phase transition theory the trial correlators are physically sensible but not the renormalized ones. Nonetheless, all conclusions about the physical behavior of the system in the critical state are founded on analyses of renormalized values. The latter means that treatment of measurable quantities requires additional analysis, when it is made on the grounds of the results obtained for renormalized mean values. This statement is especially essential for the preasymptotic region (see Bagnuls and Bervillier 1981, 1984, and 1985).

4.2 Perturbation Theory

Let us consider the main aspects in constructing perturbation series. This will be done in the simplest case with ϕ^4 interaction for the $\mathcal{O}(n)$ symmetry vector model. Notwithstanding the perceived restrictiveness of this model, it can be considered as a universal basis which allows for a number of generalizations to other physical systems. It is convenient to rewrite the Hamiltonian for this model in the following form

$$H = \frac{1}{2}\int d\boldsymbol{x}\left\{[\boldsymbol{\nabla}\phi(\boldsymbol{x})]^2 + g_2\phi^2(\boldsymbol{x}) + \frac{1}{12}g_4[\phi^2(\boldsymbol{x})]^2 + 2\boldsymbol{g}_1(\boldsymbol{x})\phi(\boldsymbol{x})\right\}, \quad (4.1)$$

where the space-dependent vertex $\boldsymbol{g}_1(\boldsymbol{x})$ has been introduced for further convenience. In the space-homogeneous case this vertex is equal to the negative value of the external field.

4.2.1 General Definitions

The introduction of the nonhomogeneous field $\boldsymbol{g}_1(\boldsymbol{x})$ turns the free energy into a generating functional for any connected correlation function. Below for the sake of simplicity we will denote vector and space variables with the help of numbers as

$$g_1(k) \equiv g_1^{\alpha_k}(\boldsymbol{x}_k)\ , \ \text{and}\ \phi(k) \equiv \phi^{\alpha_k}(\boldsymbol{x}_k)\ .$$

Let us define connected correlators as follows

$$G_c(1\dots k) = (-1)^{k-1}\frac{\delta^k F}{\delta g_1(1)\cdots\delta g_1(k)}, \tag{4.2}$$

where F is the free energy functional of the function $g_1(k)$ defined with the help of the Hamiltonian Eq. (1.1) as follows

$$F\{g_1(k)\} = -\ln\left(\prod\int\frac{d\phi_q}{\sqrt{2\pi V}}\exp[-H\{\phi, g_1(k)\}]\right).$$

From Eq. (1.2) one can obtain

$$\begin{aligned}
G_c(1) &\equiv M(1) = \langle\phi(1)\rangle,\\
G_c(12) &= G(12) - G(1)G(2),\\
G_c(123) &= G(123) - G(12)G(3) - G(13)G(2)\\
&\quad -G(23)G(1) + 2G(1)G(2)G(3),
\end{aligned} \tag{4.3}$$

where $G(1,\dots,k) = \langle\phi(1)\cdots\phi(k)\rangle$. Let us introduce a graph representation for the k-point correlation function

$$G_c(12\dots k) = \tag{4.4}$$

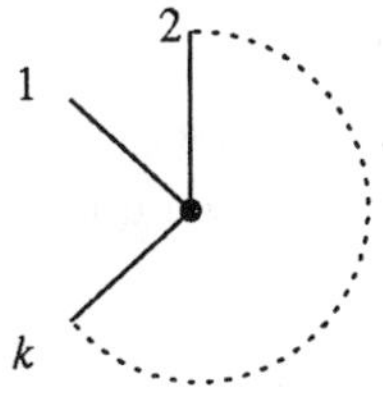

From this definition and Eq. (1.2) it follows that

$$-\frac{\delta}{\delta g_1(k)} \qquad = \qquad \tag{4.5}$$

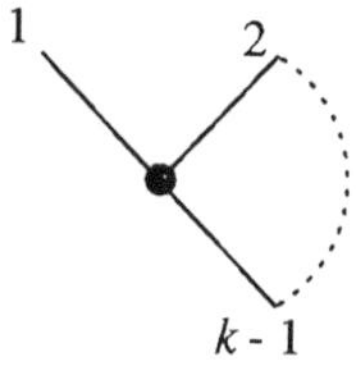

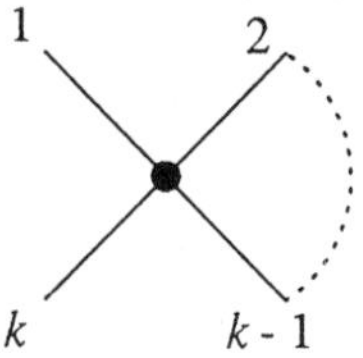

i.e. the procedure of carrying out variational derivatives from connected correlation functions adds one "leg", or simply converts a $(k-1)$-point connected correlation function into a minus k-point function. The term "connected correlators" G_c is introduced to distinguish them from the ordinary correlators G. It emphasizes the absence of unconnected diagrams in the graph representation of the g_4-power series for the G_c functions.

Along with the functions $G_c^{(k)}$, it is convenient to introduce k-point irreducible vertexes $\Gamma^{(k)}$. The perturbation series for these vertexes contain only irreducible graphs, i.e. graphs which cannot be divided in two parts by cutting only one inner line $G_c(12)$. One can also introduce a generating functional for these vertexes. Let us consider the variation of the free energy under the influence of an infinitesimal change in g_1

$$\delta F = \int_1 M(1)\delta g_1(1), \tag{4.6}$$

where $\int_1 = \sum_\alpha \int d\boldsymbol{x}$. The generating functional can be obtained from Eq. (1.6) with the help of Legendre transformation

$$\Gamma\{M\} = F - \int_1 g_1(1)M(1)\ , \qquad \delta\Gamma = -\int_1 g_1(1)\delta M(1), \tag{4.7}$$

This functional generates all irreducible vertexes as follows

$$\Gamma(1\dots k) = \frac{\delta^k\Gamma}{\delta M(1)\cdots\delta M(k)}, \tag{4.8}$$

A graph representation for the irreducible vertexes can be introduced in the same manner as has been done for correlation functions in Eq. (1.4)

$$\Gamma(12\dots k) = \tag{4.9}$$

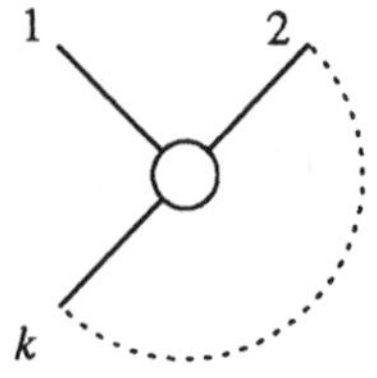

and one can also write the analog of Eq. (1.5)

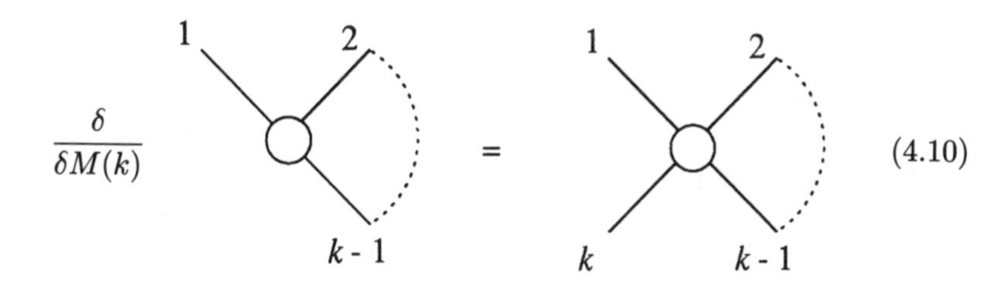

Now, we are going to use the identity

$$\frac{\delta M(1)}{\delta M(1')} \equiv \delta(1-1') \tag{4.11}$$

to prove that $G_c(12)$ and $\Gamma(12)$ are inverse operators in the sense defined by the following equation

$$G_c(12)\Gamma(21') = \Gamma(12)G_c(21') = \delta(1-1'), \tag{4.12}$$

where we have omitted the symbol $\int_2$ which will always be implied when the same number is met twice in an expression, the δ-symbol in Eq. (1.11) means

$$\delta(1-1') \equiv \delta_{\alpha_1\alpha_{1'}}\delta(\boldsymbol{x}_1 - \boldsymbol{x}_{1'}).$$

The proof can be done simply by changing the variational variables in Eq. (1.11) from $M(1)$ to $g_1(1)$ and by taking into account that $G(1) = M(1)$ and $\Gamma(1) = -g_1(1)$. From Eq. (1.12) one can get the relation $\delta G_c\Gamma = -G_c\delta\Gamma$ which leads to the following graph representation for the variational derivative with respect to M from the function G_c

$$-\frac{\delta}{\delta M(k)} \quad \text{———●———} \quad = \; - \quad \text{—●—○—●—} \tag{4.13}$$

k

The rule for the differentiation of graphs with respect to M and the identity

$$-\frac{\delta}{\delta g_1(1)} \equiv G_c(12)\frac{\delta}{\delta M(2)}, \tag{4.14}$$

help to find relations between any connected Green's function and irreducible vertexes. As an example one can find the consequent relation for the 3-point functions

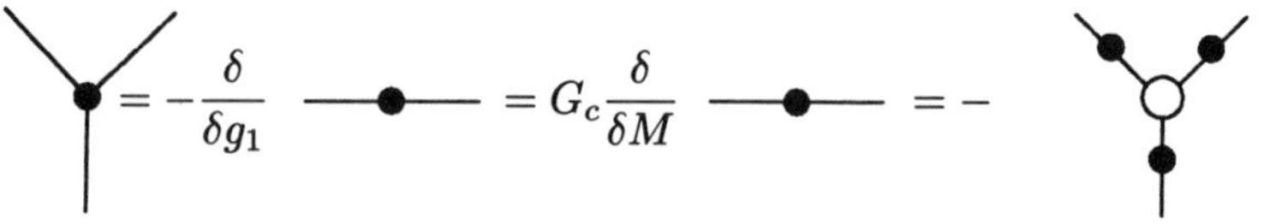

$$= -\frac{\delta}{\delta g_1} \quad = G_c \frac{\delta}{\delta M} \quad = - \qquad (4.15)$$

Another example is the 4-point correlator. It can be obtained with the help of Eq. (1.15) by carrying out the derivative with respect to g_1. This derivative must be consecutively taken from each element in Eq. (1.15) with the application the graph rules stated in Eqs. (1.10), (1.13), and the use of the identity Eq. (1.14). As a result of these actions one arrives at

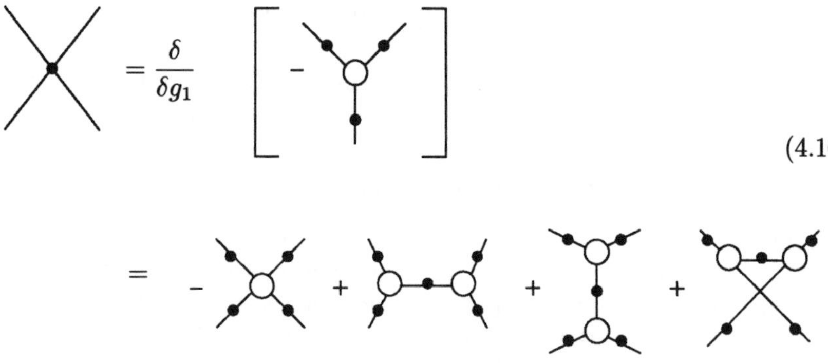

$$= \frac{\delta}{\delta g_1} \left[- \right] \qquad (4.16)$$

$$= - \quad + \quad + \quad +$$

Analogously one can find subsequent relations for the higher order correlation functions. The procedure of carrying out the derivatives with respect to g_1 can be graphically formalized. The formal rules for this procedure are quite simple:

In order to obtain the result of the variational derivative action with respect to $-g_1$ from any graph, one has to find the sum of all graphs which are created from the given one by the addition of a simple line to each black circle and one line with a black circle to each white circle. After that, all 3-point connected correlators should be expressed in terms of the consequent irreducible vertexes with the help of Eq. (1.15).

4.2.2 Graph Equations for ϕ^4-Model

The relations Eq. (1.15) and Eq. (1.16) constitute formal consequences of the initial definitions. These equations are valid for any field-statistical model independently of its Hamiltonian. This means that in order to develop a perturbation theory over the interaction constant one needs to know some additional exact equations, which should result from a particular structure of the initial model. These equations can be obtained on the basis of the following identity

$$\int D\phi \frac{\delta}{\delta\phi} e^{-H} = -\int D\phi e^{-H} \frac{\delta H}{\delta \phi} = 0.$$

Substituting the Hamiltonian Eq. (1.1) into this identity and carrying out the variational derivative with respect to ϕ one arrives at

$$(\nabla^2 - g_2)M(1) - \frac{1}{3!} g_4 \langle \phi(1)\phi^2(1)\rangle - g_1(1) = 0. \tag{4.17}$$

This equation can be written in the graph form as follows

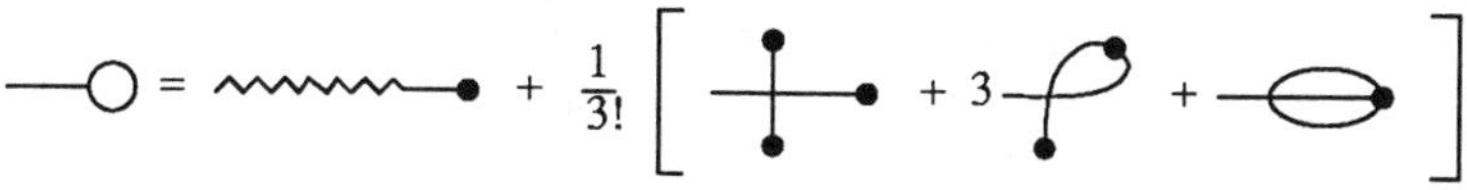

$$—\!\bigcirc = \text{∿∿∿∿}\!—\bullet + \frac{1}{3!}\left[\text{(cross graph)} + 3\,\text{(loop graph)} + \text{(bubble graph)} \right] \tag{4.18}$$

where

$$1 \;\text{∿∿∿∿∿}\; 1' = \Gamma_0^{(2)}(11'),$$

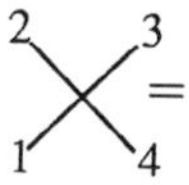

$$\times_{1\,4}^{2\,3} = g_4 e_{\alpha_1\alpha_2\alpha_3\alpha_4}\delta(x_1 - x_2)\delta(x_2 - x_3)\delta(x_3 - x_4) = \Gamma_0^{(4)}(1\ 2\ 3\ 4).$$

The introduced here notation $\Gamma_0^{(2)}$ corresponds to the zero-order vertex $\Gamma^{(2)}$ at $g_4 = 0$. The tensor $e_{\alpha_1\alpha_2\alpha_3\alpha_4}$ is completely symmetrical fourth rank tensor

$$e_{\alpha_1\alpha_2\alpha_3\alpha_4} = \frac{1}{3}(\delta_{\alpha_1\alpha_2}\delta_{\alpha_3\alpha_4} + \delta_{\alpha_1\alpha_3}\delta_{\alpha_2\alpha_4} + \delta_{\alpha_1\alpha_4}\delta_{\alpha_2\alpha_3}).$$

The derivation of Eq. (1.18) is quite simple. One has to use Eq. (1.17) in which the function $G(123)$ is expressed in terms of connected correlators with the help of Eq. (1.3).

The rules for the variational derivative action, formulated at the end of Sect. 1.2.1, constitute a basis for derivation of any expansion in power series of the exact $G_c(11')$ and exact irreducible vertexes. For instance, by carrying out the derivative from Eq. (1.18) with respect to M, one arrives at

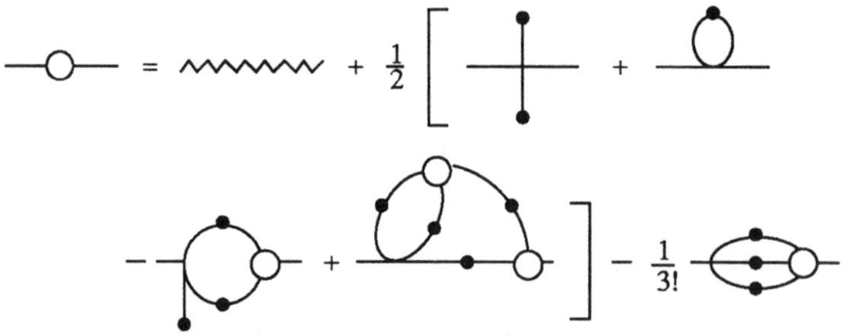

(4.19)

In the same manner, after variation of Eq. (1.18) with respect to g_1 and multiplication by the zero-order Green's function

$$G_0(11') = 1 \text{ —×— } 1'$$

one can obtain the equation for $G_c(11')$

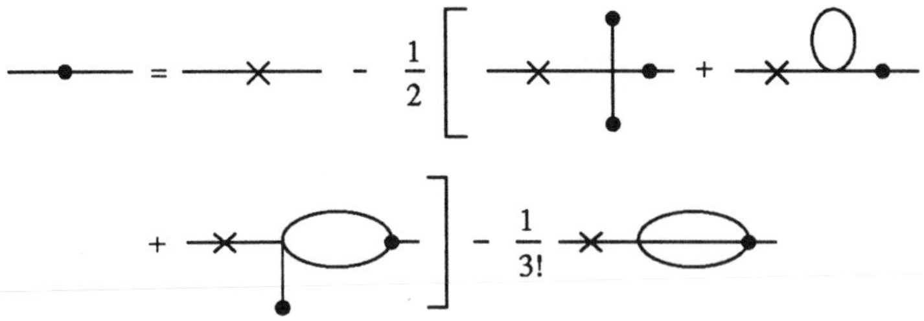

(4.20)

It is interesting to note that the construction of perturbation series with this method reveals one important feature. All higher order vertexes with $k > 4$ can be written as expansions in powers of the exact $G_c^{(2)}$, M, and exact vertexes $\Gamma^{(3)}$ and Γ^4. This feature is inherent for the ϕ^4-model. In general, any series can be obtained by consecutive iterations of equations resulting from the variations of Eq. (1.19) with respect to M. The additional characteristic feature, inherent for any model, is the appearance of one zero-order vertex ($\Gamma_0^{(4)}$ for ϕ^4-model) in each graph. For example, the expansion for the vertex $\Gamma^{(6)}$ starts (at $h = M = 0$) from the terms

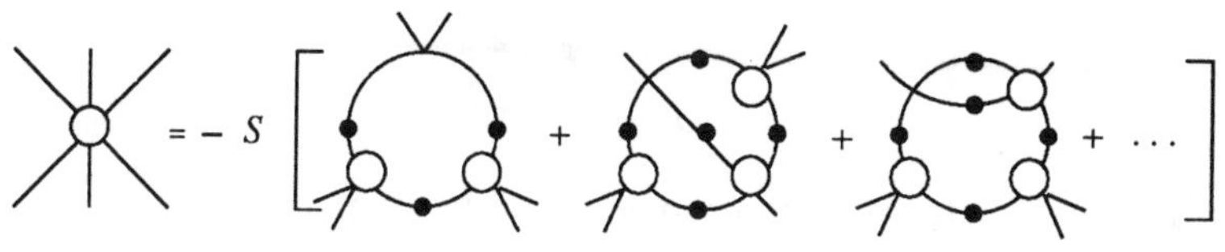

where the symbol S means symmetrization with respect to all external points.

4.2.3 Rules of Evaluation

In the following we will need expansions in powers of g_4 and zero-order two-point propagators. Such expansions can be obtained with the help of consecutive iterations of Eq. (1.20) and substitutions of respective graph series into expressions under consideration. We will also restrict deliberation to the case of translation-invariant systems ($g_1 = const$, $M = const$). As usual, in this case it is more convenient to deal with the Fourier's transforms of k-point functions in d-dimensional space

$$\delta^{(d)}\left(\sum_{i=1}^{k} \boldsymbol{q}_i\right) G^{\alpha_1 \ldots \alpha_k}(\boldsymbol{q}_1, \ldots, \boldsymbol{q}_k)$$

$$= \int d^d x_1 \ldots \int d^d x_k G^{\alpha_1 \ldots \alpha_k}(\boldsymbol{x}_1, \ldots, \boldsymbol{x}_k) \exp\left(-i \sum_{i=1}^{k} \boldsymbol{q}_i \cdot \boldsymbol{x}_i\right). \quad (4.21)$$

The rules of correspondence, with the help of which one could write analytical expressions in the given order of perturbation theory, can be derived after analysis of different series. For $\Gamma^{(k)}$ functions at $M = 0$ they can be formulated as follows:

1. All possible graphs are to be drawn with k external lines and with the number of zero-order vertexes equal to the order of the perturbation with respect to g_4.

2. In each topologically different graph one has to redistribute in all possible ways q_i momentum variables and α_i-vector indexes ($i = 1, \ldots, k$) corresponding to k external lines. To each internal line a momentum must be ascribed so as to satisfy the momentum conservation law at each vertex. If a graph has m closed loops, m internal momenta should be introduced.

3. The factor $G_0(q)$ must be written for each internal line with momentum $\boldsymbol{q}$, as well as the factor $g_4 e_{\alpha_1\alpha_2\alpha_3\alpha_4}$ for each vertex.

4. Every internal momentum must be integrated according to

$$\int_q = \int \frac{d^d q}{(2\pi)^d},$$

and every internal vector index should be summed up.

5. Any diagram with l vertexes must be multiplied by the factor

$$-\left(-\frac{1}{4!}\right)^l \frac{1}{l!} C_i,$$

where C_i is the number of different ways by which the graph of type i can be organized.

For the case of other types of vertexes the detailed description of the correspondence rules can be found in the books by Amit (1984) and Zinn-Justin (1989). In these books there are also a number of examples with which one may practice the procedure of finding the combinatorial coefficients C_i.

Everywhere below we will consider only the perturbation series for $\Gamma^{(k)}$. If one remembers that external lines in these functions do not contain as factors zero order propagators, then it is possible not to place a cross on internal lines without any misinterpretation. It is also necessary to note that construction of diagrams and their analytical counterparts is only the first step in the study of perturbation series. The next step is the procedure of apparent evaluation of different terms along with the physical interpretation. As has already been mentioned, some of expressions are divergent, if one uses for their evaluations the zero order propagator $G(q) = (q^2 + g_2)^{-1}$.

4.2.4 Regularization

Divergent diagrams in the theory of critical phenomena can be treated in the same manner as in the quantum field theory. This procedure consists in "cutting off" the divergent integrals (regularization) and then infinite expressions, which appear in the limit $\Lambda \to \infty$, are adsorbed into the definition of physical parameters and normalization factors. Usually, in the theory of critical behavior the analogy with the Schwinger's parametrical representation for Feynmann graphs is used. This representation allows us to reduce d-dimensional integrals to one-dimensional and to make natural regularization by a simple modification of the free propagator. The essence of this representation is to write the conventional Feynmann propagator $(q^2 + m^2)^{-1}$ in the form

$$\frac{1}{q^2 + m^2} = \int_0^\infty d\alpha \exp[-\alpha(q^2 + m^2)]. \tag{4.22}$$

A cutoff can be introduced in this representation with the help of integral factor $f(x)$ with $x = \alpha\Lambda^2$. The constraints on the choice of the function f can be formulated as follows. When x is large, the function f should be close to unit, but at $x \to 0$ it should decrease quicker than some power k of x ($f < Ax^k$). The value of k depends on the structure of the theory, as it defines the behavior of $G(p)$ at large momenta. For the interaction Hamiltonian of the ϕ^4-model the value of k must be higher than one (see Sect. 1.2.4). Of course, one can choose the cutoff function to ensure regularization for a quite broad class of Hamiltonians. Thus, $f = \theta(x-1)$ (where θ is Heaviside's step function) allows for the regularization of a theory having divergences of any power of Λ^2.

The representation Eq. (1.22) helps to rewrite any diagram with L loops and I internal lines as a Gaussian integral, which can be easily carried out (see Appendix A)

$$\begin{aligned}
&\int \prod_{i=1}^{I} d\alpha_i f(\alpha_i \Lambda) \prod_{k=1}^{L} d^d q_k \\
&\times \exp\left\{ -\frac{1}{2} \sum_{kj} \boldsymbol{q}_k A_{kj}(\alpha_i) \boldsymbol{q}_j + \sum_k \boldsymbol{q}_k \boldsymbol{B}_k(\alpha_i) \right\} \\
&= \int \prod_{i=1}^{I} d\alpha_i f(\alpha_i \Lambda) \frac{(2\pi)^{Ld/2}}{(\det A)^{1/2}} \\
&\times \exp\left\{ \frac{1}{2} \sum_{kj} \boldsymbol{B}_k(\alpha_i) \left[A^{-1}(\alpha_i) \right]_{kj} \boldsymbol{B}_j(\alpha_i) \right\}.
\end{aligned} \tag{4.23}$$

This type of calculations gives expressions which are apparently dependent on the space dimension d. This then allows the analytical continuation to noninteger d, which is essential for ϵ-expansion.

4.3 Renormalization

In this section we will discuss the main ideas of the renormalization theory. The impetus of the discussion is mainly on the physical aspects of this problem. One can find more rigorous considerations as well as detailed analysis of all subtleties of this theory in the review by Bresin *at al* (1976), the books by Amit (1984) and Zinn-Justin (1989).

Historically, the renormalization theory appeared as a means to cope with ultraviolet divergences in quantum electrodynamics. Now it is clear that this theory reflects a much more important symmetry aspect of field theoretical calculations, i.e. a symmetry with respect to renormalization group transformations. Nonetheless, even now, it seems quite natural to start rendition of this theory from the analysis of divergences. This analysis for the ϕ^4-model should be done taking into account the possibility of the appearance of other kinds of vertexes. The latter is essential if one would introduce counter-terms to the zero-order Hamiltonian which must be compensated by the consequent terms in the interaction part of the Hamiltonian.

4.3.1 Calculus of Divergences

Let us consider an arbitrary local vertex composed from the v_i power of the field variable ϕ on which k_i derivatives are acting. For instance, in the case of ϕ^4 interaction one has $v = 4$ and $k = 0$ $(i = 1)$, and consequently for $(\boldsymbol{\nabla}\phi)^2(\phi^2)^n$ we have $v = 2(n+1)$ and $k = 2$. It is convenient to introduce an index $\delta(\gamma)$ which is equal to the power of the increase of a diagram γ, when all momenta tend to infinity with the same rate. The rate of increase due to the integrations is equal to dm, as the number of internal integrals is equal to the number of loops m (see Sect. 1.2.3). In addition it is necessary to take into account increase caused by the derivatives in vertexes. This increase amounts to $\sum_i k_i n_i$, where n_i is a number of vertexes of the i-th kind. At last, from the whole rate of increase $dm + \sum_i k_i n_i$ we must subtract decrease caused by the zero order propagators. As the number of the propagators is equal to the number of the internal lines I and each propagator in the limit of large momentum is proportional to q^{-2}, we have decrease equal to $2I$. Therefore, as a result for the index $\delta(\gamma)$ one obtains the relation

$$\delta(\gamma) = dm + \sum_i k_i n_i - 2I. \tag{4.24}$$

This relation can be transformed in such a way that the value of $\delta(\gamma)$ will be characterized only by the types of vertexes $(k_i,\ v_i)$ and by the type of irreducible diagram, i.e. by the number of external lines. Let us consider the simplest graph with one loop and one conservation law (one vertex). If one inserts in this graph one additional conservation law, then the number of lines for the loop becomes equal to two. By continuation of this procedure one can build any diagram. However, the number of loops will always exceed by one the number of internal lines minus the number of all conservation laws in the diagram, i.e.

$$m = 1 + I - \sum_i n_i, \tag{4.25}$$

Additional relations can be obtained, if one takes into account an apparent topological property, that each internal line is connected to two vertexes while each external line is connected to only one. Therefore, the number of lines entering into all vertexes (equal to $\sum_i n_i v_i$) can be obtained from a graph if one would cut each internal line in two pieces, or count them twice. The latter gives the following relation between the number of external lines and characteristic numbers for vertexes

$$k + 2I = \sum_i n_i v_i, \tag{4.26}$$

With the use of Eqs. (1.25), and (1.26) one can find an expression for m and I and transform Eq. (1.24) as follows

$$\delta(\gamma) = d - \frac{d-2}{2}k + \sum_i n_i \delta_i, \tag{4.27}$$

where the value δ_i is characterized only by the choice of vertex and is equal

$$\delta_i = k_i + v_i \frac{d-2}{2} - d. \tag{4.28}$$

A diagram which has $\delta \geq 0$ is divergent. When $\delta < 0$ the diagram, as a whole, is convergent, as divergences in the completely integrated expression are absent. However, internal integrations can be divergent. Such divergences are related to subdiagrams. The behavior of a diagram with the order of perturbation is dependent on the sign of δ_i. When $\delta_i > 0$, δ is increasing with the increase of the order of the i-th perturbation. This type of interaction leads to a nonrenormalizable theory. A theory of this kind cannot be made convergent with the help of the introduction of a a finite number of counter terms in the Hamiltonian. Therefore, in order to have a renormalizable theory we should restrict consideration with the case $\delta_i \leq 0$. When $\delta_i = 0$ the value of δ is independent of the order of perturbation (renormalizable theory). In the case with $\delta_i < 0$ the value of δ is a decreasing function of the order of perturbation theory (super-renormalizable theory). In both cases one has $\delta \leq d - (d-2)k/2$. Hence, if $d \geq 2$ divergent diagrams have only a finite number of external lines $k \leq 2d/(d-2)$.

Let us investigate which kinds of interactions are allowed for renormalizable theories. When $d = 4$, the condition $\delta_i \leq 0$ leaves only ϕ, ϕ^2, ϕ^3, ϕ^4, and $(\nabla\phi)^2$, i.e. $(k_i + v_i) \leq 4$. When $d = 3$, to these interactions one should also add ϕ^5 and ϕ^6 $(k_i + v_i/2 \leq 3)$. In the two-dimensional case, from the condition $\delta_i \leq 0$ follows $k_i \leq 2$ and are allowed interactions ϕ^v with any v along with $(\nabla\phi)^2\phi^v$. In the case of the ϕ^4-model within the region $2 \leq d \leq 4$, only the vertexes $\Gamma^{(2)}$ and $\Gamma^{(4)}$ are divergent as a whole. All divergences of other irreducible vertexes are completely incorporated into the inner parts of diagrams with the structures of corrections to $\Gamma^{(2)}$ and $\Gamma^{(4)}$.

4.3.2 Simple Generalizations

In a more general case, we need to introduce, along with the irreducible vertexes $\Gamma^{(k)}$ and, corresponding to them, connected correlators, correlators with coinciding spatial coordinates x_i. For example, such a function occurs, if one calculates the energy correlator. In this case, even if all divergences in $\Gamma^{(k)}$ are incorporated into factors and parameter renormalization of the $\Gamma^{(2)}$ and $\Gamma^{(4)}$, we will have a divergent result in the limit when $x_i \to x_j$. This means that we need to advance an additional renormalization to find a physically reasonable result. To be consistent with the general scheme of this rendition one can introduce a generating functional to produce correlators and vertexes with the inserts $\phi^2(x)$. Such inserts will generate functions $G^{(l,k)}$ and $\Gamma^{(l,k)}$, where l is the number of the inserts of the operators $\phi^2(x)$ The generating functional can be found with the help of a simple generalization of the initial Hamiltonian Eq. (1.1). Let us assume that the vertex g_2 in the Hamiltonian is a coordinate dependent function. Then the functions $G^{(l,k)}$ and $\Gamma^{(l,k)}$ are defined with the help of the functionals $F\{g_1, g_2\}$ and $\Gamma\{M, g_2\}$ as follows

$$G^{(l,k)}(x_1 \dots x_l, 1 \dots k) = \frac{(-1)^{l+k+1}\delta^{l+k}F\{g_1, g_2\}}{\delta g_2(x_1)\cdots\delta g_2(x_l)\delta g_1(1)\cdots\delta g_1(k)}, \tag{4.29}$$

$$\Gamma^{(l,k)}(x_1 \dots x_l, 1 \dots k) = \frac{\delta^{l+k}\Gamma\{M, g_2\}}{\delta g_2(x_1)\cdots\delta g_2(x_l)\delta M(1)\cdots\delta M(k)}. \tag{4.30}$$

In full analogy with Eq. (1.4) and Eq. (1.9) one can introduce graph representations

$$G^{(l,k)}(x_1 \dots x_l, 1 \dots k) = \tag{4.31}$$

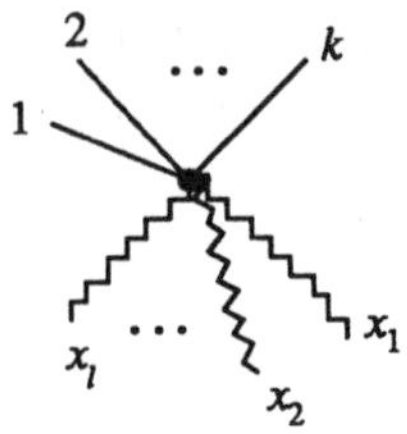

$$\Gamma^{(l,k)}(x_1 \ldots x_l, 1 \ldots k) = \tag{4.32}$$

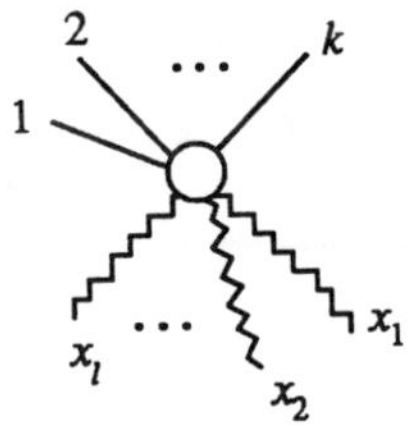

The graph relations between the functions $G^{(l,k)}$ and $\Gamma^{(l,k)}$ can be obtained in the same manner as it has been done for the functions $G_c^{(k)}$ and $\Gamma^{(k)}$ in Sect. 1.2.1. However, in this case one has to take into account that in addition to Eq. (1.14) the change of variational variables leads to the identity

$$\left.\frac{\delta}{\delta g_2(x)}\right|_{g_1} \equiv \left.\frac{\delta}{\delta g_2(x)}\right|_{M} + \frac{\delta M(1)}{\delta g_2(x)} \left.\frac{\delta}{\delta M(1)}\right|_{g_2}. \tag{4.33}$$

From the definitions of Eqs. (1.31), (1.32) and the identity, Eq. (1.33) one can find that

$$-\frac{\delta}{\delta g_2(x)} \tag{4.34}$$

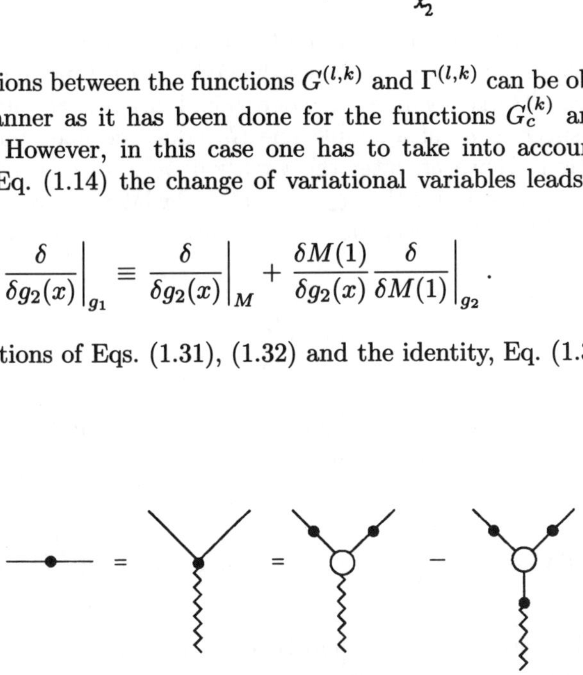

As an example let us define a relation between $G^{(1,3)}$ and $\Gamma^{(l,k)}$, where $l \leq 1$ and $k \leq 3$. This relation can be found by carrying out the derivative with respect to $g_2(x)$ from Eq. (1.8) at a fixed value of g_1. Taking also into account Eq. (1.33) one arrives at the following graph equation

$$\tag{4.35}$$

where

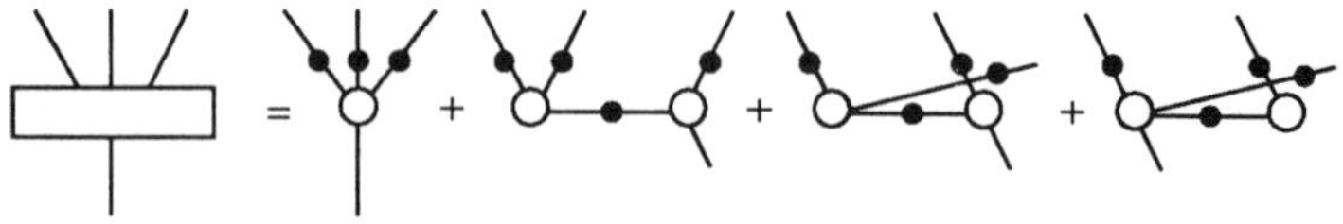

(4.36)

Analogously one can find relations for the higher order correlation functions.

Let us now consider divergency behavior of the functions $\Gamma^{(l,k)}$. In order to find the divergency index $\delta(\gamma)$ one should take into account that variation with respect to g_2 of a simple line G_0 adds one line and one additional integration. However, after the variation we have to return to the space homogeneous situation (i.e. $g_2 = const$). The latter leads to the appearance of one additional conservation of momentum law related to the zero order vertex

and therefore, the additional integration will be canceled. As a result each wavy line will decrease the value of $\delta(\gamma)$, defined by the formula, Eq. (1.27), on two units, i.e.

$$\delta(\gamma) = \Delta_{lk} + \sum_i n_i \delta_i, \qquad \Delta_{lk} = d - \frac{d-2}{2}k - 2l. \tag{4.37}$$

Using this formula one can find that in the ϕ^4-model, there are three additional divergent quantities at $d = 4$: $\Gamma^{(1,0)}$, $\Gamma^{(2,0)}$ and $\Gamma^{(1,2)}$. The vertex with $k = 0$ and $l = 1$ is a constant equal to $\langle\phi^2\rangle$. This constant is not essential in the following consideration, since the corresponding counter-term in the Hamiltonian also does not have an operator structure. The logarithmically divergent function $\Gamma^{(2,0)} = \langle\phi^2(x)\,\phi^2(0)\rangle$ does not have external lines. Hence, this diagram cannot appear as a divergent fragment in graph representations for the functions $\Gamma^{(l,k)}$ at $k \neq 0$. As a result this divergence can be canceled with the help of the subtraction not affecting any other function $\Gamma^{(l,k)}$.

4.3.3 Multiplicative Group Equations

At the point of phase transition $\Gamma^{(0,2)}(0) = 0$. This equation defines renormalization of the transition temperature, i.e. the value of g_{2c}. This value in the zero order approximation over g_4 is absent. Therefore, the equation defining the value g_{2c} as a function of the system parameters g_4, Λ in the implicit form can be written as

$$\Gamma^{0,2}(0, g_{2c}, g_4, \Lambda) = 0. \tag{4.38}$$

Let us make the following substitutions for the field variables and trial vertexes

$$\phi(x) = Z_3^{1/2} \phi_R(x), \tag{4.39}$$

$$g_4 = g Z_1 Z_3^{-2}, \tag{4.40}$$

$$g_2 = g_{2c} + \tau Z_2 / Z_3. \tag{4.41}$$

The goal of these substitutions will be cleared by what follows. The introduced new field variables ϕ_R define new irreducible vertexes $\Gamma_R^{(l,k)}$. Let us explicitly denote momentum dependence in the functions $\Gamma_R^{(l,k)}(\{\boldsymbol{p}, \boldsymbol{q}\}, \tau, g, \Lambda)$ where we have introduced two sets of momenta: $\boldsymbol{p} = (\boldsymbol{p}_1, \ldots, \boldsymbol{p}_l)$ are momenta related to wavy lines and $\boldsymbol{q} = (\boldsymbol{q}_1, \ldots, \boldsymbol{q}_k)$ are momenta related to ϕ lines. The new vertexes can be easily expressed in terms of the initial ones

$$\Gamma_R^{(l,k)}(\{\boldsymbol{p}, \boldsymbol{q}\}, \tau, g, \Lambda) = Z_3^{k/2-l} Z_2^l \Gamma^{(l,k)}(\{\boldsymbol{p}, \boldsymbol{q}\}, g_2, g_4, \Lambda). \tag{4.42}$$

This equation can be easily proved by comparison of the graph representations of perturbation series for Γ_R and Γ. The transformation $R(Z_1, Z_2, Z_3)$ from the values Γ to the values Γ_R constitutes a triparametrical group. From Eq. (1.42) one can see that this group possesses the multiplicativity property

$$R(Z_1, Z_2, Z_3) R(Z_1', Z_2', Z_3') = R(Z_1 Z_1', Z_2 Z_2', Z_3 Z_3'). \tag{4.43}$$

It is evident that the covariance of such a type is a general property of any field theory. In the case of renormalizable theories one can choose the parameters Z_1, Z_2 and Z_3 so as to have the values $\Gamma_R^{(l,k)}$ finite in the limit $\Lambda \to \infty$. The latter is possible because we need to use only a finite number of counter-terms in the Hamiltonian in order to eliminate all divergences. At $d = 4$ only the values $\Gamma_R^{(0,2)}$, $\Gamma_R^{(0,4)}$ and $\Gamma_R^{(1,2)}$ should be made finite as the other vertexes can be represented in the form of expansions over the exact $G_c^{(2)}$ and exact new vertexes $\Gamma_R^{(0,4)}$ and $\Gamma_R^{(0,2)}$ (see Sect. 1.2.1).[1]

[1] In the general case of vertexes with some number of ϕ^2 inserts, the value $\Gamma_R^{(1,2)}$ also enters into expansions.

Evidently, if one chooses Z_1, Z_2 and Z_3 in such a way that $\Gamma_R^{(l,k)}$ are finite at some fixed values of momenta, these functions will be finite at any values of momenta. Let us then fix the quantities of renormalized vertexes by some normalization conditions at $|\boldsymbol{q}| = |\boldsymbol{p}| = \mu$ and $\boldsymbol{q}_i\boldsymbol{q}_j = \mu^2(4\delta_{ij} - 1)/3$. Naturally, this choice of the values of momenta for the normalization is arbitrary. It should not be considered better than any other choice. However, it leads to some simplifications in concrete calculations and at present such a choice is widely accepted. Imposing the constraints of the normalization, one should take into account that the most divergent terms (at $d = 4$ diverging like Λ^2) must be eliminated by the change of the trial critical temperature (Eq. (1.38)). However, this condition still leaves a logarithmic divergence in $\Gamma^{(0,2)}(\boldsymbol{q}, g_{2c}, g_4, \Lambda)$. This divergence looks like $\boldsymbol{q}^2 \ln \Lambda$. It can be eliminated if one chooses the normalization as follows

$$\left. \frac{\partial \Gamma_R^{(0,2)}(\boldsymbol{q}, 0, g, \Lambda)}{\partial \boldsymbol{q}^2} \right|_{q^2=\mu^2} = 1. \tag{4.44}$$

In the power series over g_4 the logarithmic divergence first appears in the second order term from the graph

(4.45)

After the subtraction of the terms contributing to g_{2c} in the limit of large Λ, one obtains the quantity of the order of $\boldsymbol{q}^2 \ln \Lambda$. The condition Eq. (1.44) adsorbs this divergence into the value of the multiplicative constant Z_3, which can be defined with the help of Eq. (1.42) and Eq. (1.44) as

$$Z_3 = \left[\left. \frac{\partial \Gamma^{(0,2)}(\boldsymbol{q}, g_{2c}, g_4, \Lambda)}{\partial \boldsymbol{q}^2} \right|_{q^2=\mu^2} \right]^{-1}. \tag{4.46}$$

After this normalization, there remain only two different kinds of logarithmic divergences in the values $\Gamma^{(0,4)}$ and $\Gamma^{(1,2)}$. The divergence in $\Gamma^{(0,4)}$ can be absorbed into the definition of a new interaction constant g. This can be done with the help of the following condition

$$\Gamma_R^{(0,4)}(q_1, q_2, q_3, 0, g, \Lambda)|_{q_iq_j=(4\delta_{ij}-1)\mu^2/3} = g, \tag{4.47}$$

Consequently, the divergence caused by the ϕ^2 inserts can be eliminated by the normalization condition

$$\Gamma_R^{(1,2)}(\boldsymbol{p}, \boldsymbol{q}_1, \boldsymbol{q}_2, 0, g, \Lambda)|_{q_1^2=q_2^2=\mu^2;\ q_1q_2=-\mu^2; p^2=\mu^2} = 1. \tag{4.48}$$

Thus, the set, Eqs. (1.38), (1.44), (1.47), and Eq. (1.48) defines values g_{2c}, g, Z_i $(i = 1, 2, 3)$ as functions of μ, Λ and g_4.

4.3.4 Scaling Properties

In accordance with the main hypothesis of Kadanoff invariance (see Sect. 3.2) at the critical point, the system can be described independently of the value of the length scale. The transformation from one scale to another corresponds to the change of normalization. It is clear that in order to find points in the parametric space g_2, g_4 with this invariance one has to study scale covariance properties of the field theory. Naturally, the scale covariance is only an additional tool, it cannot solve the problem without consideration of the renormalization covariance of the theory.

By changing the scale of momenta in λ times, one can find the conditions describing the scale covariance of the initial Γ-functions in the form

$$\Gamma^{(l,k)}(\{p,q\},g_2,g_4,\Lambda) = \lambda^{\Delta_{lk}}\Gamma^{(l,k)}\left(\left\{\frac{p}{\lambda},\frac{q}{\lambda}\right\},\frac{g_2}{\lambda^2},\frac{g_4}{\lambda^\epsilon},\frac{\Lambda}{\lambda}\right). \tag{4.49}$$

Analogously, one can find a similar relations for the renormalized functions

$$\Gamma_R^{(l,k)}(\{p,q\},g_2,g_4,\Lambda) = \lambda^{\Delta_{lk}}\Gamma_R^{(l,k)}\left(\left\{\frac{p}{\lambda},\frac{q}{\lambda}\right\},\frac{g_2}{\lambda^2},\frac{g_4}{\lambda^\epsilon},\frac{\Lambda}{\lambda}\right). \tag{4.50}$$

The system of units used here has only one dimensional unit, the unit of length Λ^{-1} or momentum Λ. This means that dimensions of all quantities in the problem can be defined in terms of the power a of Λ. From Eq. (1.50) one can see that for the vertexes g_2 and τ the power is given by $a = 2$, while for the interaction constants g_4 and g, $a = \epsilon$. In the following we will denote dimensionless quantities normalizing initial parameters of the problem by Λ and renormalized ones by the value of the normalization momentum μ. Introducing the dimensionless quantities $u_0 = g_4\Lambda^{-\epsilon}$, $u = g\mu^{-\epsilon}$, $t = \tau\mu^{-2}$, and $\rho = \mu/\Lambda$ one can rewrite Eqs. (1.39), (1.40), and Eq. (1.41) in the form

$$\phi(x) = Z_3^{1/2}(u,\rho)\phi_R(x), \tag{4.51}$$

$$u_0 = \rho^\epsilon u Z_1(u,\rho)Z_3^{-2}(u,\rho), \tag{4.52}$$

$$\frac{g_2}{\Lambda^2} = \frac{g_{2c}}{\Lambda^2} + \rho t Z_2(u,\rho)/Z_3(u,\rho). \tag{4.53}$$

In the same manner, substituting $\lambda = \Lambda$ in Eq. (1.49) and $\lambda = \mu$ in

Eq. (1.50) one can transform Eq. (1.42) as

$$\rho^{\Delta_{lk}} Z_3^{l-k/2}(u,\rho) Z_2^{-l}(u,\rho) \overline{\Gamma}_R^{(l,k)}(\{p/\mu, q/\mu\}, t, u, \rho)$$

$$= \overline{\Gamma}^{(l,k)}(\{p/\Lambda, q/\Lambda\}, g_2/\Lambda^2, u_0), \qquad (4.54)$$

where the following notations are introduced

$$\overline{\Gamma}_R^{(l,k)}(\{p/\mu, q/\mu\}, t, u, \rho) = \Gamma_R^{(l,k)}(\{p/\mu, q/\mu\}, t, u, \rho^{-1})$$

$$\overline{\Gamma}^{(l,k)}(\{p/\Lambda, q/\Lambda\}, g_2/\Lambda^2, u_0) = \Gamma^{(l,k)}(\{p/\Lambda, q/\Lambda\}, g_2/\Lambda^2, u_0, 1)$$

It is necessary to mention that the conditions for subtraction of divergences Eqs. (1.38), (1.44), (1.47), and Eq. (1.48) turn the triparametrical renormalization group into a one-parameter group. All normalization constants Z_i can be expressed as functions of a single arbitrary parameter μ. Really, we have Eq. (1.52) and Eq. (1.53) which contain two variables ρ, μ and fixed physical parameters u_0 and Λ. The critical limit is achieved when g_2 tends to g_{2c} at fixed u_0 and Λ. Consequently, for the renormalized theory the critical point can be reached, as one can see from Eq. (1.53), when $t \to 0$ at fixed u and ρ. In this manner the renormalized theory practically solved the problem of the summation of all infrared divergences. These divergences arise as a result of the expansion in the power series with respect to u_0 or u when $g_2 \to g_{2c}$ or $t \to 0$.

4.3.5 Differential Group Equations

The dependence of all vertexes on some arbitrary parameter μ allows one to obtain an equation applicable even at the critical point. Let us use the independence of the right hand side of Eq. (1.54) on μ or ρ. After differentiation of this equation with respect to ρ, when the values g_2, u_0 and Λ are fixed, one arrives at the differential RG equations in the form

$$\left\{ \rho \frac{\partial}{\partial \rho} + w(u,\rho) \frac{\partial}{\partial u} - \frac{t}{\nu(u,\rho)} \frac{\partial}{\partial t} - \sum_{i=1}^{l} \boldsymbol{p}_i \frac{\partial}{\partial \boldsymbol{p}_i} - \sum_{i=1}^{k} \boldsymbol{q}_i \frac{\partial}{\partial \boldsymbol{q}_i} \right.$$

$$\left. + d - \frac{l + k\beta(u,\rho)}{\nu(u,\rho)} \right\} \overline{\Gamma}_R(\{p,q\}, t, u, \rho) = 0, \qquad (4.55)$$

where $\{p, q\}$ are dimensionless sets of momenta, other functions defined by the following equations

$$w(u,\rho) = \rho \left(\frac{\partial u}{\partial \rho}\right)\bigg|_{u_0}, \tag{4.56}$$

$$\nu(u,\rho) = \frac{1}{2-\eta_3(u,\rho)+\eta_2(u,\rho)}, \tag{4.57}$$

$$\beta(u,\rho) = \frac{1}{2}\nu(u,\rho)\{d-2+\eta_3(u,\rho)\}, \tag{4.58}$$

$$\eta_i(u,\rho) = \rho\,\frac{\partial \ln Z_i(u,\rho)}{\partial \rho}\bigg|_{u_0} \qquad (i=2,3). \tag{4.59}$$

Until now no approximation has been made. One is compelled to make approximations only in concrete calculations. The first essential simplification arises from the restriction to the critical region where we have inequalities $t \ll \rho^{-1}$ and $\{|p_i|, |q_i|\} \ll \rho^{-1}$. The universal character of critical behavior, i.e. its independence of the value Λ, enables one to assume that in the renormalized theory critical behavior corresponds to $\Lambda \to \infty$, when u and μ are fixed. This means, that one has to define the functions $Z_i(u,\rho)$ in the limit $\rho \to 0$ at $u = const$. These functions can only be defined in the perturbation theory over the powers of u. Therefore, it is necessary to find conditions under which such an expansion is possible (for instance $\epsilon \ll 1$). Nonetheless, some general analysis of Eq. (1.55) can be done in the critical region without performing concrete calculations.

4.4 Scaling Laws

Let us find a formal solution of Eq. (1.55) using the method of characteristics. This method leads to the following representation for vertexes

$$\overline{\Gamma}_R(\{p,q\},t,u,0) = \exp\left\{-\int_\lambda^1 \frac{dx}{x}\left[d - \frac{l+k\beta(x)}{\nu(x)}\right]\right\}\overline{\Gamma}_R(\{\bar{p},\bar{q}\},\bar{t},\bar{u},0), \tag{4.60}$$

where $\beta(x)$ and $\nu(x)$ are functions equal to $\beta(\bar{u}(x),0)$ and $\nu(\bar{u}(x),0)$. The system of characteristic equations is defined by

$$\lambda\frac{d\bar{u}}{d\lambda} = w(\bar{u},0), \tag{4.61}$$

$$\lambda\frac{d\bar{t}}{d\lambda} = -\frac{\bar{t}}{\nu(\bar{u},0)}, \tag{4.62}$$

$$\lambda\frac{d\bar{\boldsymbol{p}}_i}{d\lambda} = -\bar{\boldsymbol{p}}_i, \quad \lambda\frac{d\bar{\boldsymbol{q}}_i}{d\lambda} = -\bar{\boldsymbol{q}}_i. \tag{4.63}$$

Initial conditions for these equations should be chosen as follows: $\bar{u}(1) = u$, $\bar{t}(1) = t$, $\bar{\boldsymbol{p}}_i = \boldsymbol{p}_i$ and $\bar{\boldsymbol{q}}_i = \boldsymbol{q}_i$.

One can verify that Eq. (1.60) is a solution of Eq. (1.55) by taking the derivative with respect to λ. As the left hand side of Eq. (1.60) does not depend on λ and the evolution with λ of the functions having bars is defined by Eqs. (1.61), (1.62), and (1.63) one can obtain Eq. (1.55) in which, however, all the variables $t, u, \ldots$ are replaced by $\bar{t}, \bar{u}, \ldots$.

Let us start analysis of Eq. (1.60) from the case when all momenta are equal to zero. By putting $\bar{t} = s$, with s finite and positive one obtains from Eq. (1.62) the expression for t

$$t = s\exp\left\{-\int_\lambda^1 \frac{dx}{x\nu(x)}\right\}. \tag{4.64}$$

When $t \to 0$ the parameter λ must tend to zero too. The function $\nu(x)$ at $x \to 0$ must have a finite limit $\nu = \nu(u^*)$, where, as one can see from Eq. (1.61), u^* should be defined from condition

$$w(\bar{u})|_{\bar{u}=u^*} = 0. \tag{4.65}$$

The function $\overline{\Gamma}_R(\{0\}, t, u^*, 0)$ is a finite value. As follows from Eq. (1.64) $t \propto \lambda^{1/\nu}$. Therefore, from the solution, Eq. (1.60), one can find that at $t \to 0$

$$\overline{\Gamma}_R^{(l,k)}(\{0\}, t, u, 0) \propto t^{-l+\nu\left[d-\frac{k}{2}(d-2+\eta)\right]}, \tag{4.66}$$

where $\eta = \eta_3(u^*, 0)$. When $l = 2$ and $k = 0$ we must get the value proportional to the specific heat, i.e. $t^{-\alpha}$. Comparing this result with Eq. (1.66) one obtains $d\nu = 2 - \alpha$ (see Sect. 3.3.4, Eq. (3.77)). When $l = 0$ and $k = 2$ the vertex Γ defines inverse correlation function $(G^{(2)})^{-1}$ (see Eq. (1.12)) which is proportional to t^γ. Hence, $\gamma = (2 - \eta)\nu$, i.e. one more well known scaling law. Now, let us consider the case $t = 0$. The solutions of Eq. (1.63) have the forms $\bar{\boldsymbol{p}}_i(\lambda) = \boldsymbol{p}_i/\lambda$ and $\bar{\boldsymbol{q}}_i(\lambda) = \boldsymbol{q}_i/\lambda$. Analogously, as has been done with the t-variable, let us fix some values for $|\bar{\boldsymbol{p}}_i|$ and $|\bar{\boldsymbol{q}}_i|$ at $|\boldsymbol{p}_i|$ and $|\boldsymbol{q}_i|$ tending to zero. As a result one can find

$$\overline{\Gamma}_R^{(l,k)}(\{\lambda\bar{p}, \lambda\bar{q}\}, 0, u, 0) \propto \lambda^{\left[d-\frac{k}{2}(d-2+\eta)-\frac{l}{\nu}\right]}. \tag{4.67}$$

To obtain this equation one should also take into account that the value $\overline{\Gamma}_R^{(l,k)}(\{\bar{p},\bar{q}\},0,u^*,0)$ does not contain singularities with respect to λ at fixed $\{\bar{p},\bar{q}\}$. Now, for the case $l=0$ and $k=2$ from Eq. (1.67) it follows that

$$\chi^{-1}(q,0)=\Gamma^{(0,2)}(p,0,u,0)\propto\Gamma_R^{(0,2)}(p,0,u,0)\propto q^{2-\eta}.$$

Taking into account also that $\chi^{-1}(0,t)\propto t^{\nu(2-\eta)}$ (see Eq. (1.66)), one can derive that correlation length $\xi\propto t^{-\nu}$ in the critical region (see also Sect. 3.3.4).

Until now all the consideration has been restricted with the case $t\geq 0$. The spontaneous break of symmetry below the critical temperature leads to some modifications of the above relations. First of all, one should deal with the difference in the behavior of a one-component system ($n=1$) and systems with Goldstone massless modes ($n\geq 2$). Let us first consider the case with $n=1$. When $t<0$ the magnetization in the system $M\neq 0$, and therefore, all irreducible vertexes $\Gamma_R^{(l,k)}$ will be dependent (along with the variables $\{p,q\},t,u$) on M. This accounts for some changes in the RG equations. In order to find the consequent equations replacing Eq. (1.55), one could go along the lines of the derivation given in the Sect. 1.3. However, there is a much simpler way if one makes use of the definition $\Gamma^{(l,k)}$ from the generating functional $\Gamma\{M,g_2\}$ (see Eq. (1.30))

$$\Gamma_R^{(l,0)}(x_1,\ldots,x_l,M)=\sum_{k=0}^{\infty}\frac{M(1)\ldots M(k)}{k!}\Gamma_R^{(l,k)}(x_1,\ldots,x_l,1,\ldots,k).$$

From this equation, it follows that when $M=const$

$$\overline{\Gamma}_R^{(l,k)}(\{p,q\},t,M,u,\rho)=\sum_{m=0}^{\infty}\frac{M^m}{m!}\overline{\Gamma}_R^{(l,m+k)}(\{p,q\},t,0,u,\rho). \tag{4.68}$$

As we know equations for Γ_R when $M=0$ (see Eq. (1.55)), Eq. (1.68) helps to find equations at arbitrary value of M. Here, for the sake of simplicity, we give this equation at zero values of the variables $\{p,q\}$ and ρ

$$\left\{w(u)\frac{\partial}{\partial u}-\frac{t}{\nu(u)}\frac{\partial}{\partial t}+d-\frac{l+k\beta(u)}{\nu(u)}-\frac{\beta(u)}{\nu(u)}M\frac{\partial}{\partial M}\right\}\overline{\Gamma_R^{(l,k)}}=0. \tag{4.69}$$

The solution of this equation can be obtained in a form analogous to Eq. (1.60)

$$\overline{\Gamma}_R(\{0\},t,M,u,0)=\exp\left\{-\int_\lambda^1\frac{dx}{x}\left[d-\frac{l+k\beta(x)}{\nu(x)}\right]\right\}\overline{\Gamma}_R(\{0\},\bar{t},\bar{M},\bar{u},0), \tag{4.70}$$

where along with Eq. (1.61) and Eq. (1.62) defining the functions $\bar{u}$ and $\bar{t}$ one should introduce the following equation for $\bar{M}$

$$\lambda\frac{d\bar{M}}{d\lambda} = -\frac{\beta(\bar{u})}{\nu(\bar{u})}\bar{M} \tag{4.71}$$

with the initial condition $\bar{M}(1) = M$.

As before, let us fix $\bar{t} = \pm s$ $(s > 0)$ and put $t \to 0$. Again $|t| \propto \lambda^{1/\nu}$ and from Eq. (1.71) one can find $M \approx \bar{M}\lambda^{\beta/\nu}$ $(\bar{t} < 0)$, i.e. $M \propto |t|^{\beta}$ at fixed $\bar{M}$. The analog of Eq. (1.66) can also be easily obtained in the form

$$\overline{\Gamma}_R^{(l,k)}(\{0\}, t, M, u, 0) \approx \left|\frac{t}{s}\right|^{-l+\nu\left[d-\frac{k}{2}(d-2+\eta)\right]} \overline{\Gamma}_R^{(l,k)}(\{0\}, -s, \bar{M}, u^*, 0). \tag{4.72}$$

This equation shows that the critical exponents α, γ and ν at $T < T_c$ coincide with their counterparts at $T > T_c$. If one changes the value M at fixed $|\bar{t}| = s$ being in the critical region ($|t| \to 0$), then from Eq. (1.72) it follows that

$$\overline{\Gamma}_R^{(l,k)} = M^{\left[\frac{2d}{d-2+\eta} - \frac{l}{\beta} - k\right]} F_{\pm}^{(l,k)}(\bar{M}), \tag{4.73}$$

where

$$F_{\pm}^{(l,k)}(\bar{M}) = \bar{M}^{\left[\frac{l}{\beta} + k - \frac{2d}{d-2+\eta}\right]} \overline{\Gamma}_R^{(l,k)}(\{0\}, \pm s, \bar{M}, u^*, 0).$$

Assuming $l = 0$, $k = 1$ and taking into account the relation between M and $\bar{M}$, and recalling that $h = -g \propto \overline{\Gamma}_R^{(0,1)}$, one can obtain the equation of state in the form proposed by Widom (1965)

$$h = M^{\delta} f^{\pm}(|t|M^{-1/\beta}),$$

where the function $f_{\pm}$ is formally defined by the relation

$$f^{\pm} = (x/s)^{\beta\delta}\overline{\Gamma}_R^{(0,1)}(\{0\}, \pm s, (s/x)^{\beta}, u^*, 0).$$

It is interesting to notice that this function is an analytic function in the vicinity $x = 0$. Therefore, the notations with the two different branches f_+ when $t > 0$ and f_- when $t < 0$ are superfluous and the equation of state can be written as

$$\frac{h}{M^{\delta}} = f\left(\frac{t}{M^{1/\beta}}\right). \tag{4.74}$$

The proof of the analyticity of f can be made in the following way. Let us calculate h with the help of Eq. (1.17), but we will modify the procedure of the expansion representing the Hamiltonian in the form

$$H\{\phi\} = H\{M + \Phi\} = H(M) + \overline{H}_0\{\Phi\} + \overline{H}_{int}\{\Phi\} = H(M) + \overline{H},$$

where M is the exact value of the magnetization $\langle\phi\rangle = M$ (i.e. the value of M should be defined from the condition $\langle\Phi\rangle_{\overline{H}} = 0$), the Hamiltonian $\overline{H}_0\{\Phi\}$ is defined by

$$\overline{H}_0\{\Phi\} = \frac{1}{2}\int_{x_1}\int_{x_2}\left\{\left.\frac{\delta^2 H}{\delta\phi(x_1)\delta\phi(x_2)}\right|_{\phi=M} - g_{2c}\delta(x_1 - x_2)\right\}\Phi(x_1)\Phi(x_2).$$

As a result of the use of the modified Hamiltonian, the zero-order propagator will have the form $q^2 + g_2 - g_{2c} + g_4 M^2/2$. In all diagrams for h this propagator will not have singularities at $t \to 0$, since we have $M \neq 0$. From this follows that the function f has all derivatives with respect to x in the vicinity $x = 0$.

Let us consider peculiarities accounting for the existence of the massless modes when $n \geq 2$. In essence, these peculiarities manifest themselves even in the self-consistent approximation. However, it is important that they are conserved in all orders of the perturbation theory, since they satisfy the Ward identity. This identity for the case of the $\mathcal{O}(n)$-symmetry is reduced to the trivial statement that the function $\Gamma(M)$ is the function depending only on modulus M. As a result one has

$$h^\alpha = \frac{\partial\Gamma}{\partial M}\frac{M^\alpha}{M} = \frac{\partial\Gamma}{\partial M}n^\alpha, \tag{4.75}$$

i.e. $h^\alpha \parallel M^\alpha$ and $h = \partial\Gamma/\partial M$. By taking the derivative from Eq. (1.75) with respect to M^α, one can obtain the inverse susceptibility tensor $\chi^{-1}_{\alpha\beta} = \Gamma^{(2)}_{\alpha\beta}$ when $q = 0$

$$\chi^{-1}_{\alpha\beta} = n^\alpha n^\beta\frac{\partial^2\Gamma}{\partial M^2} + (\delta_{\alpha\beta} - n^\alpha n^\beta)\frac{\partial\Gamma}{\partial M}\frac{1}{M}. \tag{4.76}$$

From this equation one can see that at $h = 0$ below T_c the value $\Gamma^{(2)}_\perp$ is equal to zero, which shows that the transverse modes have zero mass when $T < T_c$. The transverse susceptibility, as is seen from Eq. (1.76), $\chi_\perp = M/h$, i.e. at $h \to 0$ $\chi_\perp \to \infty$.

4.5 ϵ-Expansion

In this section we show how the calculations of different renormalization factors can be performed in the series expansion over $\epsilon = 4-d$. The program of these calculations is as follows. One needs to find the function $w(u, \rho)$ as an expansion with respect to the powers of ϵ and u, which will define the value u^* according to Eq. (1.65). The critical exponents are defined with the help of Eqs. (1.57), (1.58), and (1.59). We are only interested here in the universality limit $\rho \to 0$ at a fixed value ϵ. This means that the only divergence is the one absorbed into the definition of g_{2c}. All other constants Z_i are finite in this limit. In order to find the limiting values let us represent the function $w(u,\rho)$ with the help of Eq. (1.52) in the form

$$w(u,\rho) = \rho\frac{\partial u}{\partial \rho}\bigg|_{u_0} = -\epsilon u + u\,\frac{\partial \ln Z_1^{-1} Z_3^2}{\partial u}\bigg|_{\rho}\,\rho\frac{\partial u}{\partial \rho}\bigg|_{u_0} + u\rho\,\frac{\partial \ln Z_1^{-1} Z_3^2}{\partial \rho}\bigg|_{u}.$$

This equation has the following solution when $\rho \to 0$

$$w(u,0) = -\epsilon u \left\{1 + u\frac{\partial}{\partial u}\ln[Z_1(u)Z_3^{-2}]\right\}^{-1}. \tag{4.77}$$

In the derivation of this equation the following facts were essential. The powers of $\ln\rho$, which arise in the expansions of Z_i at $\epsilon = 0$, has been replaced by the consequent powers of $\rho^\epsilon - 1$. This leads to the nullification for expressions like $\rho(\partial Z_i/\partial\rho)$ in the limit $\rho \to 0$. One more important consequence emerges from this procedure: one can expand over ϵ only those values which are finite at $d = 4$, but should not expand the normalization factors. This procedure can also be used to obtain expressions for $\eta_i(u,0) \equiv \eta_i(u)$ in the form

$$\eta_i(u) = w(u)\frac{d}{du}\ln Z_i(u). \tag{4.78}$$

The following calculations can be done differently. Below we briefly discuss two approaches.

4.5.1 Addition of Counter-Terms

In this approach expansions over u for the functions $\Gamma_R^{(l,k)}$ are made with the help of the introduction of compensating counter-terms in the Hamiltonian. Let us rewrite the Hamiltonian Eq. (1.1) at $g_1 = 0$ with the help of Eqs. (1.51), (1.52), and (1.53) as follows

$$H = \frac{1}{2}\int d\boldsymbol{x}\,\{[\boldsymbol{\nabla}\phi_R(\boldsymbol{x})]^2 + t{\phi_R}^2(\boldsymbol{x})\} + H_{int}, \tag{4.79}$$

$$H_{int} = \int d\boldsymbol{x}\left\{\frac{u}{4!}[{\phi_R}^2(\boldsymbol{x})]^2 + \frac{1}{2}\Big[(Z_3-1)[\boldsymbol{\nabla}\phi_R(\boldsymbol{x})]^2 + \bar{g}_{2c}Z_3{\phi_R}^2(\boldsymbol{x}) + (Z_2-1)t{\phi_R}^2(\boldsymbol{x}) + \frac{u}{4!}(Z_1-1)[{\phi_R}^2(\boldsymbol{x})]^2\Big]\right\}. \tag{4.80}$$

The last four terms (counter-terms) are usually represented as expansions over u

$$\begin{aligned}\sum_{n=2}^{\infty} A_n u^n &= Z_3 - 1, \\ \sum_{n=1}^{\infty} B_n u^n &= \rho^{d-2}\overline{g}_{2c} \equiv \rho^{d-2}\tfrac{g_{2c}}{\mu^2}, \\ \sum_{n=1}^{\infty} C_n u^n &= Z_2 - 1, \\ \sum_{n=1}^{\infty} D_n u^n &= Z_1 - 1.\end{aligned} \tag{4.81}$$

In Eq. (1.79) and Eq. (1.80) the values x and ϕ_R are normalized by μ^{-1} and $\mu^{(d-2)/2}$. The renormalization conditions, defining the constants A_n, B_n, C_n, and D_n, can be rewritten in the dimensionless variables as follows

$$0 = \overline{\Gamma}_R^{(0,2)}(0, 0, u, \rho), \tag{4.82}$$

$$1 = \left.\frac{\partial\overline{\Gamma}_R^{(0,2)}(q, 0, u, 0)}{\partial q^2}\right|_{q^2=1}, \tag{4.83}$$

$$u = \left.\overline{\Gamma}_R^{(0,4)}(\boldsymbol{q}_1, \boldsymbol{q}_3, \boldsymbol{q}_3)\right|_{q_i q_j=(4\delta_{ij}-1)/3}, \tag{4.84}$$

$$1 = \left.\overline{\Gamma}_R^{(1,2)}(\boldsymbol{p}, \boldsymbol{q}_1, \boldsymbol{q}_2, 0, u, 0)\right|_{p^2=q_1^2=q_2^2=1, q_1 q_2=-1/3}. \tag{4.85}$$

The condition Eq. (1.82) defines the value g_{2c} which diverges as ρ^{2-d} at $\rho \to 0$. All other terms are finite in this limit. The latter leads to the absence of ρ dependence of the coefficients A_n, B_n, C_n, and D_n.

4.5.2 An Alternative Approach

One can also succeed in calculations without the introduction of counter-terms. In this approach, Eqs. (1.38) and (1.42) are used along with the normalization conditions to obtain the following set of equations

$$0 = \overline{\Gamma}^{(0,2)}(0, g_{2c}, u_0), \tag{4.86}$$

$$\frac{1}{Z_3} = \left.\frac{\partial\overline{\Gamma}^{(0,2)}(\boldsymbol{q}, g_{2c}, u_0)}{\partial \boldsymbol{q}^2}\right|_{\rho^2=q^2}, \tag{4.87}$$

$$\frac{1}{Z_1} = \left.\frac{\overline{\Gamma}^{(0,4)}(\boldsymbol{q}_1, \boldsymbol{q}_2, \boldsymbol{q}_3, \boldsymbol{q}_4, g_{2c}, u_0)}{u_0}\right|_{q_i q_j = \frac{4\delta_{ij}-1}{3}\rho}, \tag{4.88}$$

$$\frac{1}{Z_2} = \left.\overline{\Gamma}^{(1,2)}(\boldsymbol{p}, \boldsymbol{q}_1, \boldsymbol{q}_2, g_{2c}, u_0)\right|_{q_1^2 = q_2^2 = \rho^2; q_1 q_2 = -\rho/3; p^2 = \rho^2}. \tag{4.89}$$

In this set all momenta, as well as g_{2c} are dimensionless, but the characteristic scale is Λ. The renormalization constants and g_{2c} can be calculated in this approach, in the same manner as in the first one, in the form of the power series over u. This expansion can be made with the help of Eq. (1.52). This equation enables one to expand u_0 over u with the use of Eqs. (1.86), (1.87), (1.88), and (1.89). It is necessary to mention that results for renormalization constants Z_i and the function $w(u)$ are dependent on a particular formulation of normalization conditions, the choice of the normalization point, the choice of the dimensionless scale, and so on. However, all physically observable values, for instance, critical exponents and equation of state, should be the same independent of the concrete calculational scheme.

4.5.3 Results

Here we present the results for the critical exponents calculated in the approach with the counter-term addition. The normalization constants for the calculations of the functions $w(u)$, $\eta(u)$ and $\nu(u)$ are given with the help of the diagrams, values of which are given with accuracy enough to obtain η up to the third order of ϵ and γ – the second order. The coefficients in the expansions of Eq. (1.81) can be expressed in terms of the following four graphs

$$\sigma_1 = \quad = \frac{1}{8\epsilon}\left(1 + \frac{5}{4}\epsilon\right),$$

$$\sigma_2 = \quad = -\frac{1+2\epsilon}{6\epsilon^2},$$

$$U_1 = \quad = \frac{1}{\epsilon}\left(1 + \frac{\epsilon}{2}\right),$$

$$U_2 = \quad = \frac{1}{2\epsilon^2}\left(1 + \frac{3\epsilon}{2}\right).$$

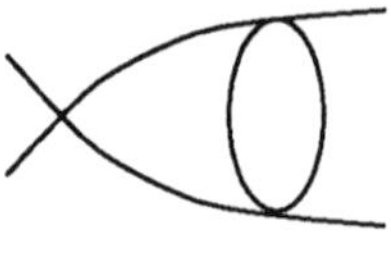

In these graphs, the factors originated from the angular integration are omitted. Each diagram with L loops has a factor $S_d^L/(2\pi)^{Ld}$, where S_d is the surface area of the unit d-dimensional sphere equal to $2(\pi)^{d/2}/\Gamma(d/2)$ (here $\Gamma(x)$ is the gamma-function). Such factors are absorbed into the redefinition of the interaction constant $u \to u(2\pi)^d/S_d$. Now, if one leaves in Z_3 terms of the order of u^3 and of the order of u^2 in the constants Z_1 and Z_2, the coefficients A_n, C_n, and D_n can be represented as follows

$$\begin{aligned} A_2 &= \tfrac{n+2}{18}\sigma_1, \quad A_3 = \tfrac{(n+2)(n+8)}{54}\left(\sigma_1 U_1 - \tfrac{1}{2}\sigma_2\right), \\ C_1 &= \tfrac{n+2}{6}U_1, \quad C_2 = \tfrac{n+2}{6}\left(\tfrac{n+8}{6}U_1^2 - U_2\right), \\ D_1 &= \tfrac{n+8}{6}U_1, \quad D_2 = \tfrac{n^2+26n+108}{36}U_1^2 - \tfrac{5n+22}{9}U_2. \end{aligned} \tag{4.90}$$

The function $w(u)$, defined by Eq. (1.77), can be obtained with the help of Eq. (1.90) and some simple transformations. Here we have to restrict the expansion over the powers of $u^p\epsilon^q$ by $p+q=3$

$$w(u) = -\epsilon u + \frac{n+8}{6}\left(1+\frac{\epsilon}{2}\right)u^2 - \frac{3n+14}{12}u^3 + o(u^4, \epsilon u^3). \tag{4.91}$$

From this equation one can obtain the value u^* at the fixed point

$$u^* = \frac{6\epsilon}{n+8}\left[1+\epsilon\left(\frac{3(3n+14)}{(n+8)^2} - \frac{1}{2}\right)\right] + o(\epsilon^3). \tag{4.92}$$

The functions η_i can be written in this approximation as follows

$$\eta_2(u) = -\frac{n+2}{6}u\left(1+\frac{\epsilon}{2}-\frac{u}{2}\right), \tag{4.93}$$

$$\eta_3(u) = \frac{n+2}{72}u^2\left(1+\frac{5}{4}\epsilon - \frac{n+8}{12}u\right). \tag{4.94}$$

One can obtain expressions for critical exponents taking into account the relations $\eta = \eta_3(u^*)$ and $\gamma = (2-\eta)\nu = (2-\eta)/[2-\eta+\eta_2(u^*)]$ (see Eq. (1.57)). Substituting Eq. (1.92) into Eq. (1.93) and Eq. (1.94) we arrive at

$$\begin{aligned} \eta &= \frac{\epsilon^2(n+2)}{2(n+8)^2}\left[1+\epsilon\left(\frac{6(3n+14)}{(n+8)^2} - \frac{1}{4}\right)\right] + o(\epsilon^4), \\ \gamma &= 1 + \frac{n+2}{2(n+8)}\epsilon + \frac{(n+2)(n^2+22n+52)}{4(n+8)^3}\epsilon^2 + o(\epsilon^3). \end{aligned}$$

4.5.4 Correction to Scaling

Let us now consider how one can obtain corrections to the main asymptotic term in the field theoretical scheme. In Wilson's theory, as is shown in Sect. 3.3.4, one has to take into account terms with an irrelevant eigenvector corresponding to a maximum (negative) eigenvalue. This leads to expansion over the powers of $|t|^{-\Delta_i}$, where Δ_i in the notations of the scaling field method is equal to the ratio $\lambda_{40}/\lambda_{20}$ (for critical behavior, i.e at $\sigma = 1$). It is evident that the analog of this expansion in the field theoretical approach is an expansion with respect to the deviation u from its fixed point value u^*. One also has to consider a more accurate approximation for the scaling function w, i.e.

$$w(\bar{u}(\lambda), 0) \approx \omega[\bar{u}(\lambda) - u^*],$$

where

$$\omega = \left.\frac{dw(u,0)}{du}\right|_{u=u^*} > 0.$$

Using this relation one can expand Eq. (1.60) in the vicinity u^*. As an example, one can see that the function $\overline{\Gamma}_R(\{0\}, t, u, 0)$ and all its derivatives with respect to $\bar{u}$ are well defined at $\bar{u} = u^*$, if $s \neq 0$. Such an expansion gives corrections to scaling as powers of $|t|^{\Delta}$, where $\Delta = \omega\nu$. In the ϕ^4-model one can prove that $\Delta = -\Delta_{40}$. The new exponent ω can be easily found from Eq. (1.91) and Eq. (1.92). It is equal to

$$\omega = \epsilon - \frac{3(3n+14)}{(n+8)^2}\epsilon^2 + o(\epsilon^3)$$

and

$$\Delta = \frac{1}{2}\epsilon - \frac{(-n^2+8n+68)}{4(n+8)^2}\epsilon^2 + o(\epsilon^3).$$

More precise calculations can be found in the articles by Baker *et al* 1978, Gorishny *et al* 1984, and Guinta *et al* 1985.

4.5.5 Improvement of Convergence

The examples of expansions over ϵ that have been given demonstrate the method, which in principle can be used to obtain higher order corrections. However, the increase of the order makes calculation very cumbersome. Bresin *et al* (1976) calculated critical exponents in the $\mathcal{O}(n)$ symmetrical model over the third order of ϵ. Wallace (1976) made such calculations up to ϵ^4. Comparison of these results with the results obtained in high temperature expansions and with experimental data shows that agreement is quite good only for terms of the order of ϵ^2. Beginning from the third order, correction terms increase with the order and the agreement worsens. Therefore, there is a problem with convergence of the ϵ-expansion. This problem results from the asymptotical nature of perturbation series of the ϵ-expansion. In 1977 Lipatov obtained a formula for the higher order terms increasing with the number of the perturbation order k as $k!$. In 1980 Le Guillou and Zinn-Justin used Lipatov's asymptotics and applied to ϵ-expansion Borel's summation formula

$$\sum_{k=0}^{\infty} k! a_k x^k = \int_0^{\infty} e^{-y} f(xy) dy,$$

where $f(x) = \sum_{k=0}^{\infty} a_k x^k$. The essence of their method is continuation of the first few expansion terms toward the asymptotics obtained by Lipatov. The critical exponents calculated with the help of this method are in better agreement with the high temperature expansions and experimental data. The accuracy of this method is essentially dependent on the number of the terms computed in the series. In principle, this method enables one to evaluate the quantities Z_i for any value of u.[2] In addition, one can also estimate sensitivity of results to parameters being introduced in the process of continuation, i.e. one can find the plausibility limits. However, all this is only correct for functions which are analytic in u. Unfortunately, we are interested in the region where $u \approx u^*$. In this region, as one can see from Eq. (1.77) and Eq. (1.78), the functions Z_i are essentially nonanalytic, i.e.

$$\begin{aligned} Z_1 &\propto (u^* - u)^{2\eta - \epsilon/\omega}, \\ Z_2 &\propto (u^* - u)^{\eta_2/\omega}, \\ Z_3 &\propto (u^* - u)^{\eta/\omega}. \end{aligned}$$

This means that true values of calculational errors cannot be defined with the help of plausibility limits. Bagnuls and Bervillier (1981) substantially simplified a modified version of this summation method.

4.6 Expansions at $d = 3$

Another very important achievement of the field theoretical approach is formulation of a theory applicable at $d = 3$. It may seem that the theory formulated in Sect. 1.3 is reliable at any dimension and the problem is to find a proper small parameter. Strictly speaking this is not the case. In 1973 Symanzik showed that such a kind of theory would inevitably lead to false infrared singularities. The graph Eq. (1.45) is proportional to $(\epsilon -$

[2]The increase of u affects only convergence of the series.

$1)^{-1} - \ln q^2$ when $\epsilon \to 1$ and $q \to 0$. In discussing this phenomenon Parisi (1980) noticed that ϵ-expansion should not lead to a reliable quantitative information if one would not use an additional hypothesis and eliminate this divergence. This problem, however, has bearing only on critical amplitudes, as the definition of critical exponents and universal combinations made from amplitudes require only consideration the limit $\epsilon \to 0$ in calculations of the constants Z_i. In order to eliminate divergences at $\epsilon \to 1$, Parisi proposed using the so-called massive theory.[3] In this theory one does not impose conditions subtracting divergences at critical point. There is also no need for the definitions of the critical value of the vertex g_{2c} as well as the use of the linear with respect to temperature scaling field t. Instead, it is necessary to introduce a value which already contains a singularity. This value is the inverse correlation length (or inverse susceptibility). The procedure defining the massive theory can be introduced as follows. Let us consider trial irreducible vertexes and introduce a change of mass $g_2 = m^2 + \delta m^2(\epsilon)$, which is defined by

$$\frac{\Gamma^{(0,2)}(0, m^2 + \delta m^2, g_4, \epsilon)}{\partial \Gamma^{(0,2)}(q, m^2 + \delta m^2, g_4, \epsilon)/\partial q^2} = m^2, \tag{4.95}$$

Due to this definition, the value m has the physical sense of the inverse correlation length. The role of the mass shift δm^2 is to eliminate all divergences arising at $\epsilon = 1$. The renormalization procedure in this case resembles the procedure used in the massless theory

$$\phi(x) = Z_3^{1/2}(g)\phi_R(x), \tag{4.96}$$

$$g_4 = m^\epsilon g Z_1(g) Z_3^{-2}(g), \tag{4.97}$$

$$\phi^2(x) = \frac{Z_2(g)}{Z_3(g)}[\phi_R(x)]_R^2. \tag{4.98}$$

Renormalized vertexes are related to trial ones by an equation similar to Eq. (1.42)

$$\Gamma_R^{(l,k)}(\{\boldsymbol{p}, \boldsymbol{q}\}, m, g, \epsilon) = Z_3^{k/2-l} Z_2^l \Gamma^{(l,k)}(\{\boldsymbol{p}, \boldsymbol{q}\}, m, g_4, \epsilon),$$

[3]An alternative approach was advanced by Tsuneto and Abrahams (1973) (ϵ-expansion, $n = 2$) and by Stephen and Abrahams (1973)($d = 2, 3$, arbitrary n). This approach was further developed by Ginzburg (1975) which showed that the use of Ward's identity helped to define a differential equation for g. That equation is a direct analog of Eq. (1.61).

where the renormalization constants Z_i are defined by the following equations

$$\frac{\partial}{\partial q^2}\Gamma_R^{(0,2)}(q,m,g,\epsilon)\bigg|_{q=0} = 1, \tag{4.99}$$

$$\Gamma_R^{(0,4)}(\{0\},m,g,\epsilon) = m^\epsilon g, \tag{4.100}$$

$$\Gamma_R^{(1,2)}(\{0\},m,g,\epsilon) = 1. \tag{4.101}$$

Two scale lengths are related in Eq. (1.97) with the help of the renormalization of $u_0 = g_4/\Lambda^\epsilon$. They are Λ^{-1} (g_4 is of the order of Λ^ϵ) and $\xi = m^{-1}$. Eq. (1.95) redefines the field h and the constant g_1 in the initial Hamiltonian (see Eq. (1.1)). In a similar fashion Eq. (1.98), introducing Z_2, renormalizes the linear scaling field $\bar{\tau} = g_2 - \delta g_2(\epsilon) - g'_{2c}$, where $\delta g_2(\epsilon)$ subtract the pole at $\epsilon = 1$ from g_{2c}, and g'_{2c} is defined by

$$\Gamma^{(0,2)}(\{0\}, g'_{2c}, g_4, \epsilon) = 0.$$

All this gives the following relation between linear $\bar{\tau}$ and nonlinear scaling fields

$$\bar{\tau} = m^2 + \lim_{\epsilon\to 1}[\delta m^2(\epsilon) - \delta g_2(\epsilon)].$$

In this manner Eqs. (1.96), (1.97) and (1.98) define renormalized physical values relating them to the scaling parameter ξ, while the trial parameters $h, g_k, \bar{\tau}$ are related to the scale Λ^{-1}.

4.6.1 Comparison of Massive and Massless Theories

It is interesting to compare the renormalization procedure of Sect. 1.3 with the massive theory. In the massless theory, the situation is described at the critical point. At this point there is no characteristic length with the exception of Λ^{-1}. The scale μ^{-1} is a completely arbitrary scale reflecting on the scaling invariance hypothesis. In the massive theory, the scaling invariance fails as the critical point is not achieved. The latter explains why the system of RG equations in the scheme with μ is more simple than in the massive theory. Nonetheless, one should remember that the critical point is a special point and one which is really only an idealization. As a result, notwithstanding that the scheme using μ gives all consequences of the fundamental theoretical ideas, it is not free from some drawbacks (for instance the discussed above pole at $\epsilon \to 1$). In this sense the massive theory is a more realistic theory, as it allows us to approach the critical point only asymptotically, which is really the case corresponding to the experimental situation.

The procedure of calculations in the massive theory resembles the one used in the scheme with fixed μ. The critical limit can be achieved when $m \to 0$ at a fixed value g_4. In this limit, Eq. (1.97) defines the renormalized interaction constant in such a way that the combination $gZ_1(g)/Z_3^2(g)$ diverges as $m^{-\epsilon}$. The main hypothesis essential in this limit is that g reaches some finite value g^*, when $m \to 0$ at fixed g_4. In the same manner, as in the case of massless theory, the value g^* is defined as a simple nontrivial zero of the function $w(g)$, given by an equation like Eq. (1.77), which at $d = 3$ has the form

$$w(g) = -\left\{\frac{d\ \ln[gZ_1(g)/Z_3^2(g)]}{dg}\right\}^{-1}.$$

Behavior of the function $w(g)$ in the vicinity of the fixed point g^* is assumed to be analytical

$$w(g) = \omega(g - g*) + o((g - g^*)^2), \tag{4.102}$$

with a positive ω. The critical exponents are defined by the relations $\eta = \eta_3(g^*)$ and $\eta_2 = \eta_2(g^*)$, where the functions $\eta_i(g)$ are given (compare with Eq. (1.78)) by

$$\eta_i = w(g)\frac{d}{dg}\ln Z_i(g)$$

Bagnuls and Bervillier (1984) calculated the first six powers of g for the functions Z_i and found in this order the functions $w(g)$ and η_i. The nature of divergency at higher orders in this theory is also defined by the Lipatov's asymptotics. The final step of calculations is application of the summation method by Le Guillou and Zinn-Justin (1980). Numerical calculations of zeros for the function $w(g)$ confirmed the hypothesis of Eq. (1.102).

It is necessary to pay attention to one additional attractive feature of the massive theory. This theory is formulated at a fixed value of temperature, which allows for a description of the preasymptotic region. Bagnuls and Bervillier (1985) made use of this obstacle and found preasymptotic critical behavior in the ϕ^4-model at $d = 3$ for $n = 1, 2, 3$.

4.6.2 Experimental Situation

Experimental studies of the scaling laws region are restricted to the measurements in a very close vicinity to T_c. When the Ginzburg parameter G_i is small, such studies have a number of different complications. It is necessary to fix and stabilize temperature with a very high degree of accuracy, take into account gravitational effects, and many others. These are the reasons why experimenters have to introduce corrections to scaling in order to fit data obtained in a region farther from T_c, where, for instance, gravitational effects are irrelevant. This means, that in order to make a proper fit, one has to find how far from T_c it is possible to use Wegner's series (see Sect. 3.3) like

$$\Gamma^{(0,2)}(\{0\},t) \propto t^{\gamma} \left\{ 1 + \sum_{k=1}^{\infty} a_k t^{\Delta_k} \right\}, \tag{4.103}$$

where for the ϕ^4-model ($\Lambda \to \infty$) $\Delta_k = k\Delta$. The fitting is usually aggravated by the fact that the series Eq. (1.103) is represented in the powers of scaling field t. This value should be related to the measurable temperature $\tau = (T - T_c)/T_c$ and field h. In the vicinity of T_c, t can be represented as $t = a\tau + o(\tau^2, \tau h^2)$, where a is some undefined constant. Substituting this expression for t in Eq. (1.103), one can see that the problem of experimental fitting in the preasymptotic region, with an accuracy better than the first nonvanishing correction, may be too complicated. This can be demonstrated on the example of the 3-dimensional Ising model. For this model, due to the small value of $\Delta \approx 0.5$, the first correction is well distinguished from others, when the value τ is quite small. The following corrections have quite close exponents. They are a term with $\Delta_2 \approx 1.05$ (see Newman and Riedel (1984)), a term with the exponent $2\Delta \approx 1$, and also the first correction from the expansion of t over τ which also gives exponent equal to 1. It is very unlikely that these corrections can be distinguished in experimental fitting. More than that, additional numbers of fitting parameters with nonuniversal amplitudes decreases accuracy in the separation of the scaling region. Nonetheless, Bagnuls and Bervillier (1984, 1985) and Bagnuls, Bervillier, and Garrabas (1984) found a way to describe the preasymptotic region along the critical isochore. They introduced a set of theoretical functions of temperature for the full description of measurable values — correlation length, susceptibility, and specific heat. These functions show crossover from the fluctuational behavior (in the T_c vicinity) to the self-consistent behavior (away from T_c). Such a description corresponds to the experimentally measured situation. However, the crossover demonstrated in these functions is very unlikely to be found in experiments, due to the noted restrictions of the used model. The authors of the last mentioned works discussed conditions which would help to improve analysis of experimental data with the help of the introduced nonasymptotic functions. The use of a minimal set of fitting parameters, which are intrinsically related to the physics of critical behavior, owing to the very high accuracy of the theoretical calculations, helps us to make estimates (within the limits of experimental accuracy) for the temperature region of plain scaling behavior and preasymptotic regime.

4.7 Results of Different Approaches

The diversity of concrete calculational schemes used in the analysis of critical behavior may naturally raise the question: Which approach is the best? Possibly, there is no single answer. Very often, simple and convenient calculational schemes are considered as such only with respect to a particular model, or only in the lower orders of perturbation theory. In higher orders they frequently become so cumbersome that initial advantages, as a rule, no longer apply. In view of the absence of a definite answer it may be instructive to compare experimental results with the theoretical ones obtained within the frameworks of different approaches. In Table 4.1 we show experimental data for a simple liquid (L) and binary solutions (BS), along with the results of calculations for $n = 1$ obtained with the help of the scaling field method (SFM), high temperature expansions (HTE), field theoretical approach (FTA), and Monte Carlo simulations (MCS). The HTE was the first successful approach to the problem of critical exponent calculations at $d = 3$. In 1955 Rushbrooke and Wood calculated the partition function and susceptibility for the Heisenberg model with an arbitrary spin in the order of T^{-7}. Later, Domby and Wood (1964) found two additional terms for the spin $S = 1/2$ and an arbitrary lattice. It is necessary to mention, that calculation of every additional term is exponentially increasing in difficulty (see Fisher's review (1974)). The results shown in the table obtained by Zinn-Justin (1981) are derived from the analysis of a series containing terms up to T^{-21} (see also Chang *et al* (1982)). The plausibility limits in the table show only the statistical errors of calculations. The accuracy of HTE is very high only for the Ising model. Unfortunately, the expansion series obtained until now for isotropic and anisotropic n-vector models are quite short and the accuracy of calculations for these models cannot be compared with the case $n = 1$.

The SFM, based on the Wilson's RG approach, was adjusted by Newman and Riedel (1984) for calculations without the use of any kind of perturbation theory. Without essential complications their method can be applied to anisotropic vector models. The main approximation in this method is the truncation of the infinite set of hierarchical equations. This truncation should be made with the help of a "principle of balance" introduced by Newman and Riedel for the occasion. The principle serves as a criterion of self-consistency for the method at the level of a fixed accuracy

System, method	ν	η	Δ	*Reference*
L	$.63 \pm .01$	$.05 \pm .01$	$.51 \pm .003$	Martinetz and Matizen (1974). Anisimov *et al* (1979). Pestak and Chan (1984).
BS	$.625 \pm .005$	$.036 \pm .007$		Beysens and Bourgon (1979). Beysens (1982).
SFM	$.626 \pm .009$	$.04 \pm .007$	$.51 \pm .04$	Newman and Riedel (1984).
HTE	$.630 \pm .0015$	$.035 \pm .003$	$.52 \pm .07$	Zinn-Justin (1981).
FTA	$.63 \pm .002$	$.031 \pm .011$	$.496 \pm .004$	Baker *et al* (1978).
	$.63 \pm .0015$	$.031 \pm .004$	$.498 \pm .02$	LeGuillou and Zinn-Justin (1980).
	$.628 \pm .001$	$.035 \pm .002$	$.50 \pm .02$	Gorishny *et al* (1984).
MCS	$.629 \pm .004$	$.031 \pm .005$	$.63 \pm .07$	Pawly *et al* (1984).

Table 4.1: Experimental and theoretical results for critical exponents obtained within the frameworks of different approaches ($n = 1$).

of calculations.

The main difference between Wilson's RG (WRG) approach and the FTA can be seen as follows. In the WRG an infinite set of vertexes is introduced, while in the FTA one deals with a finite set of a small number of lower order vertexes. However, the definition of the renormalization constants is essentially based on the perturbation theory. The perturbation series in this theory are asymptotical, and in order to obtain quite accurate results one has to sum the series using Borel's formula and the extrapolation procedure from the first few terms to the Lipatov's asymptotics. Traditionally the results obtained in the FTA for the $\mathcal{O}(n)$-symmetry ($n = 1, 2, 3$), are considered the most plausible. However, the results obtained by Newman and Riedel (1984), though with the accuracy not as high as in the FTA (see for example Gorishny *et al*), show that the WRG has hidden potential. Newman and Riedel found higher order corrections to the scaling region ($\Delta_{422} = 1.67 \pm .11$ and $\Delta_{500} = 1.5 \pm .3$) not found in other approaches.

Among the numerous numerical calculations it is necessary to mention the application of the WRG to the Ising model in Monte Carlo simulations. Pawly *et al* (1984) made calculations for the simple cubic lattice with 64^3

sites. In comparison with the two-dimensional case (see Swendsen 1982) convergence for $d = 3$ is very slow. Nonetheless, the values obtained for the exponents ν and η can be compared with the HTE results. The index $\Delta = .63 \pm .07$ is not so good which may be attributed to the bad convergence. Swendsen (1984) tried to improve the method by the inclusion of 17 vertexes into the Hamiltonian and by the use of a special optimization procedure. On the whole the use of the MC in RG calculations has not been developed to a level competitive with the other methods, especially if one considers anisotropic n-vector models.

To conclude this brief comparison of the results obtained in different approaches, one may say that the most accurate results were found with the help of HTE and field theoretical RG. However, the result of HTE for the exponent Δ, characterizing deviations from scaling, is a little higher than experimental data.

Chapter 5

Generalized RG Approach

5.1 Introduction

In Chapters 3 and 4, two ideologically different RG approaches have been considered. In Wilson's approach one deals with the Hamiltonian Eq. (3.24) which includes all imaginable vertexes complying with the symmetry of the problem. By contrast the field theoretical RG can only be effectively used with the Hamiltonian containing a small number of vertexes which do not violate renormalizability of the problem. Notwithstanding such a substantial difference in the formulation, the results of calculations in these approaches in the critical region are practically the same. Therefore, there is some property, common to both approaches, which is of crucial importance for the description of critical region. This property is Kadanoff's invariance completing the calculational schemes in both methods. Kadanoff's invariance is an intrinsic property of a system which can be found in a particular point of the parametric space of the Hamiltonian describing this system. This point may or may not exist, but the scale transformations and the consequent scale covariance is always present in any physical problem being an inherent attribute of the configurational space. Hence, one could start the study of a system from the scale transformations and then, looking for the point of Kadanoff's invariance, come to the origin from where these two different approaches merge.

5.2 Scale Transformations

In order to find the scale covariance of a system, one has to find a transformation which would not affect measurable values. This means that the change of the momentum unit should lead to the change of all values considered in the problem (field variables, coordinates, volume, and vertexes) in a way conserving measurables. In the most general case the scale transformation should be defined as follows (compare with Eq. (3.18))

$$\begin{aligned} \boldsymbol{q}' &= (1+\delta l)\boldsymbol{q}, & \Lambda' &= (1+\delta l)\Lambda, \\ V &= (1+\delta l d)V', & \phi_q &= \left(1+\delta l\, A(q')\right)\phi'_{q'}, \end{aligned} \tag{5.1}$$

where we have introduced a function of momentum $A(q)$. This function can only be defined with the help of additional assumptions. In the Wilson's approach (see Sect. 3.3) instead of $A(q)$ we have $d/2$. This value results from the additional condition $\phi_{q'}/\sqrt{V'} = \phi_q/\sqrt{V}$ (see discussion preceding Eq. (3.18)). In the case of the field theoretical RG there is no such requirement and, as is seen from Eq. (1.49), $A(q) = d/2+1$. The difference between the values of A in different approaches may raise the question: how does it happen that the scale transformations are different but the results for the critical exponents are the same? The answer to this question can be obtained via the following deliberation. In the Wilson approach the preliminary steps of the RG equation derivation do not play an important role. In the final RG equation (see Eq. (3.20)) the value $d/2$ comes in the combination $d/2+h(q)$, where the function $h(q)$ is to be chosen in a way ensuring the presence of the fixed point. For the Gaussian Hamiltonian the critical fixed point corresponding to the Landau approximation $H_0 = \int_q q^2|\phi_q|^2/2$ is only possible at $h(q) = 1$. Therefore, in this case we have the same value as in the FTA $A = d/2+1$. One can verify that this value leads to the conservation of the initial form for the Gaussian critical Hamiltonian, or we may say that this Hamiltonian is invariant under the scale transformation with $A = d/2+1$. We also need to introduce regularization (see Sect. 1.2.4) into this Hamiltonian in a way conserving the invariance. The latter can be done if one chooses H_0 in the form

$$H_0 = \frac{1}{2}\int_q q^2 S^{-1}\left(\frac{q^2}{\Lambda^2}\right)|\phi_q|^2, \tag{5.2}$$

where the function $S(x)$ provides the necessary momentum cutoff in the theory. For the case with the smooth cutoff $S(x)$ is monotonous with $S(0) = 1$

and $\lim_{x\to\infty} S(x)x^k = 0$, for any k. The sharp cutoff can be introduced with the help of the Heaviside step function $S(x) = \Theta(1-x)$.

The next step consists in the derivation of rules describing the transformations of vertexes for the given interaction Hamiltonian H_I. In a quite general case H_I can be taken as in Eq. (3.24). We will require for the interaction Hamiltonian only covariant properties, which can be presented as follows

$$\sum_{\alpha_1}\int_{q_1}\cdots\sum_{\alpha_k}\int_{q_k} g_k^{\alpha_1\cdots\alpha_k}(\boldsymbol{q}_1\cdots\boldsymbol{q}_k)\delta\left(\sum_{i=1}^{k}\boldsymbol{q}_i\right)\prod_{i=1}^{k}\phi_{q_i}^{\alpha_i}$$
$$=\sum_{\alpha_1}\int_{q_1'}\cdots\sum_{\alpha_k}\int_{q_k} g_k'^{\alpha_1\cdots\alpha_k}(\boldsymbol{q}_1'\cdots\boldsymbol{q}_k')\delta\left(\sum_{i=1}^{k}\boldsymbol{q}_i'\right)\prod_{i=1}^{k}\phi_{q_i'}'^{\alpha_i}, \tag{5.3}$$

and such a condition should hold for any k. From Eq. (5.3), with the help of Eq. (5.1), one can obtain the transformation properties for vertexes $\hat{g}_k \equiv g_k^{\alpha_1\dots\alpha_k}$

$$\hat{g}_k(\boldsymbol{q}_1\cdots\boldsymbol{q}_k) = \left\{1+\delta l\left[(k-1)d - \sum_{i=1}^{k} A(q_i')\right]\right\}\hat{g}_k'(\boldsymbol{q}_1'\cdots\boldsymbol{q}_k'). \tag{5.4}$$

5.2.1 Differential Form of Scale Covariance

The set of Eqs. (5.1) and (5.4) defines the transformation of all quantities under an infinitesimal change of the momentum scale. With the help of this set, one can derive differential equations reflecting scale covariance for any measurable value. Using the traditional definition for the k-point correlation function

$$(2\pi)^d\delta\left(\sum_{i=1}^{k}\boldsymbol{q}_i\right)G_k^{\alpha_1\cdots\alpha_k}(\boldsymbol{q}_1\cdots\boldsymbol{q}_k) = \left\langle\prod_{i=1}^{k}\phi_{q_i}^{\alpha_i}\right\rangle, \tag{5.5}$$

one can find the transformation of the function $\hat{G}_k \equiv G_k^{\alpha_1\cdots\alpha_k}$ by substituting $\boldsymbol{q}$, Λ, V, $\boldsymbol{\phi}_q$, and g_k from Eqs. (5.1) and (5.4) into Eq. (5.5)

$$\delta\left(\sum_{i=1}^{k}\boldsymbol{q}_i\right)\hat{G}_k(\{\boldsymbol{q}_i\},\Lambda,g_k\{\boldsymbol{q}_i\}) = \left(1+\delta l\sum_{i=1}^{k}A(q_i')\right)$$
$$\times\delta\left((1-\delta l)\sum_{i=1}^{k}\boldsymbol{q}_i'\right)\hat{G}_k(\{\boldsymbol{q}_i'\},\Lambda',g_k'\{\boldsymbol{q}_i'\}). \tag{5.6}$$

Now, expanding the right-hand side of this equation in the powers of δl, one can find a differential equation reflecting the scale covariance in the system. As a result of this expansion we arrive at

$$\left\{ \sum_{i=1}^{k} \left(A(q_i) + \boldsymbol{q}_i \frac{\partial}{\partial \boldsymbol{q}_i} \right) - d - d\, V \frac{\partial}{\partial V} + \frac{\partial \hat{H}_I}{\partial l} \frac{\delta}{\delta \hat{H}_I} + \Lambda \frac{\partial}{\partial \Lambda} \right\} \hat{G}_k = 0, \quad (5.7)$$

where the operator $\partial/\partial V$ is essential when the function $\hat{G}_k$ evidently depends on the volume.

In Eq. (5.7) we introduced the operator $\frac{\partial \hat{H}_I}{\partial l} \frac{\delta}{\delta \hat{H}_I}$ which means that one has to take variational derivatives with respect to vertexes and then the derivative from the primed vertexes with respect to l (see Eq. (5.4)). The latter gives the apparent definition

$$\frac{\partial \hat{H}_I}{\partial l} \frac{\delta}{\delta \hat{H}_I} \equiv \sum_{k=1}^{\infty} \int_{q_1 \cdots q_k} \left\{ \left(\epsilon_k + \sum_{i=1}^{k} \boldsymbol{q}_i \frac{\partial}{\partial \boldsymbol{q}_i} \right) \hat{g}_k\{\boldsymbol{q}_i\} \right\} \frac{\delta}{\delta \hat{g}_k}, \quad (5.8)$$

where $\epsilon_k = d + k(d-2)/2$.

5.2.2 Two Ways of Utilizing Scale Covariance

Relations of the type of Eq. (5.7) can also be obtained for any k-point irreducible vertex. These equations are quite general as they reflect only the scale covariance of the physical system. There are two ways to use of these equations in the phase transitions theory. In the first, one should restrict consideration only to renormalizable Hamiltonians H_I. This case will substantially simplify the operator Eq. (5.8) as we usually consider local, i.e. momentum independent vertexes. After rescaling Λ and introduction of renormalized values (see Eq. (1.42)) one arrives (when $A(q) = (d+2)/2$) at Eq. (1.55). Thus, this approach leads to the field theoretical RG. The other way consists of the elimination of the partial derivative with respect to Λ from Eq. (5.7). The latter is only possible for the general type of Hamiltonians containing all vertexes consistent with the symmetry of the problem. Such a consideration leads to a formulation of a first order, quasi-linear equation in partial derivatives. The characteristic set for this equation can be transformed to a single equation in variational derivatives in which one can recognize the RG equation of Wilson's approach. In the following section we will show how to obtain the equation which is a generalized form of Wilson's equation. This form is generalized in the sense that it is lacking the most essential drawback of the traditional approximation — the presence of redundant eigenvectors in the linear RG equation (see Sect. 3.3.3)).

5.3 Scale Equations

We will call the equation which will be obtained after the elimination of the derivative with respect to Λ, the "scale equation" (see Ivanchenko *et al* 1990 (a)). First of all, it is necessary to notice that the definition of the scale transformations given in the previous section is superfluous. In the zero order Hamiltonian H_0 we have the function $G_0^{-1}(q) = q^2 S^{-1}(q^2/\Lambda^2)$ as a factor at $|\phi_q|^2$. On the other hand, in the interaction Hamiltonian we also have the term with the second order vertex $g_2(\boldsymbol{q})$. As the function $A(q)$ so far has not been defined, one can use it to prevent the appearance of the nonlocality in the vertex g_2 under the influence of the scale transformations. This condition leads to the following definition of the function $\eta(q) = d + 2 - 2A(q)$ (see Ivanchenko *et al* 1990 (b))

$$\eta(q) = \eta(0) + \frac{D(q) - D(0) - \eta(0)G_0^{-1}(q)}{G_0^{-1}(q) + g_2}, \tag{5.9}$$

where the following notations have been introduced

$$\begin{aligned} D(q) &= Q(q) - 2g_2^2 h(q), \quad \eta(0) = \frac{1}{2d} \left. \frac{\partial^2 D(q)}{\partial \boldsymbol{q}^2} \right|_{q=0}, \\ Q(q) &= \int_p h(p) \left\{ n g_4(-\boldsymbol{p}, \boldsymbol{p}, \boldsymbol{q}, -\boldsymbol{q}) + 2g_4(\boldsymbol{p}, \boldsymbol{q}, -\boldsymbol{p}, -\boldsymbol{q}) \right\}. \end{aligned} \tag{5.10}$$

The function $h(q)$ is defined by the initial choice of the regularization procedure, i.e. $h(q) = \partial G_0(q, \Lambda)/\partial \ln \Lambda^2$.

5.3.1 Conventional Approach

Now we are going to briefly describe the procedure for the elimination of the derivative $\partial/\partial\Lambda$. The conventional approximation consists in the choice $A(q) = (d+2)/2$. Let us find the deviation of the function G_k under the influence of the infinitesimally small change in Λ. The following considerations are based on the fact that the averaging over a Gaussian field ϕ can be replaced by two independent averages over Gaussian fields ϕ_1 and ϕ_2, providing $\phi = \phi_1 + \phi_2$ and the sum of the correlators $G_{01}(q, \Lambda) = \langle |\phi_1(\boldsymbol{q})|^2 \rangle_0$ and $G_{02}(q, \Lambda) = \langle |\phi_2(\boldsymbol{q})|^2 \rangle_0$ is equal to the correlator of the initial field

$$G_0(q, \Lambda) = \langle |\phi_q|^2 \rangle_{0,\Lambda} = G_{01}(q, \Lambda) + G_{02}(q, \Lambda).$$

Let us choose ϕ_1 so that $G_{01}(q, \Lambda) = G_0(q, (1 - \delta l)\Lambda)$, where $\delta l \to 0$. Then the value G_{02} is of the order of δl, i.e.

$$G_{02}(q, \Lambda) = G_0(q, \Lambda) - G_0(q, (1 - \delta l)\Lambda) \approx 2\delta l h(q). \tag{5.11}$$

Now one can find the change of the averaged value of a functional $W\{\phi\}$. At the first stage we consider averaging only over the Gaussian ensemble defined by the Hamiltonian Eq. (5.2). The procedure of averaging can be done with the help of the equality $\langle W\rangle_0 = \langle W\rangle_{0,12}$. Due to the smallness of δl the averaging over the field ϕ_2 can be performed in the first nonvanishing order of δ. As a result one arrives at

$$\langle W\rangle_{0,\Lambda} = \langle (1+\delta\,\hat{L})W\rangle_{0,(1-\delta l)\Lambda}, \tag{5.12}$$

where the operator $\hat{L}$ has the form

$$\hat{L} = \int_q h(q)\frac{\delta^2}{\delta\phi_q\delta\phi_{-q}}.$$

From Eq. (5.12) immediately follows

$$\Lambda\frac{\partial}{\partial\Lambda}\langle W\rangle_0 = \langle \hat{L}W\rangle_0. \tag{5.13}$$

This relation enables one to find the consequent equation for the averaged value $\langle W\rangle$ with the full Hamiltonian $H = H_0 + H_I$. Let us rewrite this quantity as $\langle W\rangle \equiv \langle e^{-H_I}W\rangle_0/\langle e^{-H_I}\rangle_0$. Applying to this equality the operator $\Lambda\partial/\partial\Lambda$ one can find with the help of Eq. (5.13) the expression

$$\begin{aligned}\Lambda\frac{\partial}{\partial\Lambda}\langle W\rangle = \int_q h(q)\left\langle \frac{\delta^2 H_I}{\delta\phi_q\delta\phi_{-q}} - \frac{\delta H_I}{\delta\phi_q}\frac{\delta H_I}{\delta\phi_{-q}}\middle| W\right\rangle \\ +\langle\hat{L}W\rangle - 2\int_q h(q)\left\langle\frac{\delta H_I}{\delta\phi_q}\frac{\delta W}{\delta\phi_{-q}}\right\rangle,\end{aligned} \tag{5.14}$$

where $\langle\cdots|\cdots\rangle$ means a connected average.

This equation can be further simplified with the help of the identity

$$\begin{aligned}0 &= Z^{-1}\int_q h(q)\prod\int\frac{d\phi_p}{\sqrt{2\pi V}}\frac{\delta}{\delta\phi_q}\left\{e^{-H}\frac{\delta W}{\delta\phi_{-q}}\right\} \\ &= -\int_q h(q)\left\langle\frac{\delta H}{\delta\phi_q}\frac{\delta W}{\delta\phi_q}\right\rangle + \langle\hat{L}W\rangle,\end{aligned} \tag{5.15}$$

and evident representation for the functional W. In order to obtain the scale for the k-point correlation function, we have to make the following choice

$$W = \prod_{i=1}^{k}\phi_{q_i}^{\alpha_i}. \tag{5.16}$$

Using the identity Eq. (5.15) one can transform the last two terms in Eq. (5.14) as follows

$$\langle \hat{L} W \rangle - 2 \int_q h(q) \left\langle \frac{\delta H_I}{\delta \phi_q} \frac{\delta W}{\delta \phi_q} \right\rangle = \sum_{i,j} \delta_{\alpha_i \alpha_j}$$

$$\times \int_q h(q) \delta(\boldsymbol{q}_i - \boldsymbol{q}) \delta(\boldsymbol{q}_j + \boldsymbol{q}) \left\langle \prod_{l \neq i,j}^{k} \phi_{q_l}^{\alpha_l} \right\rangle + \langle W \rangle \sum_i^k G_0^{-1}(\boldsymbol{q}_i) h(\boldsymbol{q}_i). \quad (5.17)$$

Inserting this expression into Eq. (5.14) one gets, in the definition of the derivative with respect to Λ, average values containing terms with different powers of field operators ϕ. These terms can be reduced to the action of some new operator with the structure $\hat{\Phi} = \sum_k \hat{U}_k[\hat{g}_m] \delta / \delta \hat{g}_k$. The operator of this kind is already present in Eq. (5.7) (see Eq. (5.8)). Therefore, it is natural to include the operator Eq. (5.8) into the operator $\hat{\Phi}$ by modifying the definition of $\hat{U}_k$. It is interesting to note that the action of the operator Eq. (5.8) can be also represented as

$$\frac{\partial \hat{H}_I}{\partial l} \frac{\delta}{\delta \hat{H}_I} \langle W \rangle = \left\langle \int_q \left(\frac{d+2}{2} \phi_q + \boldsymbol{q} \frac{\partial \phi_q}{\partial \boldsymbol{q}} \right) \frac{\delta H_I}{\delta \phi_q} \right| W \Big\rangle$$

so that the resulting operator can be generated by the form

$$\begin{aligned} \frac{\partial H_I}{\partial l} \equiv & \int_q \left[\frac{d+2}{2} \phi_q + \boldsymbol{q} \frac{\partial \phi_q}{\partial \boldsymbol{q}} \right] \frac{\delta H_I}{\delta \phi_q} \\ & + \int_q h(\boldsymbol{q}) \left[\frac{\delta^2 H_I}{\delta \phi_q \delta \phi_{-q}} - \frac{\delta H_I}{\delta \phi_q} \frac{\delta H_I}{\delta \phi_{-q}} \right]. \end{aligned} \quad (5.18)$$

With the help of this equation one can obtain the scale equation for the function $\hat{G}_k$ in the form

$$\begin{aligned} & \left\{ k \frac{d+2}{2} - d + \sum_{i=1}^{k} \left(\boldsymbol{q}_i \frac{\partial}{\partial \boldsymbol{q}_i} + 2 G_0^{-1}(\boldsymbol{q}_i) h(\boldsymbol{q}_i) \right) + \hat{\Phi}[\hat{g}_m] \right\} \\ & \times \hat{G}_k\{\boldsymbol{q}_i\} = \sum_{j=1}^{k} \sum_{l \neq j}^{k} \delta(\boldsymbol{q}_j + \boldsymbol{q}_l) h(\boldsymbol{q}_j) \delta_{\alpha_j \alpha_l} \hat{G}_{k-2}^{(j,l)}\{\boldsymbol{q}_i\}, \end{aligned} \quad (5.19)$$

where the function $\hat{G}_k^{(j,l)}$ is a k-point correlation function not depending on momenta $\boldsymbol{q}_j$ and $\boldsymbol{q}_l$. For the Hamiltonian containing only even powers of

ϕ, the operator $\hat{\Phi}$ is defined with the help of

$$\hat{U}_k[\hat{g}_m] = U_k^{\alpha_1\cdots\alpha_{2k}} = \left(\epsilon_k + \sum_{i=1}^{2k} \boldsymbol{q}_i \frac{\partial}{\partial \boldsymbol{q}_i}\right) \bar{g}_k + \int_q h(\boldsymbol{q}) \{ (k+1)$$

$$\times (2k+1) g_k^{\gamma\gamma\alpha_1\cdots\alpha_{2k}}(\boldsymbol{q}, -\boldsymbol{q}, \boldsymbol{q}_1 \cdots \boldsymbol{q}_{2k}) - \sum_{m=1}^{k+1} 2m(k-m+1) \quad (5.20)$$

$$\times \hat{S}\left[g_m^{\gamma\alpha_1\ldots\alpha_{2m-1}}(\boldsymbol{q}, \boldsymbol{q}_1 \cdots \boldsymbol{q}_{2k}) g_{k-m+1}^{\gamma\alpha_{2m}\cdots\alpha_{2k}}(-\boldsymbol{q}, \boldsymbol{q}_{2m} \cdots \boldsymbol{q}_{2k})\right] \}$$

The operator $\hat{S}$ in Eq. (5.20) symmetrizes the product $\hat{g}_m \hat{g}_{m-k+1}$ with respect to the permutations of the variables $\boldsymbol{q}_i$ and $\boldsymbol{q}_j$ (with the simultaneous permutations of the indexes α_i and α_j). In Eq. (5.19), after the elimination of the derivative $\partial/\partial\Lambda$ the value Λ has been set to unity, which means that momentum variables are measured in the units of Λ and the cutoff parameter is also equal to unity.

5.3.2 Generalized Approach

The scale equation in the form given by Eq. (5.19) belongs to the class of quasilinear equations in partial derivatives. The theory for such equations is well developed. The traditional approach to the analysis of equations in partial derivatives is to study the characteristic set of equations. If one restricts consideration only by the characteristic set for the operator $\hat{\Phi}$

$$\frac{d\hat{g}_k(\boldsymbol{q}_1 \ldots \boldsymbol{q}_{2k})}{dl} = \hat{U}_k[\hat{g}_m] \quad k = 1, 2, \ldots, \quad (5.21)$$

then it is possible to prove (see Sect. 5.4) that this set is equivalent to the RG formulation similar to Wilson's equation. Therefore, this method contains all the drawbacks of the conventional RG approach, namely the problem of redundant eigenvectors. In order to avoid the confusion caused by redundant eigenvectors we consider a generalized approach using the unique definition of the field scaling parameter $A(q) = (d + 2 - 2\eta(q))/2$. As is discussed at the beginning of Sect. 5.3, the introduction of the function $\eta(q)$, defined from Eqs. (5.9) and (5.10) conserves the initial form of the Hamiltonian H_0 under the scale transformation. The latter changes the definition of the operation $\partial H_I/\partial l$ (see Eq. (5.18)) not only in the explicit way by the exchange of $(d+2)/2$ with $(d + 2 - 2\eta(q))/2$ but also by the addition of compensating terms linear in η resulting in

$$\frac{\partial \tilde{H}_I}{\partial l} = \overline{\frac{\partial H_I}{\partial l}} + \frac{1}{2} \int_q \eta(\boldsymbol{q}) \left\{ V - G_0^{-1} |\phi_q|^2 - \phi_q \frac{\delta H_I}{\delta \phi_q} \right\}, \quad (5.22)$$

where the operator $\overline{\frac{\partial H_I}{\partial l}}$ is defined by Eq. (5.18) with $(d+2)/2$ replaced by the function $(d+2-2\eta(q))/2$. As will be shown later, the function $\eta(q)$ is directly related to the critical exponent η.

5.4 RG Equations

In this section we will apparently derive the RG equation in the framework of the considered scale transformations (see Ivanchenko and Lisyansky 1992, Ivanchenko *et al* (1992) (b)). In order to find this equation we need to briefly discuss the traditional approach due to Wilson (see Sect. 3.3.1). Wilson's equation can be written schematically as

$$\frac{\partial H}{\partial l} = \hat{R}\{H\}, \tag{5.23}$$

where $\hat{R}$ stands for some nonlinear operator performing RG transformations. The Hamiltonian H at the critical point of a phase transition is determined by

$$\hat{R}\{H^*\} = 0, \tag{5.24}$$

and is called a *fixed-point Hamiltonian.* This term, however, has a much wider meaning because not every fixed point Hamiltonian describes critical behavior. The term *critical Hamiltonian* is more restrictive. Every critical Hamiltonian belongs to the set of fixed-point Hamiltonians, but the converse is not true. One can move from the critical point by choosing $H = H^* + \Delta H$ using two distinct ways. If ΔH has nonzero projection on the relevant direction (see Sect. 3.3.2), the transformation Eq. (5.23) will lead away from the critical point with increasing l. In the other case ΔH will eventually tend to zero. The latter case introduces a subset of Hamiltonians which, after undergoing an RG transformation, will turn into critical ones. Traditionally, this subset is called the *critical surface.* The term is usually interpreted as a surface in the space defined by all the parameters needed to close an RG transformation. For instance, in the Hamiltonian Eq. (3.24) the vertices $\hat{g}_k(\mathbf{q}_1 \cdots \mathbf{q}_k)$ are the requisite space defining parameters. Let us assume that $d > 4$ Then any of the Hamiltonians of the form described in Eq. (5.2) with different choices of the function S are critical by their physical definitions.[1] However, Eq. (3.20) treats them differently, choosing only one particular form as critical. Hence there should be such a renormalization procedure for which Eq. (5.24) is satisfied independently of a particular choice of the function $S(x)$. Now, let us decrease the dimension to $d < 4$; then Hamiltonians as in Eq. (5.2) are no longer critical, but they

[1]The dangerous mode is the mode with $q = 0$.

are still fixed-point Hamiltonians. Therefore, when trying to find critical Hamiltonians for $d < 4$, one should write instead of Eq. (5.23) the following relation

$$\frac{\partial(H_0 + H_I)}{\partial l} \equiv \frac{\partial H_I}{\partial l} = \hat{R}\{H_0 + H_I\} \equiv \overline{R}\{H_I\}. \qquad (5.25)$$

Until now the operator $\hat{R}$ has only been defined in Wilson's approach (see Sect. 3.3). For practical purposes the definition of $\hat{R}$ is not important. Below we will find the operator $\overline{R}$ in the framework of the transformations developed in this chapter.

5.4.1 Conventional Approach

In order to derive an RG equation let us follow the formal Wilson's scheme considered in Chapter 3. As the first step of an RG transformation one has to perform an integration over short-wave modes in the partition function. For this purpose it is convenient to rewrite the partition function in the form

$$Z = \int D\phi \exp\left(-H\{\phi\}\right) = \langle \exp\left(-H_I\{\phi\}\right)\rangle_{0,\Lambda} \equiv Z_0 \langle \rho_I \rangle_{0,\Lambda}, \qquad (5.26)$$

where

$$Z_0 = \int D\phi \exp\left(-H_0\{\phi\}\right)$$

and the averaging $\langle \cdots \rangle_{0,\Lambda}$ is performed with the Gaussian functional at a given value of Λ.

In the following consideration we will use the procedure with two Gaussian fields, developed in Sect. 5.3.1. As a result of the application of this procedure to Eq. (5.26) one can find

$$Z = Z_0 \langle (1 + \delta l \hat{L}) \rho_I \rangle_{0,(1-\delta l)\Lambda}. \qquad (5.27)$$

The operator $\hat{L}$ in Eq. (5.27) is the same as in Eq. (5.13). The RG transformation will be completed by the change of the momentum scale in order to restore the initial value of the parameter Λ. The latter can be done with the help of the scale transformation Eq. (5.1) in which one should use conventional expression $A(q) = (d + 2)/2$. By rewriting the transformation of the field in the form

$$\phi_q = \left(1 + \delta l \frac{d+2}{2}\right) \phi'_{q'} = \left\{1 + \delta l \left(\frac{d+2}{2} + \boldsymbol{q} \frac{\partial}{\partial \boldsymbol{q}}\right)\right\} \phi'_q$$

one arrives at the following equation

$$\rho'_I\{\phi\} = \left\{1 + \delta l (\hat{L} + \hat{L}_\phi + \hat{L}_V)\right\} \rho_I\{\phi\}, \qquad (5.28)$$

where the operators $\hat{L}_\phi$ and $\hat{L}_V$ are defined by

$$\hat{L}_\phi = \left[\frac{d+2}{2}\phi_q + \boldsymbol{q}\frac{\partial \phi_q}{\partial \boldsymbol{q}}\right]\frac{\delta}{\delta\phi_q} \quad \text{and} \quad \hat{L}_V = dV\frac{\partial}{\partial V}$$

and they can be identified as the operators realizing a dilation transformation (see Eq. (3.19)). Inserting the evident form $\rho = \exp(-H_I)$ in Eq. (5.28) we obtain

$$\frac{\partial H_I}{\partial l} = dV\frac{\partial H_I}{\partial V} + \int_q \left[\frac{d+2}{2}\phi_q + \boldsymbol{q}\frac{\partial \phi_q}{\partial \boldsymbol{q}}\right]\frac{\delta H_I}{\delta \phi_q}$$
$$+ \int_q h(\boldsymbol{q})\left[\frac{\delta^2 H_I}{\delta\phi_q \delta\phi_{-q}} - \frac{\delta H_I}{\delta\phi_q}\frac{\delta H_I}{\delta\phi_{-q}}\right]. \tag{5.29}$$

This equation is an exact RG equation written in the form of Eq. (5.25). It differs from the equation obtained in Chapter 3, even if one rewrites Eq. (3.20) in terms of H_I. Nonetheless, one can obtain all known results of RG theory from this equation. However, this equation is not convenient for practical uses. As with Wilson's equation, it contains redundant eigenvectors that should be properly excluded to obtain results having physical meaning.

5.4.2 Generalized Approach

Though the transformation Eq. (5.28) restores the initial form of the Hamiltonian H_0, strictly speaking, one should not consider this transformation as conserving the zero-order Hamiltonian. Terms having the structure of H_0 appear in the interaction Hamiltonian under the influence of this transformation. Therefore, the operator $\overline{R}$ is not defined in Eq. (5.29). This equation generates terms with the vertex $g_2(\boldsymbol{q})$ that must be incorporated into the Hamiltonian H_0. As a consequence, the effective value of the zero order Hamiltonian changes under the RG transformation Eq. (5.29), while the choice of H_0 should define the operator $\overline{R}$ and must not undergo any changes with the RG transformation. The latter can be achieved if instead of restoring the functional H_0 one reduces the functional $H_0' + \delta H_0$ to H_0, at each step of the RG process with the help of some scale transformation of the field ϕ_q. A few words should be said about the addition δH_0. This term should include only momentum-dependent parts of the vertex $g_2(\boldsymbol{q})$ because the terms with $g_2(0) \equiv g_2$ contains a projection on the relevant direction and, therefore, it cannot be incorporated into the functional H_0. Using the notation of Sect. 5.3.2 let us rewrite the function $A(q)$ as

$$A(q) = \frac{d+2}{2} - \eta(q). \tag{5.30}$$

Then instead of Eq. (5.28) we have

$$\rho_I'\{\phi\} = \left\{ 1 + \delta l \left(\hat{L} + \hat{L}_\phi' + \hat{L}_V - \frac{1}{2} \int_q \eta(q) G_0^{-1}(q) |\phi_q|^2 \right.\right.$$
$$\left.\left. - n \frac{V}{2} \int_q \eta(q) \right) \right\} \rho_I\{\phi\}, \qquad (5.31)$$

where the operator $\hat{L}_\phi'$ is defined as

$$\hat{L}_\phi' = \left[\frac{d+2-2\eta(q)}{2} \phi_q + \boldsymbol{q} \frac{\partial \phi_q}{\partial \boldsymbol{q}} \right] \frac{\delta}{\delta \phi_q}$$

Using Eq. (5.31) and Eq. (5.26) one can find

$$\frac{\partial H_I}{\partial l} = dV \frac{\partial H_I}{\partial V} - n \frac{V}{2} \int_q \eta(q) - \frac{1}{2} \int_q \eta(q) G_0^{-1}(q) |\phi_q|^2$$
$$+ \int_q \left[\frac{d+2-2\eta(q)}{2} \phi_q + \boldsymbol{q} \frac{\partial \phi_q}{\partial \boldsymbol{q}} \right] \frac{\delta H_I}{\delta \phi_q}$$
$$+ \int_q h(\boldsymbol{q}) \left[\frac{\delta^2 H_I}{\delta \phi_q \delta \phi_{-q}} - \frac{\delta H_I}{\delta \phi_q} \frac{\delta H_I}{\delta \phi_{-q}} \right] \equiv \hat{R}_g\{H_I\}. \qquad (5.32)$$

Again, we have obtained an exact RG equation. This equation is a generalized form of Eq. (5.29), which enables us to carry out the reduction of $H_0' + \delta H_0$ to the Hamiltonian H_0. Eq. (5.32) contains a, so far, undefined, arbitrary function $\eta(q)$ that accounts for this. The following steps are of crucial importance for the definition of the generalized RG, though they can be easily anticipated. First of all, let us separate terms of zeroth and first orders in ϕ^2 in the functional H_I,

$$H_I\{\phi\} = bV + \frac{1}{2} \int_q g_2(\boldsymbol{q}) |\phi_q|^2 + H_I'. \qquad (5.33)$$

Now, using Eq. (5.32) we obtain a simple equation for a renormalization of the constant b,

$$\frac{db}{dl} = db + \frac{1}{2} \int_q \eta(q) + n \int_q h(\boldsymbol{q}) g_2(\boldsymbol{q}). \qquad (5.34)$$

The equation for $g_2(q)$ has the form

$$\frac{dg_2(q)}{dl} = Q(q) - 2h(q) g_2^2(q) - \eta(q) G_0^{-1} + \left(2 - \eta(q) - 2q^2 \frac{\partial}{\partial q^2} \right) g_2(q), \qquad (5.35)$$

where the function $Q(q)$ is given by Eq. (5.10). Now let us separate the part dependent on momenta from the vertex $g_2(q) = g_2 + g_2'(q)$. Equations for g_2 and $g_2'(q)$ can be written in the form

$$\begin{aligned} \frac{dg_2}{dl} &= [2-\eta(0)]g_2 + Q(0) - 2h(0)g_2^2, \\ \frac{\partial g_2'(q)}{\partial l} &= -\eta(q)G_0^{-1}(q) - [\eta(q)-\eta(0)]g_2 + Q(q) - Q(0) \\ &\quad -2[h(q)-h(0)]g_2^2 + \left(2-\eta(q)-2q^2\frac{\partial}{\partial q^2}\right)g_2'(q) \\ &\quad -2h(q)\left[g_2'^2(q) + 2g_2g_2'(q)\right]. \end{aligned} \tag{5.36}$$

Our objective is to eliminate the generation of the q-dependent part of the vertex $g_2(q)$ in the RG equation, provided that its initial value does not depend on momentum.[2] Using Eq. (5.35) one can easily find that the function $\eta(q)$ defined by Eq. (5.9) ensures such behavior. This completes the procedure of reducing $H_0' + \delta H_0$ to the initial value of the Hamiltonian H_0 thus forcing it to be an invariant of the RG transformation. Now the function $\eta(q)$ depends on all higher order vertices and has clear physical meaning: at a stable fixed point (critical behavior) the value $\eta^*(0)$ is equal to the exponent η, i.e. at the critical point the correlation function $G(q) \propto q^{-2+\eta^*(0)}$.[3]

5.4.3 RG as a Characteristic Set of Scale Equations

In the previous section, we obtained the operator $\overline{R}$ as an operator $\hat{R}_g\{H_I\}$ but with the function $\eta(q,l)$ defined by Eqs. (5.9) and (5.10). Insofar as the Hamiltonian H_0 is given, this RG transformation is unique and, therefore, it does not contain any redundant eigenvectors. In comparison with the traditional RG equations in the Wilson approach, the developed scheme may seem to be a little cumbersome. This is not really the case because the traditional approach, having a simpler definition of the RG transformation (see Eq. (5.29)), transfers the difficulties to the problem of elimination of redundant eigenvectors. In this section we are going to show some additional insight into the problem of RG transformation. Namely, *the RG transformations are completely defined by the chosen scale transformation.* The latter can be proven if one consider characteristic sets of scale equations

---

[2]This means that we require $g_2'(q,l) = 0$ at any l.

[3]Proof of this statement is given in Sect. 5.5.1.

(SE). The derivation of SE in Sect. 5.3 was given in a form that permits us to find this connection easily. The characteristic set for the operator $\hat{\Phi}$ in the form of Eq. (5.21) can be reduced to one equation in variational derivatives. This can be done by multiplying each k-th equation of the set on the product of $2k$ field operators containing the same momenta as present in the derivative $d\hat{g}_k(\boldsymbol{q}_1 \cdots \boldsymbol{q}_{2k})/dl$. After that one needs to integrate each term over all momenta, as in Eq. (5.3) and add all equations with factors defining the definition of the Hamiltonian H_I in the form of Eq. (3.24). The right hand side of Eq. (5.21) can be transformed with the help of variational derivatives to the operator $\hat{R}_g\{H_I\}$. This proof can also be made starting from Eq. (5.32). In that case, one must insert into this equation the representation of H_I in the form of Eq. (3.24) and, by equalizing powers of field operators, reduce the RG equation to the set Eq. (5.21).

In conclusion it is interesting to compare some formal aspects of the Wilson approach and the generalized scheme developed in this chapter. In Wilson's approach one has to deal with some arbitrary function defining an RG transformation. By elimination some of the redundant eigenvectors, one may restore this function.[4] In the generalized scheme, the RG procedure is defined uniquely but this definition depends on the choice of the function $S(x)$. Therefore, in this scheme all eigenvalues λ_i of the linear RG operator (see Sect. 3.3.2) are functionals of $S(x)$ and universality is lost. The latter accounts for the introduction of a new dimensionless scale x_0. This is the scale on which the function $S(x)$ goes to zero with increasing x. The loss of universality is not a big problem because it can be immediately restored simply by letting $x_0 \to \infty$. In this limit all the RG defining Hamiltonians give the same results. Nonetheless, this argument shows that, for example, if one calculates exactly the critical exponents for $d = 2$ and $n = 1$ they should not coincide with the exponents of the Ising model. This accounts for the fact that Ising model keeps H_0 (or x_0) fixed, while traditionally we put $x_0 \to \infty$.

5.5 Applications of Generalized SE

In this section we consider some of applications of the generalized approach bearing principal features of the method. Applications to particular models are considered in Chapter 6.

The scale equations are inhomogeneous. In the right-hand side of the SE for the k-point correlation function we have a linear combination of $(k-2)$-point correlation functions. The latter means that at $k = 2$ we have as a inhomogeneous term the function $h(q)$ and therefore, one could expect that

[4] In the general case we have an infinite number of redundant eigenvectors.

under the proper boundary conditions for the function $G(q)$, this function can be restored exactly in the critical region. So far this program has not been realized, though behavior of the correlator at the critical point was found by Ivanchenko *et al* in 1990 (b). Below, for the sake of simplicity, we put the cutoff parameter $\Lambda = 1$, which means that all momenta are measured in units of Λ.

5.5.1 Structure of the Correlation Function at the Transition Point

The behavior of correlation functions at the critical point can be found owing to the fact that this point is characterized by the equality $\hat{\Phi} = 0$. The latter condition substantially reduces the number of variables in SE. In the simplest case of the two-point correlation function

$$\delta_{\alpha\beta} G(\boldsymbol{q}) = \frac{1}{V} \langle \phi_q^\alpha \phi_{-q}^\beta \rangle$$

one can obtain from Eq. (5.19), taking into account Eq. (5.22), the following equation

$$\left\{ 2 - \eta(\boldsymbol{q}) + 4q^2 h(q) S^{-1}(q) + \boldsymbol{q}\frac{\partial}{\partial \boldsymbol{q}} \right\} G(\boldsymbol{q}) = 2h(q). \tag{5.37}$$

The integral surface of this equation containing the point $(\boldsymbol{q}_o,\ G(\boldsymbol{q}_o))$ is defined by

$$G(\boldsymbol{q}) = \frac{S^2(q^2)}{q^2} \exp\{-\kappa(q^2, q_0^2)\} \left\{ q_0^2 G(\boldsymbol{q}_o) S^{-2}(q_0^2) - \int_{q^2}^{q_0^2} dt \frac{dS^{-1}(t)}{dt} \exp\{\kappa(t, q_0^2)\} \right\}, \tag{5.38}$$

where

$$\kappa(t, q_0^2) = \frac{1}{2} \int_t^{q_0^2} d\tau \frac{\eta(\tau)}{\tau}.$$

In the derivation of Eq. (5.38) the relation $h(t) = -dS(t)/dt$ has been taken into account. Formally, Eq. (5.38) defines the function $G(\boldsymbol{q})$ provided $G(\boldsymbol{q}_o)$ is known. With respect to the function $G(\boldsymbol{q})$ we know its asymptotic behavior $G(\boldsymbol{q}) \to G_0(q)$ at $q \to \infty$. Therefore, by setting $q_0 \to \infty$ and choosing $G(\boldsymbol{q}_o) = q_0^{-2} S(q_0^2)$ we can get the solution of Eq. (5.38) corresponding to the real physical situation

$$G(\boldsymbol{q}) = \frac{S(q^2)}{q^2} \exp\{-\kappa(q^2)\} \left\{ 1 - S(q^2) \int_{q^2}^{\infty} dt \frac{dS^{-1}(t)}{dt} [\exp\{\kappa(t)\} - 1] \right\}, \tag{5.39}$$

where $\kappa(q^2) \equiv \kappa(q^2, \infty)$. The integral over t in the right-hand side of Eq. (5.39) is convergent at $q \to \infty$, as the function $\eta(q)$ according to Eq. (5.9) is proportional to $G_0(q)$ and at large momenta it falls quite rapidly with the increase of q. One can also see that Eq. (5.39) at $q \to \infty$ gives the right asymptote $G_0(q)$. $q \to 0$ is another physically important limit. In this limit $\kappa(q^2) \to -\eta(0) \ln q$ and simple transformations lead to

$$G(q) = f(q) q^{-2+\eta(0)}, \tag{5.40}$$

where the function $f(q)$ is defined as

$$f(q) = \exp\left\{-\frac{1}{2}\int_{q^2}^{\infty} d\tau \frac{\eta(\tau)-\eta(0)}{\tau}\right\}\left\{1-\int_{q^2}^{\infty} dt \frac{dS^{-1}(t)}{dt}[\exp(\kappa(t))-1]\right\}. \tag{5.41}$$

When $q \to 0$ the function $f(q)$ has a finite limit, which means that $f(0)$ is a preexponential factor in the asymptote $G(q \to 0) \propto q^{-2+\eta(0)}$. Thus, as follows from Eq. (5.40), the value $\eta(0)$, calculated at all vertexes taken at the fixed point, is the exponent η. It is important to notice that Eq. (5.39) not only defines the critical asymptote but also gives the behavior of the two-point correlation function at any value of momenta q. Naturally, this function is not a universal function as it essentially depends on the choice of the cutoff function $S(q^2)$. In the case of a sharp cutoff, when $S(q^2) = \Theta(1-q^2)$, Eq. (5.39) contains divergences and is no longer applicable. The latter is a consequence of the assumption used in the derivation of Eq. (5.39). All functions have been assumed to have finite derivatives. Hence, the case when $S(q^2)$ is Heaviside's step function and $h(q)$ is Dirac's δ-function ($h(q) = \delta(1-q^2)$) should be considered as special. In this case instead of Eq. (5.37) we have

$$\left\{\Theta(1-q^2)\left[1-\frac{1}{2}\eta(q)q^2\frac{d}{dq^2}\right] + 2q^2\delta(1-q^2)\right\} G(q) = \delta(1-q^2)$$

The solution of this equation should have the form $G(q) = \Theta(1-q^2)\overline{G}(q)$. By separating singular terms one finds

$$q^2\overline{G}(q)\delta(1-q^2) = \delta(1-q^2),$$

which means that $\overline{G}(1) = 1$. Nonsingular terms give the following equation

$$\left\{1-\frac{1}{2}\eta(q) + q^2\frac{d}{dq^2}\right\}\overline{G}(q) = 0,$$

from which the solution for $G(q)$ can be obtained in the final form as

$$G(q) = \frac{\Theta(1-q^2)}{q^2}\exp\left\{-\frac{1}{2}\int_{q^2}^{1}\frac{d\tau}{\tau}\eta(\tau)\right\}. \tag{5.42}$$

Now in the place of the limit $q \to \infty$ we have $q^2 \to 1$, where evidently $G(q) \to \Theta(1-q^2)/q^2$, i.e. $G(q)_{q^2\to 1} \to G_0(q)$. In the opposite limit one can get $G(q) \to f(0)q^{-2+\eta(0)}$ with

$$f(0) = \exp\left\{-\frac{1}{2}\int_0^1 d\tau \frac{\eta(\tau)-\eta(0)}{\tau}\right\}.$$

Finally, Eqs. (5.39) and (5.42) define the two-point correlation functions for the cases of smooth and sharp cutoffs. However, these equations will give explicit expressions only if the function $\eta(q)$ is known. In the general case finding this function can be extremely difficult. In the following section, this problem is considered in the ϵ-expansion in the first nonvanishing approximation. Nonetheless, even when the exact expressions for $\eta(q)$ are unknown, Eqs. (5.39) and (5.42) contain valuable information about the correlation function's structure. In these equations $\eta(q)$ enters in the integrated forms. As this function, by definition, is a slowly changing monotonic function, one can approximate it by any smooth function which coincides with the experimental value of the exponent η at $q = 0$ and as $q \to \infty$ tends to $G_0(q)$. Such an approximation should give very good results as the integrations will further decrease errors. Hence, the structure of the correlation function can be restored at any value of q as a nonuniversal function depending on the cutoff procedure.

5.5.2 Function $\eta(q)$ at the Fixed Point

The function $\eta(q)$ is defined by the nonlocal structure of the vertex $g_4(\boldsymbol{q}_1, \boldsymbol{q}_{1'}, \boldsymbol{q}_2, \boldsymbol{q}_{2'})$ (see Eqs. (5.9), (5.10)) and the local vertex g_2. According to the general consideration of Chapter 3, the fixed point can be defined with the help of Eq. (5.21) as $d\hat{g}_k/dl = 0$. Therefore, the equations from which one can get values g_2 and $g_4\{\boldsymbol{q}_i\}$ should be written as

$$U_k[g_m(\boldsymbol{q}_1 \cdots \boldsymbol{q}_m)] = 0, \tag{5.43}$$

where U_k as well as vertexes g_m for the considered here case of $\mathcal{O}(n)$ symmetry are defined as $U_k^{\alpha_1 \cdots \alpha_k} = (\delta_{\alpha_1\alpha_2} \cdots \delta_{\alpha_{k-1}\alpha_k} + permutations)U_k$. The set from Eq. (5.43) is a formally exact set and one can choose any particular approximation scheme to obtain its solutions. Here we consider as an example calculation of $\eta(q)$ in the first nonvanishing order in the ϵ-expansion scheme. An analysis of the set of Eq. (5.43) shows that the vertexes g_2 and g_4 are of the order of ϵ, g_6 is of the order of ϵ^2, while other vertexes have higher orders of ϵ. Hence, in the lowest orders we only have the following three equations

$$g_2 = -\frac{1}{2}(n+2)\int_q h(q), \tag{5.44}$$

$$\left\{ \epsilon - \hat{R}_2\{\boldsymbol{q}_i\} - 2g_2 \sum_{i=1}^{2} [h(q_i) + h(q_{i'})] \right\}$$
$$\times g_4(\boldsymbol{q}_1, \boldsymbol{q}_{1'}; \boldsymbol{q}_2, \boldsymbol{q}_{2'}) + Q_2(\boldsymbol{q}_1, \boldsymbol{q}_{1'}; \boldsymbol{q}_2, \boldsymbol{q}_{2'}) = 0, \tag{5.45}$$

$$\left\{ 2 + \hat{R}_3\{\boldsymbol{q}_i\} \right\} g_6(\boldsymbol{q}_1, \cdots, \boldsymbol{q}_{3'}) + 8 \int_p h(p)$$
$$\times \hat{S}[g_4(\boldsymbol{p}, \boldsymbol{q}_1; \boldsymbol{q}_2, \boldsymbol{q}_{2'}) g_4(-\boldsymbol{p}, \boldsymbol{q}_{1'}; \boldsymbol{q}_3, \boldsymbol{q}_{3'})] = 0. \tag{5.46}$$

Here the operators $\hat{R}_k$ and the function $Q_2\{\boldsymbol{q}_i\}$ are defined by formulae

$$\begin{aligned} \hat{R}_k\{\boldsymbol{q}_i\} &= \textstyle\sum_{i=1}^{k} \left(\boldsymbol{q}_i \frac{\partial}{\partial \boldsymbol{q}_i} + \boldsymbol{q}_{i'} \frac{\partial}{\partial \boldsymbol{q}_{i'}} \right), \\ Q_2\{\boldsymbol{q}_i\} &= 3 \textstyle\int_p h(p) \left\{ \frac{n}{2} g_6(-\boldsymbol{p}, \boldsymbol{p}; \boldsymbol{q}_1, \cdots, \boldsymbol{q}_{3'}) \right. \\ &+ \left. 2\hat{S} g_6(\boldsymbol{p}, \boldsymbol{q}_1; -\boldsymbol{p}, \boldsymbol{q}_{1'}; \boldsymbol{q}_2, \boldsymbol{q}_{2'}; \boldsymbol{q}_3, \boldsymbol{q}_{3'}) \right\} \end{aligned}$$

Let us separate g_4, independent on momenta part, from the vertex $g_4\{\boldsymbol{q}_i\}$. This part can be defined from

$$\epsilon g_4 - 8 g_2 g_4 h(0) + Q_2\{\boldsymbol{q}_i{=}0\} = 0. \tag{5.47}$$

The function $\bar{g}_4\{\boldsymbol{q}_i\} = g_4\{\boldsymbol{q}_i\} - g_4$ satisfies the equation that can be obtained from Eq. (5.45) with the help of Eq. (5.47)

$$\hat{R}_2\{\boldsymbol{q}_i\} \bar{g}_4\{\boldsymbol{q}_i\} = -[\chi\{\boldsymbol{q}_i\} - \chi\{0\}], \tag{5.48}$$

where

$$\chi\{\boldsymbol{q}_i\} = 2 g_2 g_4 \sum_{i=1}^{2} [h(q_i) + h(q_{i'})] - Q_2\{\boldsymbol{q}_i\}.$$

From Eqs. (5.47) and (5.48) one can see that g_4 is of the order of ϵ, while $\bar{g}_4\{\boldsymbol{q}_i\}$ is of the order of ϵ^2. As a result, we can reduce Eq. (5.46) to

$$\left\{ 2 + \hat{R}_3\{\boldsymbol{q}_i\} \right\} g_6\{\boldsymbol{q}_i\} = -g_4^2 H\{\boldsymbol{q}_i\}, \tag{5.49}$$

where

$$H\{\boldsymbol{q}_i\} = \frac{2}{3} \sum_{i \neq j}^{3} [h(\boldsymbol{q}_i + \boldsymbol{q}_j + \boldsymbol{q}_{i'}) + (j \rightarrow j')].$$

The solutions of Eqs. (5.48) and (5.49) can be formally written as integrals

$$\bar{g}_4\{\boldsymbol{q}_i\} \quad = \quad -\int_0^1 \frac{dx}{x}\left[\chi(\boldsymbol{q}_i x) - \chi(0)\right], \tag{5.50}$$

$$g_6\{\boldsymbol{q}_i\} \quad = \quad -g_4^2 \int_0^1 dx x H\{\boldsymbol{q}_i x\}, \tag{5.51}$$

From Eq. (5.51) one can define the function $Q\{\boldsymbol{q}_i\}$ which in turn defines (see definition for χ) $\bar{g}_4\{\boldsymbol{q}_i\}$. In particular, we find that

$$Q_2\{\boldsymbol{q}_i = 0\} = -4g_4^2 \int_p h(q) \int_0^1 dx x[(n+8)h(px) - 2(n+2)h(0)],$$

and as a result from Eq. (5.47) we arrive at

$$g_4 = \epsilon \left[4(n+8) \int_p h(p) \int_0^1 dx x h(px)\right]^{-1}, \tag{5.52}$$

By taking into account that $h(p) = -dS/dp^2$, $S(0) = 1$, and $S(\infty) = 0$ we can calculate the integrals in Eq. (5.52) exactly and, as a result, the value g_4 can be expressed in the traditional form as $g_4 = 2\epsilon/K_4(n+8)$, where $K_4 = 1/8\pi^2$.

According to Eqs. (5.9) and (5.10) the function $\eta(q)$ can be expressed in terms of the difference $Q(q) - Q(0)$. After trivial, though somewhat cumbersome calculations, we obtain this difference as

$$\begin{aligned} Q(q) - Q(0) = {} & -4g_2 g_4(n+2) \int_0^1 \frac{dx}{x}[h(qx) - h(0)] \int_p h(p) \\ & -4g_4^2 \int_{k,p} h(k)h(p) \int_0^1 \frac{dx}{x} \int_0^1 dy y \Big\{ (n+2)^2[h(qxy) - h(0)] \\ & + 6(n+2)[h((\boldsymbol{p} + (\boldsymbol{k}+\boldsymbol{q})x)y) - h((\boldsymbol{p}+\boldsymbol{k}x)y)]\Big\}. \end{aligned} \tag{5.53}$$

By inserting Eq. (5.53) into Eqs. (5.9) and (5.10) one obtains the function $\eta(q)$. Further calculations can only be done for some evident form of the cutoff function $S(q^2)$ (or $h(q)$).

Let us now find the critical exponent η. According to Eq. (5.10) η is related to the function $Q(q)$ as follows

$$\eta = \frac{1}{8}\frac{\partial^2}{\partial \boldsymbol{q}^2}[Q(q) - g_2^2 h(q)]\Big|_{q=0}. \tag{5.54}$$

Inserting the function $Q(q)$ in Eq. (5.54) in the form of Eq. (5.53) and making use of Eq. (5.44) we obtain

$$\eta = 24(n+2)g_4^2 \int_{k,p} h(k)h(p) \int_0^1 dxx \int_0^1 dyy^3 \frac{d^2h(\boldsymbol{p}y+\boldsymbol{k}xy)}{d(\boldsymbol{p}y+\boldsymbol{k}xy)^2}. \quad (5.55)$$

In principle the exponent η defined by Eq. (5.55) could be dependent on the cutoff procedure and thus being a nonuniversal value. In this case, however, this is not true. The integral over y can be easily carried out and Eq. (5.55) simplifies to

$$\eta = \frac{3(n+2)\epsilon^2}{2(n+8)^2 K_4} \int_{k,p} h(k)h(p) \int_0^1 dxx \frac{1}{|\boldsymbol{p}+\boldsymbol{k}x|} \frac{dh(\boldsymbol{p}+\boldsymbol{k}x)}{d|\boldsymbol{p}+\boldsymbol{k}x|}. \quad (5.56)$$

In this equation the evident form g_4 (Eq. (5.52)) has been used. The formula Eq. (5.56) coincides with the expression obtained by Rudnick (1975) for the exponent η in Wilson's approach. The combination of integrals in Eq. (5.56) does not depend on a particular choice of $h(q)$, provided $h(q) = -dS/dq^2$, and is equal to $K_4^2/48$. Hence, for η one has the usual expression $\eta = \epsilon^2(n+2)/2(n+8)^2$ (see Eq. (3.115)) arising in the main order of ϵ. Such universal behavior may not necessarily be reproduced in higher orders. On the contrary as is clear from the discussion presented in Sect. 5.4.3, results must be dependent on the choice of S being universal only in the limit $x_0 \to \infty$.

5.6 η-Expansion

From the moment of its appearance at the beginning of the seventies, the RG approach has been the main mathematical tool for analysis of critical phenomena in phase transitions. Although very good results have been achieved with the help of this technique it is not well founded since in all known versions it is based on a perturbation theory using an expansion parameter which is not small. The search for small parameters in the RG theory has been going on since the first successful applications made by Wilson and Fisher in 1972. Among the parameters considered one should mention $\epsilon = 4-d$, inverse number of components of the vector order parameter (see Abe 1972, Ma 1972); small coupling constants in the three dimensional space (see Ginzburg 1975, Colot *et al* 1975);[5] use of long range interaction potentials $V(r)$ such that $V(q) - V(0) \propto q^\sigma$ with $\sigma < 2$ as a

[5]The accuracy of these methods, strictly speaking, is equivalent to the extrapolation ϵ to unit.

zero order approximation for the following expansion in the difference $2-\sigma$ (see Fisher *et al* 1972); and so on. Drawbacks of all these methods lie in the facts that they are either suitable for a very narrow class of systems or require expansions in values which actually are not small.

The other direction of search was based on the exact nature of Wilson's equation. Its essence is an abandonment of the perturbation theory in the common sense of the word. Riedel *et al* 1985 (see also Golner 1986) reformulated the functional RG equation in the form of an infinite system of equation for scaling fields (see Sect. 3.3.6). They also formulated the "principle of balance" connecting the method of truncating the infinite system with the optimization of physical results. This approach led to a very good convergence of numerical values for critical exponents. Though the "principle of balance" cannot be strictly proved, nonetheless results of computer stimulations are very promising, giving numerical values close to experimental ones.

Nonetheless, the existence of the exact closed scheme in the form of the generalized RG equation allows one to seek new initial approximations of the perturbation theory. These may be some solutions different from the trivial Gaussian solution with the following expansion in a small parameter also different from the expansion in the coupling constants and the ϵ-expansion. Tempting candidates for the role of this parameter are the numerically small critical exponents α and η. Their appearance is closely related to the structure of the RG equation and, on the other hand, their physical (renormalized) values are experimentally small. Taking into account the fact that the determination of one of the exponents (namely ν and, with the help of it, α) can only be done by calculation of eigenvalues of the linearized RG in the vicinity of a fixed point, one could draw the conclusion that the only practical candidate for the role of the small parameter is the exponent η. The version of the theory based on the perturbations over $\eta^{1/2}$ is presented below (Ivanchenko *et al* 1990).

5.6.1 General Scheme of the Method

Returning to Eq. (5.32), let us rewrite it in a somewhat different form. This form is more convenient for the further calculation. We will separate zero and first order in ϕ^2 terms in the functional H_I

$$H_I = bV + \frac{1}{2}g_2 \int_q |\phi_q|^2 + \mathcal{H}. \tag{5.57}$$

For the constants b and g_2 one can obtain, with the help of Eqs. (5.9) and (5.10) the following relations (see also Eqs. (5.34) and (5.36))

$$\frac{db}{dl} = db + \frac{1}{2}\int_q \eta(q) + ng_2 \int_q h(\boldsymbol{q}) \tag{5.58}$$

and

$$\frac{dg_2}{dl} = [2 - \eta(0)]g_2 + Q(0) - 2h(0)g_2^2, \tag{5.59}$$

while the functional $\mathcal{H}$ satisfies the equation

$$\frac{\partial \mathcal{H}}{\partial l} = \int_q \left[\frac{d+2-2\eta(q)}{2}\boldsymbol{\phi}_q + \boldsymbol{q}\frac{\partial \boldsymbol{\phi}_q}{\partial \boldsymbol{q}}\right]\frac{\delta \mathcal{H}}{\delta \boldsymbol{\phi}_q}$$
$$+ \int_q h(\boldsymbol{q})\left[\frac{\delta^2 \mathcal{H}}{\delta \boldsymbol{\phi}_q \delta \boldsymbol{\phi}_{-q}} - \frac{\delta \mathcal{H}}{\delta \boldsymbol{\phi}_q}\frac{\delta \mathcal{H}}{\boldsymbol{\phi}_{-q}} - 2g_2\boldsymbol{\phi}_q\frac{\delta \mathcal{H}}{\delta \boldsymbol{\phi}_{-q}}\right]. \tag{5.60}$$

Let us isolate in Eq. (5.60) the following linear operator

$$\hat{L} = \hat{L}_1 + \hat{L}_2 = \int_q \left[\frac{3}{4}\boldsymbol{\phi}_q + \boldsymbol{q}\frac{\partial \boldsymbol{\phi}_q}{\partial \boldsymbol{q}}\right]\frac{\delta}{\delta \boldsymbol{\phi}_q} + \int_q h(\boldsymbol{q})\frac{\delta^2}{\delta \boldsymbol{\phi}_q \delta \boldsymbol{\phi}_{-q}}. \tag{5.61}$$

The action of this operator on the fourth order local form

$$\mathcal{H}_1 \equiv \mathcal{H}_{41} = \frac{1}{8}g_{41}\int_q \delta\left(\sum_i \boldsymbol{q}_i\right)(\boldsymbol{\phi}_{q_1}\boldsymbol{\phi}_{q_2})(\boldsymbol{\phi}_{q_3}\boldsymbol{\phi}_{q_4}) \tag{5.62}$$

leads to the trivial result

$$\hat{L}\mathcal{H}_1 = 0. \tag{5.63}$$

However, it should be mentioned that this equation holds since a second variation of the functional $\mathcal{H}_1$ proportional to $|\boldsymbol{\phi}_q|^2$ is automatically included in Eq. (5.59) (i.e. omitted in Eq. (5.63)). If one considers that in some approximation the RG equation can be reduced to the action of the operator $\hat{L}$, then the local fourth order Hamiltonian H_1 is the zero order fixed point solution of Eq. (5.60). Therefore, one should treat the remaining terms as small in orders of a so far undefined small parameter. The terms not included in the operator $\hat{L}$ contain nonlinear parts. The latter means that they should be treated in a way permitting the existence of the zero order solution $\mathcal{H}_1$ and hence the solvability condition should define the arbitrary constant g_{41}. Let us now consider the remaining terms in Eq. (5.60). First of all the form acting as a factor at $\phi\delta\mathcal{H}/\delta\phi$ can be represented as

$$\frac{d+2-\eta(q)}{2} = \frac{3d}{4} + \frac{\epsilon}{4} - \frac{\eta(q)}{2}. \tag{5.64}$$

The last two terms in the right-hand side of Eq. (5.64) possess, at $d = 3$, a certain smallness with respect to the first one. At $q \to 0$ the limit of the function $\eta(q)$ is the exponent η which is quantitatively small. At $q \to \infty$ the correlation function $G(q)$ turns into $G_0(q)$ so that $\eta(q) \to 0$ when $q \to \infty$. The second term at $d = 3$ has a magnitude smaller by the order of $1/(3d)$ with respect to the first one. This is rather good for possible application of the perturbation theory. In the situation when $d \to 4$ this smallness is obviously enhanced.

Let us assume that the perturbation theory can be formulated in terms of a single small parameter (denoted as x). It seems natural to present x^2 by $\eta(0) = x^2$. As the renormalized value η at the fixed point coincides with the exponent η the introduced quantity x can be considered as a truly small parameter. Using an *a priori* estimate for $\epsilon/3d \sim 10^{-1}$ and $\eta \sim 10^{-2}$ one can suppose that $\epsilon/3d \approx x + o(x^2)$ and reexpand it in powers of $\sqrt{\eta}$:

$$\frac{\epsilon}{3d} = \frac{1}{3}\sum_{k=1}^{\infty} \alpha_k x^k.$$

For the function $\eta(q)$ we also should write expansion like

$$\eta(q) = \sum_{k=2}^{\infty} \eta_k(q) x^k$$

The existence of such expansions is verified below.

The last three terms in Eq. (5.60) are either nonlinear or describe contributions to the renormalization of vertices g_k at the expense of higher order vertices g_{k+2}. It seems quite natural to suppose that $\mathcal{H}_1$ at the fixed point is of the order of x (i.e. $g_{41} \sim x$) and that the three terms lead to a regular increase of the power x in the higher orders of the perturbation theory. This assumption should also be verified.

Let us denote the contributions to the functional $\mathcal{H}$ of the order of x^k by means of $\mathcal{H}_k$, i.e.

$$\mathcal{H} = \sum_{k=1}^{\infty} \mathcal{H}_k x^k, \tag{5.65}$$

and terms in $\mathcal{H}_k$ having powers ϕ^m by means of $\mathcal{H}_{mk}$. Using the introduced notations one can rewrite Eq. (5.60) at the fixed point as

$$\begin{aligned} \hat{L}\mathcal{H}_k &= \sum_{p=1}^{k-1} \int_q \left\{ \left[2g_{2p}h(q) - \frac{1}{4} d\alpha_p \right] \phi_q \frac{\delta \mathcal{H}_{k-p}}{\delta \phi_q} + h(q) \frac{\delta \mathcal{H}_p}{\delta \phi_q} \frac{\delta \mathcal{H}_{k-p}}{\delta \phi_{-q}} \right\} \\ &+ \frac{1}{2} \sum_{p=2}^{k-1} \int_q \eta_p(q) \phi_q \frac{\delta \mathcal{H}_{k-p}}{\delta \phi_q} \equiv \Psi_k. \end{aligned} \tag{5.66}$$

In the lowest order, Eq. (5.66) coincides with $\hat{L}\mathcal{H}_1 = 0$. The important property of this equation is the fact that its solution $\mathcal{H}_1$ in the form of Eq. (5.62) is the local functional and preserves only a single constant g_{41} undefined. For the following this property is of crucial importance. In the next order ($k = 2$) we have

$$\hat{L}\mathcal{H}_2 = \int_q \left\{ \left[2g_{21}h(q) - \frac{1}{4}d\alpha_1 \right] \phi_q + h(q)\frac{\delta \mathcal{H}_1}{\delta \phi_q} \right\} \frac{\delta \mathcal{H}_1}{\delta \phi_q} = \Psi_2. \tag{5.67}$$

Substituting $\mathcal{H}_2 = \mathcal{H}_{42} + \mathcal{H}_{62}$ one gets a closed (naturally with equation $dg_2/dl = 0$) set

$$\hat{L}_1\mathcal{H}_{42} + \hat{L}_2\mathcal{H}_{62} = \Psi_{42}, \qquad \hat{L}_1\mathcal{H}_{62} = \Psi_{62}, \tag{5.68}$$

where Ψ_{mk} are terms in Ψ_l containing only m-th powers of field operators. The second equality in Eq. (5.68) determines $\mathcal{H}_{62}$ through g_{41}, and the first one defines the constant g_{41} through $\hat{L}_1\mathcal{H}_{42}\{g_{42}\} = 0$. In this order the constant g_{42} in its turn becomes indefinite, i.e. the situation with a single unknown constant is restored.

Let us consider one more order (up to x^3) of the RG set Eq. (5.66). In this order one has the set

$$\hat{L}_1\mathcal{H}_{43}+\hat{L}_2\mathcal{H}_{63} = \Psi_{43}, \qquad \hat{L}_1\mathcal{H}_{63}+\hat{L}_2\mathcal{H}_{84} = \Psi_{63}, \qquad \hat{L}_1\mathcal{H}_{83} = \Psi_{83}. \tag{5.69}$$

The third equation determines $\mathcal{H}_{83}$, the second $\mathcal{H}_{63}$, and the first, in addition to $\mathcal{H}_{43}$, determines the constant g_{42} leaving g_{43} unknown.

The following procedure is obvious. At each step, a term of the type $\mathcal{H}_{mk}$ is added to $\mathcal{H}$ so that $m = 2(k+1)$ and an additional equation of the form $\hat{L}_1\mathcal{H}_{2(k+1),k} = \Psi_{2(k+1),k}$ appears. The equation for $\Psi_{4,k}$ of the type $\hat{L}_1\mathcal{H}_{4,k} + \hat{L}_2\mathcal{H}_{6,k} = \Psi_{4,k}$ introduces, in its turn, the new constant $g_{4,k}$. In order to complete this regular procedure one should supplement it with the system of equations arising from the definition of $\eta(q)$:

$$x^2 \equiv \eta(0) = \sum_{k=2}^{\infty} \eta_k(0)x^k.$$

Denoting coefficients at x^k, in the expansion of $D(q)$ over x, as $D_k(q)$ one has

$$1 = \left.\frac{dD_2(q)}{dq^2}\right|_{q=0}, \quad 0 = \left.\frac{dD_k(q)}{dq^2}\right|_{q=0}, \quad k = 3, \cdots. \tag{5.70}$$

The system of Eqs. (5.70) defines the set of coefficients α_k thus closing the set of equations. Now the value $\eta = x^2$ is to be calculated by inversion of the formula $\sum_{k=1}^{\infty} \alpha_k x^k = \epsilon/d$.

5.6.2 Solution of Equations

In the first order of x, the set of equations is quite simple. From Eq. (5.59) one has

$$\frac{d}{2}g_{21} + Q(0) = 0, \qquad \hat{R}_2 g_{41}\{\boldsymbol{q}_i\} = 0.$$

The second equation yields the constant g_{41} which can only be defined from the second approximation. This constant enters into the definition $Q(0)$ (see Eq. (5.10)) according to the relation

$$g_{21} = -\frac{2}{d}\int_q h(q)[ng_{41} + 2g_{41}] = -\frac{2(n+2)}{d}g_{41}\bar{h}, \tag{5.71}$$

where $\bar{h} = \int_q h(q)$. All momentum integrals can be normalized by the value $\bar{h}$. Therefore, no misunderstanding will arise if for the sake of simplicity, one sets $\bar{h} = 1$. Such a property can also be introduced with the help of normalization of the function $h(q)$ by rescaling parameter Λ. In the second approximation we have the set

$$\begin{aligned} l\frac{d}{2}g_{22} + Q_{12}(0) + \frac{d}{2}g_{21}\alpha_1 - 2h(0)g_{21}^2 &= 0, \\ -\hat{R}_2 g_{42}\{\boldsymbol{q}_i\} + Q_{22}\{\boldsymbol{q}_i\} + dg_{41}\alpha_1 - 8\hat{S}h(\boldsymbol{q}_i)g_{41}g_{21} &= 0, \\ \left(\frac{d}{2} + \hat{R}_3\right) g_{62}\{\boldsymbol{q}_i\} + 8\hat{S}h(\boldsymbol{q}_1 + \boldsymbol{q}_2 + \boldsymbol{q}_{2'})g_{41}^2 &= 0, \end{aligned} \tag{5.72}$$

where we introduced notation Q_{1k} according to

$$Q(q) = \sum_{k=2}^{\infty} Q_{1k}x^k.$$

In order to close this set one has to use Eq. (5.70). Let us denote the symmetrized combination as

$$\hat{S}h(\boldsymbol{q}_1 + \boldsymbol{q}_2 + \boldsymbol{q}_{2'}) = H(\boldsymbol{q}_1, \boldsymbol{q}_{1'}; \boldsymbol{q}_2, \boldsymbol{q}_{2'}; \boldsymbol{q}_3, \boldsymbol{q}_{3'}) \tag{5.73}$$

and integrate the last equation of the set Eq. (5.72). The integral of this equation can be represented as

$$g_{62}\{\boldsymbol{q}_i\} = -8g_{41}^2 \int_0^1 dy y^{\frac{d-2}{2}} H\{y\boldsymbol{q}_i\}. \tag{5.74}$$

The contribution from g_{62} to the quantity Q_{22} is given by

$$Q_{22}\{\boldsymbol{q}_i\} = \frac{3}{2}\int_p h(p)\left[ng_{62}(\boldsymbol{p}, -\boldsymbol{p}; \{\boldsymbol{q}_i\}) + 4\hat{S}g_{62}(\boldsymbol{p}, \boldsymbol{q}_1; -\boldsymbol{p}, \boldsymbol{q}_{1'}; \boldsymbol{q}_2, \boldsymbol{q}_{2'})\right]. \tag{5.75}$$

Let us separate the part independent on $\boldsymbol{q}_i$ from the second equation of the set Eq. (5.72)

$$Q_{22}(0) + dg_{41}\alpha_1 - 8h(0)g_{41}g_{21} = 0. \tag{5.76}$$

This equation can be used to eliminate the factor α_1. Then the part dependent on $\boldsymbol{q}_i$ can be integrated with the help of the same method as used in Eq. (5.74). As a result one arrives at

$$g_{42}\{\boldsymbol{q}_i\} = \int_0^1 \frac{dy}{y}\left[\Big(Q_{22}\{y\boldsymbol{q}_i\} - Q(0)\Big) - 8\hat{S}g_{42}g_{21}(h(yq_1) - h(0))\right]$$

$$\equiv \tilde{g}_{42}\{\boldsymbol{q}_i\} + \bar{g}_{42}\{\boldsymbol{q}_i\}. \tag{5.77}$$

Our aim at this stage is to calculate the contribution of $g_{42}\{\boldsymbol{q}_i\}$ to the value $D_2(q)$, i.e. $Q_{12}(q)$. Direct substitution of the term $\tilde{g}_{42}$ defined by Eq. (5.77) gives

$$\tilde{Q}_{12} = -12g_{41}^2 \int_{kp} h(k)h(p) \int_0^1 \int_0^1 \frac{dz}{z} dy y^{\frac{d-2}{2}} B(y(\boldsymbol{p}, z(\boldsymbol{q}, \boldsymbol{k}))), \tag{5.78}$$

where the function B is given by

$$B = n^2 H(y((\boldsymbol{p}, -\boldsymbol{p}); z(\boldsymbol{k}, -\boldsymbol{k}; \boldsymbol{q}, -\boldsymbol{q}))) + 2nH(y(z(\boldsymbol{q}, -\boldsymbol{q}); \boldsymbol{p}, z\boldsymbol{k}; -\boldsymbol{p}, -z\boldsymbol{k}))$$

$$+(\boldsymbol{q} \leftrightarrow \boldsymbol{k}, \boldsymbol{p}) + 8H(y(\boldsymbol{p}, z\boldsymbol{k}; -\boldsymbol{p}, z\boldsymbol{q}; z(-\boldsymbol{q}, -\boldsymbol{k}))).$$

This function can be transformed after the elementary but cumbersome symmetrization imposed in the definition of H (see Eq. (5.73)). As a result of the transformation one arrives at

$$B = \frac{n+2}{3}\Big\{(n+2)\Big[h(yzq) + (\boldsymbol{q} \leftrightarrow \boldsymbol{k}, \boldsymbol{p})\Big] + 6h(y(\boldsymbol{p} + z(\boldsymbol{q} + \boldsymbol{k})))\Big\}. \tag{5.79}$$

The contribution to Q_{12} from $\bar{g}_{42}$ (i.e. $\bar{Q}_{12}$) can be obtained quite easily and is equal to

$$\bar{Q}_{12}(q) = \frac{8(n+2)^2 g_{41}^2}{d} \int_k h(k) \int_0^1 \frac{dy}{y} \{h(zq) + h(zk)\}. \tag{5.80}$$

In the derivation of this formula we take into account Eq. (5.71) relating g_{21} and g_{41}.

5.6.3 Evaluation of the η-Exponent

In order to conclude calculations in this order of x, we need to use formula $dD_2/dq^2\big|_{q=0} = 1$ defining the factor α_1. Considering particular q-dependence of the function $h(y(\boldsymbol{p} + z(\boldsymbol{k} + \boldsymbol{q})))$ it is more convenient to use another form of the equality $d^2 D_2/d\boldsymbol{q}^2\big|_{q=0} = 2d$. After differentiation, one arrives at

$$4g_{41}^2 \left\{ \frac{2(n+2)^2\epsilon}{d^2(d+4)} h'(0) + \frac{3(n+2)}{d} \int_{kp} h(k)h(p) \int_0^1 dzdyzy^{\frac{d-2}{2}} \frac{\partial^2 h}{\partial \rho^2} \right\} = 1, \tag{5.81}$$

where $\rho = y|\boldsymbol{p} + z\boldsymbol{k}|$. Now using Eqs. (5.71) and (5.76) we can express g_{41} as a function of α_1 and after substitution it into Eq. (5.81) determine $x^2 = \eta$. The value $Q_{22}(0)$ in Eq. (5.76) can be determined with the help of Eq. (5.75)

$$Q_{22}(0) = -\frac{8}{d} g_{41}^2 \left\{ 2(n+2)h(0) + (n+8)A \right\}, \tag{5.82}$$

where the value A is defined by

$$A = \frac{d}{2} \int_0^1 dyy^{\frac{d-2}{2}} \int_p h(p)h(yp). \tag{5.83}$$

Substituting $Q_{22}(0)$ into Eq. (5.76) we get

$$\alpha_1 = \frac{8(n+8)}{d^2} A g_{41}, \tag{5.84}$$

Finally one arrives at

$$\eta = x^2 = \left[\frac{\epsilon d}{4(n+8)A} \right]^2 \left\{ \frac{3(n+2)I}{d} + \frac{2h'(0)\epsilon(n+2)}{d(d+4)} \right\}, \tag{5.85}$$

where

$$I = \int_{kp} h(k)h(p) \int_0^1 \int_0^1 dzzdyy^{\frac{d-2}{2}} \frac{\partial^2 h(\rho)}{\partial \rho^2}.$$

The simplest way to evaluate integrals in Eq. (5.85) is to use the ϵ-expansion. In the lowest order this yields

$$\begin{aligned} \eta &= 6(n+2) \left(\tfrac{\epsilon}{2(n+8)A} \right)^2 \int_{kp} h(k)h(p) \int_0^1 dzz \left. \tfrac{\partial h}{\partial (\rho^2)} \right|_{y=1} \\ &= \tfrac{n+2}{2} \left(\tfrac{\epsilon}{n+8} \right)^2, \end{aligned} \tag{5.86}$$

which coincides with the result of the conventional calculations (see Chapter 3). In a more general case, one can see from Eq. (5.85) that result is essentially dependent on the choice of the cutoff function. It seems quite probable that most of physically accepted choices should have $h'(0) = 0$, leaving only the first term in Eq. (5.85). For the sharp cutoff with $d = 3$ one can calculate directly all the constants in this equation. It is also necessary to recall that all momentum integrals in this derivation have been normalized by the value $\bar{h} = \int_q h(q)$. By taking this normalization into account one can obtain $A = 3/8$ and $I = (1 - 1/\sqrt{2})/4$. Hence the exponent η is $\eta = (n+2)/\sqrt{2}(n+8)^2$. This value is $\sqrt{2}$ times bigger than the extrapolation $\epsilon \to 1$ of the ϵ-expansion result Eq. (5.86) (η is equal to approximately 0.03 for physically interesting n) and is closer to the experimental estimate $\eta \approx 0.04$.

5.6.4 Evaluation of the ν-Exponent

Due to the existence of the scaling laws (see Chapters 1,3, and 4), in order to find all critical exponents one only needs to find two of them. In the previous section the procedure for calculation of the exponent η was presented. In this section, we calculate the correlation-length exponent ν. In accordance with the general theory this exponent is equal to the inverse value of the highest (positive) eigenvalue of the linear RG in the vicinity of the stable fixed point ($\nu = 1/\lambda_1$). Let us linearize the set Eq. (5.72) near this point. For the sake of simplicity we denote vertex values at the fixed point as g^*_{mk} and deviations from them as g_k. Then, restricting consideration to the first order of x we have the set defining eigenvalues λ as

$$\begin{aligned}
\lambda g_1 &= \frac{d}{2}(1 + x\alpha_1) + \Delta Q_1 - 4xg^*_{21}g_1, \\
\lambda g_2 &= -(\hat{R}_2 - xd\alpha_1)g_2 + \Delta Q_2 - 2x(g^*_{21}g_2 + g^*_{41}g_1)\sum_i h(q_i), \\
\lambda g_3 &= -\left(\hat{R}_3 + \frac{d}{2}(1 - 3x\alpha_1)\right) g_3 - xg^*_{41}g_2 H\{q_i\}.
\end{aligned} \tag{5.87}$$

First of all we should verify that the fixed point solution, considered in the previous section, is stable. Let us solve the set Eq. (5.87) using the following system as a zero-order approximation

$$\begin{aligned}
\lambda g_1 &= \frac{d}{2} g_1, \\
\lambda g_2 &= 0, \\
\lambda g_3 &= -\frac{d}{2} g_3.
\end{aligned} \tag{5.88}$$

one can see that the set Eq. (5.88) gives the following set of eigenvalues: $\lambda_1 = d/2$, $\lambda_2 = 0$, and $\lambda_3 = -d/2$. These eigenvalues can be considered as the beginning of the operator $\hat{L}$ eigenvalue series $\lambda_k = d(2-k)/2$

($k = 1, 2, 3, \ldots$). In the zero-order approximation we have the marginal situation when $\lambda_2 = 0$. Therefore, the stability of the fixed point can only be provided by the appearance of a negative contribution of the order of x. Now considering $\lambda_2 \propto x$ one can find relation defining g_3 from Eq. (5.87) as

$$\left(\hat{R}_3 + \frac{d}{2}\right) g_3 + x g_{41}^* g_2 H\{\boldsymbol{q}_i\} = 0. \tag{5.89}$$

This equation can be solved in the same manner as before (see Eq. (5.74)). For the purposes of this section it is sufficient to write only the contribution from g_3 to $\Delta Q_2(0)$

$$\Delta Q_2(0) = -4x g_2 g_{41}^* \int_0^1 dy y^{\frac{d-2}{2}} \int_p h(p) \left\{2(n+2)h(yp) + 4(n+2)h(0)\right\}. \tag{5.90}$$

Note also that, according to Eq. (5.71) $g_{21}^* = -2(n+2)g_{41}^*/d$ and consequently

$$-2x(g_{21}^* g_2 + g_{41}^* g_1) \sum_{i=1}^{2} \{h(q_i + h(q_{i'})\} \approx 16 g_{41}^* g_2 h(0)$$

one can transform the second equation of the set Eq. (5.87) as

$$\lambda g_2 = x\alpha_1 d g_2 - \frac{16}{d}(n+8) A g_{41}^* g_2. \tag{5.91}$$

Now taking into account Eq. (5.84) one arrives at $\lambda_2 = -x\alpha_1 d$. It is also interesting to note that $x\alpha_1 d = \epsilon$ and therefore the expression for λ_2 coincides with the consequent expression in the ϵ-expansion (see Ma 1976). Now let us return to the calculation of the exponent ν. For this purpose λ_1 should be sought for in the form $\lambda_1 = d/2 + o(x)$. Neglecting values of the second order of x from the second equation of the set Eq. (5.87) we get

$$\left(\hat{R}_2 + \frac{d}{2}\right) g_2 + 2x g_{41}^* g_1 \sum_{i=1}^{2} (h(q_i) + h(q_{i'})) = 0. \tag{5.92}$$

It is interesting to note that the factor at g_2 in Eq. (5.92) is exactly the same as in Eq. (5.89) so that solution can be obtained by analogy with Eq. (5.90)

$$g_2 = -2x g_{41}^* g_1 \int_0^1 dy y^{\frac{d-2}{2}} \sum_{i=1}^{2} (h(yq_i) + h(yq_{i'})), \tag{5.93}$$

and the corresponding contributions from g_2 to ΔQ_1 is

$$\Delta Q_1 = -4xg_{41}^* g_1 \int_p \int_0^1 dyy^{\frac{d-2}{2}} \left[h(yp) + h(0) \right]. \tag{5.94}$$

The value λ_1 now can be determined with the help of the first equation of the set Eq. (5.87) which can be rewritten as

$$\lambda_1 = \frac{d}{2}(1 + x\alpha_1) - x\frac{8}{d}Ag_{41}^* g_1, \tag{5.95}$$

from where it follows that

$$\lambda_1 = \frac{d}{2} + x\alpha_1 \frac{d(4-n)}{2(n+8)}. \tag{5.96}$$

Finally, recalling that $x\alpha_1 d = \epsilon$ one obtains that $\nu = \lambda_1^{-1}$ is equal to $\nu = 2\left\{1 - (4-n)\epsilon/d(n+8)\right\}/d$. As an example, at $d = 3$ and $n = 1$ $\nu = 16/27 = 0.593$ which is better than the estimate of the ϵ-expansion $\nu = \left\{1 + (n+2)/2(n+8)\right\}/2 = 7/12 = 0.583$. It is interesting to note that the eigenvalue itself $\lambda_1 = d/2 + (4-n)\epsilon/2(n+8) = 2\left\{1 - (n+2)\epsilon/2(n+8)\right\}$ coincides with the consequent value in ϵ-expansion. The coincidence of λ_1 with the value in the ϵ-expansion is also found in calculations of critical exponents at the three dimensional case (see Ginzburg 1975 and Colot *et al* 1975) in the lowest order over the interaction constant. In the higher orders of the perturbation theory, however, this coincidence disappears.

5.7 Comparison of Different RG Approaches

In Sect. 5.4.3 we have shown that there is a homomorphism between SE and RG transformations. Scale equations always correspond to some particular RG transformation. The phenomenon which uniquely defines SE and therefore an RG equation is a type of scale transformations. As far as we have chosen a set of scale transformations, we automatically define the set of SE and corresponding to this set RG equation. The generalized scheme makes the choice of scale transformations unique. From this point of view, the field theoretical RG is also defined by the choice of scale transformations, but, as the number and type of vertexes are fixed by the condition of renormalizability of the Hamiltonian, we cannot require the invariance of critical Hamiltonian H_0 in addition to it (see Sect. 5.3). A substantially stronger condition is implied in the renormalizability. The invariance condition can be valid only for the case of nonrenormalizable Hamiltonians and can be considered as a weak substitute for the former renormalizability condition.

There is also some common feature in both approaches. The limit $\rho \to 0$ (see Sect. 1.3.5) in the field theoretical approach corresponds to the limit $x_0 \to \infty$ for the approach considered in this chapter. In these limits both methods lead to the same universal behavior. Nonetheless, behavior in the prescaling region, and especially corrections to scaling laws, are different. In the generalized approach critical exponents depend on the choice of the cutoff function and this dependence can be seen as resulting from the non-renormalizable nature of the Hamiltonian. In reality, we always deal with nonrenormalizable Hamiltonians. As is shown in Chapter 2, even for the simplest models like the Ising model, we obtain a Ginzburg-Landau functional of a general kind. The latter means that we should always deal with some hints of nonuniversal behavior in experiment. In such a case, there are two possible ways in treating corrections to scaling. The first one is to use Wegner's series (see Sect. 3.3), like the one for the two-point correlation function

$$G(0,t) \propto t^{-\gamma} \left\{ 1 + \sum_{k=1}^{\infty} a_k t^{\Delta_k} \right\}, \tag{5.97}$$

with exponents being functionals of $S(x)$. The other way is to consider exponents (for instance γ, Δ_k) in the limit $x_0 \to \infty$, but to deal, instead of an expansion like in Eq. (5.97), with series containing logarithmical terms. From these two choices the first one, conserving expansion Eq. (5.97) with nonuniversal exponents, may be simpler for experimental purposes. It is also necessary to mention that the presence of a dimensionless scale x_0 can lead to essential corrections of experimental results applicable in the Gaussian region. In 1968, Aslamasov and Larkin calculated Gaussian contributions to electrical conductivity (paraconductivity) in the vicinity of the superconducting phase transition. Their result was successfully tested in a number of traditional superconductors. However, in 1988 Freitas *et al* found essential deviations from the Aslamasov-Larkin formula for high-T_c $YBaCuO$ systems. They explained these deviations by the influence of the short-wavelength cutoff. Later, in 1995 Gauzzi and Pavuna, analyzing experimental results, concluded that the universal paraconductivity relation is not valid for high-T_c $YBaCuO$ films.

A few words should also be said about comparison of the traditional ϵ-expansion with the η-expansion. Difference between these expansions can be seen from the table in which starting, zero-order exponents are compared.

ϵ-*Expansion*	η-*Expansion*
$\beta = .5$	$\beta = (d-2)/d = .333$
$\gamma = 1$	$\gamma = 4/d = 1.333$
$\delta = 3$	$\delta = (d+2)/(d-2) = 5$
$\alpha = 0$	$\alpha = 0$
$\eta = 0$	$\eta = 0$

As is seen from this table, even in the zero-order approximation, η-expansion gives results which are essentially different from the mean field approach. These results are quite close to experimental ones. It is interesting to note that for ϵ-expansion the zero order exponent $\nu = \gamma/2 = 1/2$ is less than the experimental value, while for the η-expansion $\nu = 0.666$ which is a little higher. The situation with the η-expansion is further improved in the first order approximation. The comparison of the results for the first order corrections are given below for the value ν at different numbers of order parameter components.

	ϵ-*Expansion*	η-*Expansion*	*Experiment*
$n = 1$	.583	.593	.625
$n = 2$	.600	.622	.66
$n = 3$	.613	.646	.70
$n = 4$	.625	.667	

As one can see from this table η-expansion always gives better magnitudes for the exponents ν. From Eq. (5.85) one can obtain $\eta \approx 0.026$ for $n = 1$. This value can be compared with the experimental value $\eta = 0.036 \pm 0.007$ for binary solutions (see Sect. 1.7). Even for $d = 2$, the formula Eq. (5.85) gives $\eta \approx 0.1$ which is not very far from the exact value for the Ising model $\eta = 0.125$. Thus η-expansion, even in the lowest orders of $\eta^{1/2}$ leads to quite good results for all critical exponents.

Chapter 6

RG Study of Particular Systems

Any of the general methods introduced in the previous chapters can be applied to analyses of particular systems. In this chapter we consider application of the SE method to some simple examples. The central point of the SE approach is consideration of the characteristic set for the $\hat{\Phi}$ operator (see Sect. 5.3.2). As the set should define vertexes $\hat{g}_k$, which are multivariable functions, first of all, we should find representations for this functions reducing equations in partial derivatives to algebraic equations. The general approach to this problem is considered in the following section.

6.1 Reduction of the Characteristic Set

The simplest way to reduce a system like Eq. (5.21) to an algebraic set is to expand the functions $\hat{g}_k(\boldsymbol{q}_1, \ldots, \boldsymbol{q}_k)$ in power series of momenta $\boldsymbol{q}_i$. As a result, the functions $\hat{g}_k\{\boldsymbol{q}_i\}$ will be represented as a series over polynomials $P^{(k)}_{r\kappa_r}$ which have the symmetry properties implied by the symmetry of the initial Hamiltonian. Hence, we can represent vertexes in the form

$$\hat{g}_k\{\boldsymbol{q}_i\} = \sum_{r=0}^{\infty} \sum_{\kappa_r} \hat{g}_{kr\kappa_r} P^{(k)}_{r\kappa_r}\{\boldsymbol{q}_i\}, \tag{6.1}$$

where $\hat{g}_{kr\kappa_r}$ are c-number tensors in the isotopic space. In order not to overcomplicate the problem, we consider the case where there is no particular vector direction in the momentum space. That means that vertexes are dependent only on magnitudes of momenta $\boldsymbol{q}_i^2$ and their dot products $\boldsymbol{q}_i\boldsymbol{q}_j$, i.e. $\hat{g}_k\{\boldsymbol{q}_i\} \equiv \hat{g}_k\{\boldsymbol{q}_i^2, (\boldsymbol{q}_i\boldsymbol{q}_j)\}$. Thus, the values $P^{(k)}_{r\kappa_r}$ can be represented as homogeneous polynomials of the $2r$-th order

$$P^{(k)}_{r\kappa_r}\{\lambda\boldsymbol{q}_i\} = \lambda^{2r}P^{(k)}_{r\kappa_r}\{\boldsymbol{q}_i\}.$$

Each of these polynomials can be expressed in terms of integer powers of independent polynomial invariants, constituting full rational basis of invariants for the Ginzburg-Landau functional of a given symmetry (see, for instance, Spencer 1974). In our case the subscript κ_r numbers each combination of such invariants. At a given r, polynomials with different κ_r are linearly independent. Moreover, they can be orthogonolized, for example, with the help of the conditions

$$P^{(k)}_{r\kappa_r}(\boldsymbol{\nabla}_{q_1}, \ldots, \boldsymbol{\nabla}_{q_k})P^{(k)}_{r\kappa'_r}(\boldsymbol{q}_1, \ldots, \boldsymbol{q}_k) = 0$$

which hold when $\kappa_r \neq \kappa'_r$. The representation Eq. (1.1) helps to reduce the SE from the functional type Eq. (5.19) to equations in partial derivatives. Consequently, the characteristic set at the fixed point will be reduced from equations in partial derivatives to algebraic equations. Let us consider this reduction for the example of the $\mathcal{O}(n)$ symmetric Ginzburg-Landau functional containing only even powers of field operators. The tensors $\hat{g}_{r\kappa_r}$ in this case can be represented as

$$\hat{g}_k \equiv g^{\alpha_1\cdots\alpha_{2k}}_{2k,r\kappa_r} = g_{kr\kappa_r}R^{\alpha_1\cdots\alpha_{2k}}_k, \tag{6.2}$$

where

$$R^{\alpha_1\cdots\alpha_{2k}}_k = \frac{1}{(2k-1)!!}\{\delta_{\alpha_1\alpha_2}\delta_{\alpha_3\alpha_4}\ldots\delta_{\alpha_{2k-1}\alpha_{2k}} + \mathit{permutations}\}.$$

With the help of Eq. (1.2), we obtain the following equation for the case of the two-point correlation function

$$\left\{2 - 4G_0^{-1}(\boldsymbol{q})h(\boldsymbol{q}) + \boldsymbol{q}\frac{\partial}{\partial\boldsymbol{q}} + \hat{\Phi}[g_{kr\kappa_r}]\right\}G(\boldsymbol{q}, g_{kr\kappa_r}) = 2h(\boldsymbol{q}), \tag{6.3}$$

where the operator $\hat{\Phi}$ is now reduced to the ordinary operator in partial derivatives

$$\hat{\Phi}[g_{kr\kappa_r}] = \sum_{k=1}^{\infty}\sum_{r=0}^{\infty}\sum_{\kappa_r}\left\{(\epsilon_{2k} - 2r)g_{kr\kappa_r}\frac{\partial}{\partial g_{kr\kappa_r}}\right.$$

$$+(k+1)g_{k+1,r\kappa_r}\sum_{l=0}^{r}\sum_{\kappa_l}B^{(k)}_{\kappa_r\kappa_l}\frac{\partial}{\partial g_{kl\kappa_l}} - 2\sum_{m=1}^{k}m(k-m+1)$$

$$\left.\times\sum_{l=0}^{\infty}\sum_{s=r+l}^{\infty}\sum_{\kappa_l\kappa_l}g_{mr\kappa_r}g_{k-m+1,l\kappa_l}A^{(k,m)}_{\kappa_r\kappa_l\kappa_s}\frac{\partial}{\partial g_{ks\kappa_s}}\right\}. \tag{6.4}$$

The factors B and A in Eq. (1.4) are defined by the equations

$$\sum_{l=0}^{r}\sum_{\kappa_l} B^{(k)}_{\kappa_r\kappa_l} P^{(k)}_{l\kappa_l}\{\boldsymbol{q}_i,\boldsymbol{q}_{i'}\} = \int_p \bar{h}(\boldsymbol{p})$$
$$\times\left\{\frac{n}{2}P^{(k+1)}_{r\kappa_r}(-\boldsymbol{p},\boldsymbol{p};\boldsymbol{q}_1,\boldsymbol{q}_{1'};\ldots;\boldsymbol{q}_k,\boldsymbol{q}_{k'})\right.$$
$$\left.+\,k\hat{S}P^{(k+1)}_{r\kappa_r}(-\boldsymbol{p},\boldsymbol{q}_1;\boldsymbol{p},\boldsymbol{q}_{1'};\boldsymbol{q}_2,\boldsymbol{q}_{2'};\ldots;\boldsymbol{q}_k,\boldsymbol{q}_{k'})\right\}, \tag{6.5}$$

and

$$\sum_{s=l+r}^{\infty}\sum_{\kappa_s} A^{(k,m)}_{\kappa_r\kappa_l\kappa_s} P^{(k)}_{s\kappa_s}\{\boldsymbol{q}_i,\boldsymbol{q}_{i'}\} = \int_p \bar{h}(\boldsymbol{p})$$
$$\times\hat{S}\left\{\delta\left(\boldsymbol{q}_1-\boldsymbol{p}+\sum_{s=2}^{m}(\boldsymbol{q}_s+\boldsymbol{q}_{s'})\right)P^{(m)}_{r\kappa_r}(-\boldsymbol{p},\boldsymbol{q}_1;\boldsymbol{q}_2,\boldsymbol{q}_{2'};\ldots;\boldsymbol{q}_m,\boldsymbol{q}_{m'})\right.$$
$$\left.\times\, P^{(k-m+1)}_{l\kappa_l}(\boldsymbol{p},\boldsymbol{q}_{1'};\boldsymbol{q}_{m+1},\boldsymbol{q}_{m'+1};\ldots;\boldsymbol{q}_k,\boldsymbol{q}_{k'})\right\}. \tag{6.6}$$

In these equations we have introduced the notation $\bar{h}(p) = h(p)/\int_p h(p)$. In all approximations that do not need high accuracy and especially, when only qualitative feature are important, it is possible to restrict consideration to the first numbers of r. In these cases, coefficients in the operator $\hat{\Phi}$ can be found quite easily. As an example, let us calculate the factors $B_{\kappa_r\kappa_l}$ for $r = 0; 1$. In this case we have only three types of independent polynomials $P^{(k)}_{1\kappa_1}$ satisfying the orthogonality condition Eq. (1.2)

$$\begin{aligned}
P^{(k)}_{11}\{\boldsymbol{q}_i,\boldsymbol{q}_{i'}\} &= \sum_{i=1}^{k}(\boldsymbol{q}_i^2+\boldsymbol{q}_{i'}^2),\\
P^{(k)}_{12}\{\boldsymbol{q}_i,\boldsymbol{q}_{i'}\} &= 2\sum_{i=1}^{k}\boldsymbol{q}_i\boldsymbol{q}_{i'},\\
P^{(k)}_{12}\{\boldsymbol{q}_i,\boldsymbol{q}_{i'}\} &= \left\{\sum_{i=1}^{k}(\boldsymbol{q}_i+\boldsymbol{q}_{i'})\right\}^2-\sum_{i=1}^{k}(\boldsymbol{q}_i+\boldsymbol{q}_{i'})^2.
\end{aligned} \tag{6.7}$$

With the help of the orthogonality condition and the use of Eq. (1.5) we may get

$$\begin{aligned}
B^{(k)}_{\kappa_0\kappa_0} &= \frac{n}{2}+k,\\
B^{(k)}_{\kappa_1\kappa_0} &= d\left[\left(\frac{n}{2}+k\right)\delta_{\kappa_1 1}-\frac{n}{2}\delta_{\kappa_1 2}-k\delta_{\kappa_1 3}\right],\\
B^{(k)}_{\kappa_1\kappa_1'} &= \left(\frac{n}{2}+k\right)\delta_{\kappa_1\kappa_1'}+(\delta_{\kappa_1 3}-\delta_{\kappa_1 2})\delta_{\kappa_1' 2}.
\end{aligned}$$

Following the same method, the factors $A^{(k,m)}_{\kappa_r\kappa_e\kappa_s}$ can also be obtained. Thus, as an example, we have

$$\begin{aligned} A^{(k,m)}_{\kappa_0\kappa_0\kappa_0} &= 1, \\ A^{(k,1)}_{\kappa_0\kappa_0\kappa_1} &= -\frac{1}{2k}\delta_{\kappa_1 1}, \\ A^{(k,m>1)}_{\kappa_0\kappa_0\kappa_1} &= -\frac{3}{2k}\delta_{\kappa_1 1} - \frac{1}{k}\delta_{\kappa_1 2} - \frac{1}{k(k-1)}\delta_{\kappa_1 3}. \end{aligned}$$

Analogously, one can calculate all the other factors. In order to see how the representation Eq. (1.1) reduces the fixed point problem to an algebraic one, let us restrict consideration of Eq. (1.3) to $\boldsymbol{q} = 0$ using the function $h(p) = \exp(-\boldsymbol{p}^2)$. We also assume that the correlation function G can be found for the case of a local Ginzburg-Landau functional ($r = 0$). In this case we do not have vertexes with $r \neq 0$ in Eq. (1.3) and this equation can be reduced to

$$\left\{2 + \sum_{k=1}^{\infty} U_k[g_k]\frac{\partial}{\partial g_k}\right\} G(g_k) = 2, \tag{6.8}$$

where

$$\begin{aligned} U_k[g_k] = \epsilon_{2k} g_k &+ \frac{1}{2}(k+1)(2k+n)g_{k+1} \\ &-2\sum_{m+1}^{k} m(k-m+1)g_m g_{k-m+1}. \end{aligned}$$

Now the set of characteristic equations at the fixed point $U_k[g_k] = 0$ is a nonlinear algebraic set.

6.2 The Gaussian Model

Before treating more complex models, let us consider the critical behavior for the Gaussian model with the help of the SE method. This can be solved exactly. For this model all vertexes with $k > 1$ in the functional Eq. (3.24) are absent. As we are treating all models in the generalized scheme (see Sect. 5.3.2), the vertex $g_1(\boldsymbol{q})$ does not depend on momentum,[1] i.e. all $g_{1r\neq 0}$ are equal to zero. The vertex g_{10} in the critical region is proportional to the temperature. As the proportionality coefficient can always be absorbed

[1] It is necessary to remember that as the consideration is restricted only to the Hamiltonians even in powers of field operators we use notations introduced in Sect. 6.1 in Eq. (1.2), i.e. instead of g_{2k} we write g_k.

in the normalization of the field variables we will write the functional as

$$H = \frac{1}{2} \int_q [\tau + G_0^{-1}(q)] |\phi_q|^2. \tag{6.9}$$

This functional only defines a partition function for $\tau > 0$. If $\tau < 0$ the partition function is divergent. This means that it is impossible to obtain the exponent β with the help of this functional without some additional reasoning. This exponent ($\beta = 1/2$) can be obtained from ϕ^4-model considered in the Gaussian approximation. The same consideration should also apply to the exponent δ, because the addition of the term $-\boldsymbol{h}\boldsymbol{\phi}_{q=0}$, to the Hamiltonian Eq. (1.9) again leads to the divergence of the partition function at $\tau = 0$. Nonetheless, the study of the model Eq. (1.9) with the SE method is very instructive and we consider it for the correlation function Eq. (1.3). First of all let us write the characteristic set for this equation

$$\begin{aligned}
\frac{dg_{10}}{dl} &= 2g_{10} + 2B^1_{\kappa_0\kappa_0} g_{20} + 2\textstyle\sum_{l=0}^{\infty} \sum_{\kappa_l} B^{(1)}_{\kappa_l\kappa_0} g_{2l\kappa_l} - 2g_{10}^2, \\
\frac{dg_{kr\kappa_r}}{dl} &= (\epsilon_{2k} - 2r) + (k+1) \textstyle\sum_{l=r}^{\infty} \sum_{\kappa_r} B^{(k)}_{\kappa_l\kappa_r} g_{k+1,l\kappa_l}
\end{aligned}$$

$$-2 \sum_{m=1}^{k} m(k-m+1) \sum_{s=0}^{r} \sum_{l=0}^{r-s} \sum_{\kappa_s\kappa_l} A^{(k,m)}_{\kappa_s\kappa_l\kappa_r} g_{ms\kappa_s} g_{k-m+1,l\kappa_l}. \tag{6.10}$$

From this set one can see that when the initial value of any vertex $g_{k>1}$ is not equal to zero, every other vertex will be created during the change of l. The only vertex which does not generate higher order vertexes is g_{10}. This means that if the initial values for the vertexes with $k > 1$ are equal to zero, then this vertexes will never be generated by changing l. As a result the system Eq. (1.10) has only one equation for $g_{10} \equiv \tau$

$$\frac{d\tau}{dl} = 2\tau(1 - \tau).$$

This equation has two fixed points: $\tau^* = 0$ and $\tau^* = 1$. The eigenvalues corresponding to these points are $\lambda = 2$ and $\lambda = -2$. The critical state only corresponds to the point $\tau^* = 0$, as the second point has a negative eigenvalue. The latter point, with $\tau^* = 1$, sometimes is called the "high temperature fixed point" (see for instance Ma 1976). Using the scaling laws Eq. (3.77) one can obtain that the exponent α at the critical Gaussian point ($\lambda = 2$) is equal to $2 - d/2$. In order to determine the exponent γ one has to find the value η^*. From Eq. (5.10) it follows that $\eta^* = 0$ and hence $\gamma = 1$. Of course, these values for the critical exponents α, γ, and η can be obtained without the SE method, from the exact solution for the Gaussian model. The equation for the two-point correlation function (Eq. (1.3)) can also be solved for this model. In the critical region $\tau \to 0$ and $\boldsymbol{q} \to 0$ instead of Eq. (1.3) we get

$$\left(1 + q^2 \frac{\partial}{\partial q^2} + \tau \frac{\partial}{\partial \tau}\right) G(\tau, q) = 0.$$

The solution of this equation has the explicit form $G(\tau, q) \propto (\tau + q^2)^{-1}$.

6.3 The ϕ^4 Model

The model which contains the vertex $\hat{g}_2\{\boldsymbol{q}\}$, in addition to g_{10}, in the Ginzburg-Landau functional is traditionally called as ϕ^4-model. As is seen from Eq. (1.10) the vertex g_2 generates all other even vertexes. An analysis of the characteristic set Eq. (5.21) shows that odd vertexes are generated if any of them are present in the initial Hamiltonian. The appearance of all higher-order vertexes in the fixed-point Hamiltonian aggravates calculations especially because vertexes are, in the general case, of the same order as the vertex g_2 which originates them. This is the reason why so much effort was spent in search of a small parameter for the perturbation theory (see discussions in Sect. 5.6).

6.3.1 General Consideration

First of all, let us show that all vertexes are the same order of value. For this purpose we consider the simplest, $\mathcal{O}(n)$ symmetry case. From Eq. (1.10) one can find the following equations for the vertexes g_{20} and g_{30}

$$\frac{dg_{20}}{dl} = (4-d)g_{20} + 3B^{(2)}_{\kappa_0\kappa_0}g_{30} + 3\sum_{r=1}^{\infty}\sum_{\kappa_r} B^{(2)}_{\kappa_r\kappa_0}g_{3r\kappa_r} - 8A^{(2,1)}_{\kappa_0\kappa_0\kappa_0}g_{20}g_{10},$$

$$\frac{dg_{30}}{dl} = 2(3-d)g_{30} + 4\sum_{r=0}^{\infty}\sum_{\kappa_r} B^{(3)}_{\kappa_r\kappa_0}g_{4r\kappa_r} - 12A^{(3,1)}_{\kappa_0\kappa_0\kappa_0}g_{30}g_{10} - 8A^{(2,2)}_{\kappa_0\kappa_0\kappa_0}g_{20}^2 \tag{6.11}$$

From the first equation of this set one can see that if at the fixed point $g_{10} \sim g_{20}$, then when $4 - d \sim o(1)$ there are no reasonable solutions except $g_{20}, g_{30} \sim o(1)$. One can continue the chain of Eqs. (6.11) and see that all other vertexes (with $r = 0$, as well as with $r \neq 0$) are also of the order of unity. In such a situation it is impossible to apply a perturbation theory to the analysis of the set Eqs. (6.11) (as well as to the RG equation). This is the reason why Wilson and Fisher (1972) introduced analysis of phase transitions in a fictitious space with the dimension $d = 4 - \epsilon$ with $\epsilon \ll 1$. In this case the stable fixed-point solutions of Eqs. (1.10) and (6.11) have $g_{10}, g_{20} \sim \epsilon$, and $g_{30} \sim \epsilon^2$, while all other vertexes have higher orders in ϵ, namely: $g_{2r>0,\kappa_r}, g_{3r\kappa_r} \sim \epsilon^2$, $g_{4r\kappa_r} \sim \epsilon^3$ and so on. Now, it may look like it is possible to restrict consideration to the analysis of only two vertexes g_{10} and g_{20}. Unfortunately, this is not the case. There are vertexes $g_{3r\kappa_r}$ in the right hand side of the first equation in the set Eq. (6.11). These vertexes have the same order of value as the other terms. Consequently, in the first equation of the set Eq. (1.10) one has to take into account the terms of the order of g_{10}^2 and hence, the terms with $g_{2l>0,\kappa_l}$ which have the same order of value, ϵ^2.

In order to overcome these difficulties, let us return to equations with nonlocal vertexes $\hat{g}_k(\boldsymbol{q}_1, \ldots, \boldsymbol{q}_k)$. As before we assume that $\hat{g}_1, \hat{g}_2 \sim \epsilon$, $\hat{g}_3 \sim \epsilon^2$ and so on. Let us also restrict calculations to the first order in ϵ. In this case we have to take into account the term containing $\hat{g}_3$ in the equation for the vertex $\hat{g}_2$ as it has the same order of value as all other terms. In the equation defining evolution of the vertex $\hat{g}_3$, on the contrary, the term containing $\hat{g}_4$ has order ϵ^3, while all other terms are of the order of ϵ^2. The latter means that in the lowest order of ϵ the first three equations form a closed set, which we write below for the general case of an arbitrary anisotropic system

$$\begin{aligned}
\frac{dg_{10}^{\alpha_1\alpha_2}}{dl} &= 2g_{10}^{\alpha_1\alpha_2} + 3\int_p h^{\beta\gamma}(\boldsymbol{p}) g_{20}^{\beta\gamma\alpha_1\alpha_2} - 2h^{\beta\gamma}(0) g_{10}^{\beta\alpha_1} g_{10}^{\gamma\alpha_2} \\
\frac{dg_2^{\alpha_1\ldots\alpha_4}\{\boldsymbol{q}_i\}}{dl} &= \hat{\epsilon}_4\{\boldsymbol{q}_i\} g_2^{\alpha_1\ldots\alpha_4}\{\boldsymbol{q}_i\} + \frac{15}{2}\int_p h^{\beta\gamma}(\boldsymbol{p}) g_3^{\beta\gamma\alpha_1\ldots\alpha_4}(-\boldsymbol{p}, \boldsymbol{p}; \{\boldsymbol{q}_i\}) \\
&\quad - \frac{16}{15}\hat{S} h^{\beta\gamma}(\boldsymbol{q}_1) g_{10}^{\beta\alpha_1} g_2^{\gamma\alpha_2\ldots\alpha_4}(-\boldsymbol{p}, \{\boldsymbol{q}_{i>1}\}), \\
\frac{dg_3^{\alpha_1\ldots\alpha_6}\{\boldsymbol{q}_i\}}{dl} &= \hat{\epsilon}_6\{\boldsymbol{q}_i\} g_3^{\alpha_1\ldots\alpha_6}\{\boldsymbol{q}_i\} - 8\int_p h^{\beta\gamma}(\boldsymbol{p}) \\
&\quad \times \hat{S} g_2^{\beta\alpha_1\alpha_2\alpha_3}(\boldsymbol{p}, \{\boldsymbol{q}_{i\le 3}\}) g_2^{\gamma\alpha_4\alpha_5\alpha_6}(-\boldsymbol{p}, \{\boldsymbol{q}_{i>3}\}),
\end{aligned} \tag{6.12}$$

where

$$\hat{\epsilon}_k\{\boldsymbol{q}_i\} = \epsilon_k - \sum_{l=1}^{k} \boldsymbol{q}_l \frac{\partial}{\partial \boldsymbol{q}_l}.$$

The equations for the higher order vertexes ($\hat{g}_{k>1}$), within the accepted approximation, have a diagonal form

$$\frac{d\hat{g}_k\{\boldsymbol{q}_i\}}{dl} = \hat{\epsilon}_{2k}\{\boldsymbol{q}_i\}\hat{g}_k\{\boldsymbol{q}_i\}, \tag{6.13}$$

or in the $P-r$ representation Eq. (1.1)

$$\frac{\hat{g}_{kr\kappa_r}}{dl} = (\epsilon_{2k} - 2r)\hat{g}_{kr\kappa_r}. \tag{6.14}$$

The values $\epsilon_{2k} = d + k(2-d) = 4 - 2k$ are negative for $k \geq 3$, therefore, Eqs. (6.13) and (6.14) have a fixed point $g^*_{kr\kappa_r} = 0$ with negative eigenvalues. This means that the vertexes with $k > 3$ do not contribute to the critical behavior in this approximation.

Let us now consider the last equation from the set Eq. (1.12). This equation can be simplified if one cancels the generating term $\hat{g}_2 \times \hat{g}_2$ with the help of the substitution

$$g_3^{\alpha_1 \ldots \alpha_6}\{\boldsymbol{q}_i\} = u_3^{\alpha_1 \ldots \alpha_6}\{\boldsymbol{q}_i\} + 8\hat{\epsilon}_6^{-1}\{\boldsymbol{q}_i\} \int_{\boldsymbol{p}} h^{\beta\gamma}(\boldsymbol{p})$$

$$\times\, \hat{S} g_2^{\beta\alpha_1\alpha_2\alpha_3}(\boldsymbol{p}, \{\boldsymbol{q}_{i\leq 3}\}) g_2^{\gamma\alpha_4\alpha_5\alpha_6}(-\boldsymbol{p}, \{\boldsymbol{q}_{i>3}\}). \tag{6.15}$$

As a result, in the accepted approximation, one obtains $d\hat{u}_3/dl = -2\hat{u}_3\{\boldsymbol{q}_i\}$, which means that $\hat{u}_3^*\{\boldsymbol{q}_i\} = 0$. However, before making the substitution Eq. (6.15), let us find an explicit expression for the formal notation $\hat{\epsilon}_6^{-1}\hat{g}_2 \times \hat{g}_2$. It is convenient to represent the derivative $\boldsymbol{q}_l \partial/\partial \boldsymbol{q}_l$ in the operator $\hat{\epsilon}_k$ as

$$\boldsymbol{q}_l \frac{\partial}{\partial \boldsymbol{q}_l} = 2\sum_{\alpha=1}^{d} q_{l\alpha}^2 \frac{\partial}{\partial q_{l\alpha}^2} = 2\sum_{\alpha=1}^{d} \frac{\partial}{\partial \xi_{l\alpha}^2} \qquad (\xi_{l\alpha} = \ln q_{l\alpha}^2)$$

and to use in equation for $\hat{g}_3$ Fourier transformation with respect to $\xi_{l\alpha}$

$$\frac{d\hat{g}_3\{\rho_{l\alpha}\}}{dl} = -\left(2 + 2i\sum_{l=1}^{6}\sum_{\alpha=1}^{4} \rho_{l\alpha}\right)\hat{g}_3\{\rho_{l\alpha}\} + \hat{F}\{\rho_{l\alpha}\}, \tag{6.16}$$

where

$$\hat{g}_3\{\rho_{l\alpha}\} = \int \left(\prod_{l\alpha} d\xi_{l\alpha} e^{-i\rho_{l\alpha}\xi_{l\alpha}}\right) \hat{g}_3\{\xi_{l\alpha}\}$$

$\hat{F}\{\rho_{l\alpha}\}$ is a Fourier transform of $\hat{g}_2 \times \hat{g}_2$ term. Now Eq. (6.11) can be rewritten as

$$\hat{g}_3\{\xi_{l\alpha}\} = \hat{u}_3\{\xi_{l\alpha}\} + \frac{1}{(2\pi)^{24}} \int \left[\prod_{l\alpha} d\rho_{l\alpha} \exp(i\rho_{l\alpha}\xi_{l\alpha})\right]$$

$$\times \hat{F}\{\rho_{l\alpha}\} \int_0^1 dx x^{(1+2i\sum_{l\alpha}\rho_{l\alpha})},$$

from which we get

$$g_3^{\alpha_1\ldots\alpha_6}\{\boldsymbol{q}_i\} = u_3^{\alpha_1\ldots\alpha_6}\{\boldsymbol{q}_i\} - 8\int_p h^{\beta\gamma}(\boldsymbol{p})$$

$$\times \int_0^1 dxx\hat{S}g_2^{\beta\alpha_1\alpha_2\alpha_3}(\boldsymbol{p},\{x\boldsymbol{q}_{i\leq 3}\})g_2^{\gamma\alpha_4\alpha_5\alpha_6}(-\boldsymbol{p},\{x\boldsymbol{q}_{i>3}\}). \tag{6.17}$$

It is necessary to remember that the operator $\hat{S}$ in the second right-hand side term symmetrizes the expressions following it, with respect to permutations of the variables q_i and indexes α_i. As a whole we have $6!/2!(3!)^2 = 10$ permutations.

As we are only interested in the vicinity of the fixed point, the first equation of the set Eq. (1.12) gives

$$g_{10}^{\alpha_1\alpha_2} = -\frac{3}{2}\int_p h^{\beta\gamma}(\boldsymbol{p})g_{20}^{\beta\gamma\alpha_1\alpha_2}. \tag{6.18}$$

Substituting Eqs. (6.17) and (6.18) into the equation defining evolution of $\hat{g}_2$ (see Eq. (1.12)) and taking into account that $\hat{u}_3^* = 0$ one arrives at

$$\frac{dg_2^{\alpha_1\ldots\alpha_4}\{\boldsymbol{q}_i\}}{dl} = \hat{\epsilon}_4\{\boldsymbol{q}_i\}g_2^{\alpha_1\ldots\alpha_4}\{\boldsymbol{q}_i\} - 60\int_p h^{\beta\gamma}(\boldsymbol{p})\int_q h^{\delta\rho}(\boldsymbol{q})$$

$$\times \int_0^1 dxx\hat{S}g_2^{\beta\gamma\delta\alpha_1}(\boldsymbol{p},-x\boldsymbol{q};x\boldsymbol{q},x\boldsymbol{q}_1)g_2^{\rho\alpha_2\ldots\alpha_4}(-\boldsymbol{p},\{x\boldsymbol{q}_{i>1}\})$$

$$+ 24\int_p h^{\beta\gamma}(\boldsymbol{p})\hat{S}h^{\delta\rho}(\boldsymbol{q}_1)g_{20}^{\beta\gamma\delta\alpha_1}g_2^{\rho\alpha_2\ldots\alpha_4}(-\boldsymbol{q}_1,\{\ \boldsymbol{q}_{i>1}\}). \tag{6.19}$$

The operator $\hat{S}$ functions in the second and the third terms of the right-hand side expression in a little different way. The first time, as it follows from Eq. (6.17), it symmetrizes with respect to the permutations of the variables $-\boldsymbol{q},\boldsymbol{q},\boldsymbol{q}_1,\boldsymbol{q}_2,\boldsymbol{q}_3,\boldsymbol{q}_4$ and consequently the exponents $\beta,\gamma,\alpha_1,\alpha_2,\alpha_3,\alpha_4$. As it has already been said, there are ten permutations of such a kind. The second time it makes the expression invariant with respect to the permutations of variables $\boldsymbol{q}_1,\boldsymbol{q}_2,\boldsymbol{q}_3,\boldsymbol{q}_4$ and indexes $\alpha_1,\alpha_2,\alpha_3,\alpha_4$. If one takes into account all possible permutation of the variables $-\boldsymbol{q}$, $\boldsymbol{q}$ (and the indexes β,γ), then the product $\hat{S}\hat{g}_2 \times \hat{g}_2$ can be represented as follows

$$\frac{1}{10}\left\{ 6\hat{S}g_2^{\delta\beta\alpha_1\alpha_2}(\boldsymbol{p},-\boldsymbol{q},\{x\boldsymbol{q}_{i\leq 2}\})g_2^{\rho\gamma\alpha_3\alpha_4}(-\boldsymbol{p},x\boldsymbol{q},\{x\boldsymbol{q}_{i>2}\})\right.$$

$$\left. + 4\hat{S}g_2^{\delta\beta\gamma\alpha_1}(\boldsymbol{p},-x\boldsymbol{q},x\boldsymbol{q},x\boldsymbol{q}_1)g_2^{\rho\alpha_2\alpha_3\alpha_4}(-\boldsymbol{p},\{x\boldsymbol{q}_{i>1}\})\right\},$$

where the symmetrization procedure is exactly the same as in the third term of Eq. (6.19), which now can be rewritten as

$$\frac{dg_2^{\alpha_1\ldots\alpha_4}\{\boldsymbol{q}_i\}}{dl} = \left(\epsilon_4 - \sum_{l=1}^{4} \boldsymbol{q}_l \frac{\partial}{\partial \boldsymbol{q}_l}\right) g_2^{\alpha_1\ldots\alpha_4}\{\boldsymbol{q}_i\} - 36 \int_p h^{\beta\gamma}(\boldsymbol{p}) \int_q h^{\delta\rho}(\boldsymbol{q})$$

$$\times \int_0^1 dx x \hat{S} g_2^{\delta\beta\alpha_1\alpha_2}(\boldsymbol{p}, -\boldsymbol{q}, \{x\boldsymbol{q}_{i\le 2}\}) g_2^{\rho\gamma\alpha_3\alpha_4}(-\boldsymbol{p}, x\boldsymbol{q}, \{x\boldsymbol{q}_{i>2}\})$$

$$+ \int_p h^{\beta\gamma}(\boldsymbol{p}) \int_q h^{\delta\rho}(\boldsymbol{q}) \hat{S} \Big\{ (2\pi)^4 \delta(\boldsymbol{q} - \boldsymbol{q}_1) g_{20}^{\beta\gamma\delta\alpha_1} g_2^{\rho\alpha_2\ldots\alpha_4}(-\boldsymbol{q}_1, \{\boldsymbol{q}_{i>1}\})$$

$$- 2 \int_0^1 dx x g_2^{\delta\beta\gamma\alpha_1}(\boldsymbol{p}, -x\boldsymbol{q}, x\boldsymbol{q}, x\boldsymbol{q}_1) g_2^{\rho\alpha_2\alpha_3\alpha_4}(-\boldsymbol{p}, \{x\boldsymbol{q}_{i>1}\}) \Big\}. \tag{6.20}$$

From this equation one can see, that the part of the vertex $\hat{g}_2$ (i.e. $\hat{g}_{20}$) independent on momenta is of the order of ϵ in the vicinity of the fixed point, while the momentum dependent part is of the order of ϵ^2. Therefore, it is quite sufficient for our purposes to restrict consideration only with the vertex $\hat{g}_{20}$. Thus, we have to substitute in Eq. (6.20) $\hat{g}_2\{\boldsymbol{q}_i\} = (2\pi)^d \delta(\boldsymbol{q}_1 + \cdots + \boldsymbol{q}_4)\hat{g}_{20}$ and after carrying out necessary integrations put $\boldsymbol{q}_i = 0$. This leads to the compensation of the terms in braces in Eq. (6.20). As a result one arrives at

$$\frac{dg_{20}^{\alpha_1\ldots\alpha_4}}{dl} = \epsilon g_{20}^{\alpha_1\ldots\alpha_4} - 3K^{\beta\gamma\delta\rho} \Big[g_{20}^{\beta\delta\alpha_1\alpha_2} g_{20}^{\gamma\rho\alpha_3\alpha_4}$$

$$+ g_{20}^{\beta\delta\alpha_1\alpha_3} g_{20}^{\gamma\rho\alpha_1\alpha_3} + g_{20}^{\beta\delta\alpha_1\alpha_4} g_{20}^{\gamma\rho\alpha_2\alpha_3} \Big], \tag{6.21}$$

where

$$K^{\alpha_1\ldots\alpha_4} = 4\hat{S} \int_p h^{\alpha_1\alpha_2}(\boldsymbol{p}) \int_0^1 dx x h^{\alpha_3\alpha_4}(x\boldsymbol{p}). \tag{6.22}$$

In the general case the zero-order anisotropic propagator can be represented as

$$G_0^{\alpha_1\alpha_2} = \frac{1}{q^2} Q^{\alpha_1\alpha_2}(\boldsymbol{q}/\Lambda) S(q^2/\Lambda^2), \tag{6.23}$$

where $S(x)$ has the same sense as in Eq. (5.2). The function $\hat{h}$ is defined as before (see Sect. 5.6.2)

$$h^{\alpha_1\alpha_2}(\boldsymbol{q}) = \frac{1}{q^2} \Lambda^2 \frac{d}{d\Lambda^2} \Big[Q^{\alpha_1\alpha_2}(\boldsymbol{q}/\Lambda) S(q^2/\Lambda^2) \Big] \tag{6.24}$$

Using Eqs. (6.23) and (6.24) one can transform Eq. (6.22) to

$$K^{\alpha_1 \ldots \alpha_4} = \hat{S} \frac{d}{d\Lambda^2} \int_q G_0^{\alpha_1\alpha_2}(\boldsymbol{q}, \Lambda) G_0^{\alpha_3\alpha_4}(\boldsymbol{q}, \Lambda) \bigg|_{\Lambda=1} \tag{6.25}$$

This relation holds for any dimension of space. In the case when we deal with an isotropic propagator at $d = 4$ the integration in Eq. (6.22) can be carried out and as a result we have

$$K^{\alpha_1 \ldots \alpha_4} = (\hat{S}\delta_{\alpha_1\alpha_2}\delta_{\alpha_3\alpha_4}) \frac{K_4}{2}, \tag{6.26}$$

where the constant K_d is defined for arbitrary dimension of space d as the area of the surface of a unit sphere divided by $(2\pi)^d$, i.e.

$$K_d = \frac{S_d}{(2\pi)^d} = \frac{1}{2^{d-1}\pi^{d/2}\Gamma(d/2)}, \tag{6.27}$$

where $\Gamma(x)$ is Euler's gamma function.

Thus, as a result we have obtained an equation describing the evolution of the vertex $\hat{g}_{20}$ in the vicinity of the fixed point. Such an equation, obtained in the framework of the field theoretical approach, is usually called the "one-loop approximation" (see Brezin *et al* 1976). In a number of cases the tensor factor $\hat{K}_4$ can be absorbed into the definition of $\hat{g}_{20}$, then Eq. (6.21) can be simplified to

$$\frac{dg_{20}^{\alpha_1\ldots\alpha_4}}{dl} = \epsilon g_{20}^{\alpha_1\ldots\alpha_4} - \frac{3}{2}(g_{20}^{\beta\delta\alpha_1\alpha_2} g_{20}^{\beta\delta\alpha_3\alpha_4} + g_{20}^{\beta\delta\alpha_1\alpha_3} g_{20}^{\beta\delta\alpha_1\alpha_3} + g_{20}^{\beta\delta\alpha_1\alpha_4} g_{20}^{\beta\delta\alpha_2\alpha_3}). \tag{6.28}$$

It is interesting to note that Eq. (1.28) has been derived without an explicit representation for the function $S(x)$. This means that the factors in this equation do not depend on the momentum cutoff procedure. Naturally, the renormalization of the critical temperature should be essentially dependent on the cutoff function. The latter can be seen from the first equation of the set Eq. (1.12), which for the case of the isotropic zero order propagator can be written as

$$\frac{dg_{10}^{\alpha_1\alpha_2}}{dl} = 2g_{10}^{\alpha_1\alpha_2} + \frac{6}{K_4}\int_p h(\boldsymbol{p}) g_{20}^{\beta\beta\alpha_1\alpha_2} - 2h(0) g_{10}^{\beta\alpha_1} g_{10}^{\beta\alpha_2}. \tag{6.29}$$

If one chooses the function $h(q) = S(q^2) = e^{-q^2}$, Eq. (6.29) takes the form

$$\frac{dg_{10}^{\alpha_1\alpha_2}}{dl} = 2g_{10}^{\alpha_1\alpha_2} + \frac{3}{2} g_{20}^{\beta\beta\alpha_1\alpha_2} - 2g_{10}^{\beta\alpha_1} g_{10}^{\beta\alpha_2}. \tag{6.30}$$

In conclusion, we present the set of equations in the one-loop approximation for the case when the Ginzburg-Landau functional has N interacting fields. The approach described above should be somewhat modified due to the appearance of additional subscripts (s_i) in the definitions of vertexes. These subscripts denote interacting fields in the Hamiltonian. Taking into account this difference one can obtain the set

$$\begin{aligned}
\frac{dg^{\alpha_1\alpha_2}_{10,s_1s_2}}{dl} &= 2g^{\alpha_1\alpha_2}_{10,s_1s_2} \\
&+\sum_{s=1}^{N}\left\{3\int_p h^{\beta\gamma}(\mathbf{p})g^{\beta\gamma\alpha_1\alpha_2}_{20,sss_1s_2} - 2h^{\beta\gamma}(0)g^{\beta\alpha_1}_{10,ss_1}g^{\gamma\alpha_2}_{10,ss_2}\right\}, \\
\frac{dg^{\alpha_1\ldots\alpha_4}_{20,s_1\ldots s_4}}{dl} &= \epsilon g^{\alpha_1\ldots\alpha_4}_{20,s_1\ldots s_4} \\
&-3K^{\beta\gamma\delta\rho}\sum_{s=1}^{N}\sum_{s'=1}^{N}\hat{S}(g^{\beta\delta\alpha_1\alpha_2}_{20,ss's_1s_2}g^{\gamma\rho\alpha_3\alpha_4}_{20,ss's_3s_4}).
\end{aligned} \tag{6.31}$$

Thus, in the first order of ϵ, equations describing critical behavior can be obtained for a quite general case. These equations can be used for calculations of critical exponents. It is important to notice that according to Eq. (5.10) the exponent η is proportional to the fixed values of the vertexes $g_{2r\kappa_r}$ and g^2_{10} and, hence, it has the order of ϵ^2. Two different approaches for calculations of η demonstrated in Chapter 5 can also be generalized on the anisotropic case and interacting fields.

6.3.2 $\mathcal{O}(n)$ Symmetry

Let us apply the above method to the isotropic system with the n-component order parameter. Using the representation Eq. (1.2) one can reduce Eqs. (1.28) and Eq. (6.30) to the following set

$$\begin{aligned}
\frac{dg_{10}}{dl} &= 2g_{10} - 2g^2_{10} + \frac{n+2}{2}g_{20}, \\
\frac{dg_{20}}{dl} &= \epsilon g_{20} - \frac{n+8}{2}g^2_{20}.
\end{aligned} \tag{6.32}$$

As a next step we have to find fixed points for this set and then it is necessary to determine the eigenvalues of the sets linearized in the vicinity of these points. It can be easily found that in addition to the trivial fixed point (Gaussian point with $g^*_{10} = g^*_{20} = 0$) the set Eq. (6.32), in the first approximation with respect to ϵ, has one more fixed point with

$$g_{10}^* = -\frac{n+2}{4} g_{20}^*; \qquad g_{20}^* = \frac{2\epsilon}{n+8}. \tag{6.33}$$

Now, we have to consider that the vertexes g_{10} and g_{20} in the vicinity of the fixed point can be written in the form $g_{i0} = g_{i0}^* + \delta g_i$ ($\delta g_i \ll g_{i0}^*$) and linearize the set Eq. (6.32) with respect to the small deviations δg_i. This leads to the following equations

$$\begin{aligned} \frac{d\delta g_1}{dl} &= 2\left[1 - \frac{n+2}{2} g_{20}^*\right] \delta g_1 + \frac{n+2}{2} \delta g_2, \\ \frac{d\delta g_2}{dl} &= \left[\epsilon - (n+8)\, g_{20}^*\right] \delta g_2. \end{aligned} \tag{6.34}$$

The diagonalization procedure leads to the equation determining eigenvalues

$$\begin{vmatrix} 2-(n+2)g_{20}^* - \lambda & (n+2)/2 \\ 0 & \epsilon - (n+8)g_{20}^* - \lambda \end{vmatrix} = 0. \tag{6.35}$$

For the trivial fixed point, the set Eq. (6.34) gives the values $\lambda_1 = 2$ and $\lambda_2 = \epsilon$, for the point Eq. (6.33) one can obtain from Eq. (6.35)

$$\lambda_1 = 2 - \frac{n+2}{n+8}\epsilon, \qquad \lambda_2 = -\epsilon. \tag{6.36}$$

As is shown in Chapter 3, the critical behavior is defined by the fixed point with the only one positive eigenvalue λ_1 which defines exponent ν. Using the scaling laws and taking into account that in this approximation $\eta = 0$, one arrives at the following values of critical exponents

$$\begin{aligned} \nu &= \frac{1}{\lambda_1} = \frac{1}{2} + \frac{n+2}{4(n+8)}\epsilon, & \eta &= 0, \\ \gamma &= 1 + \frac{n+2}{2(n+8)}\epsilon, & \delta &= 3+\epsilon, \\ \beta &= \frac{1}{2} - \frac{3\epsilon}{2(n+8)}, & \alpha &= \frac{(4-n)\epsilon}{2(n+8)}. \end{aligned} \tag{6.37}$$

When $d > 4$ ($\epsilon < 0$) both eigenvalues in Eq. (6.36) are positive and the critical behavior will be defined by the Gaussian fixed point which gives $\lambda_1 = 2$ and $\lambda_2 = -|\epsilon|$ and we return to the Gaussian description considered in Sect. 6.2.

For the case of $\mathcal{O}(n)$- symmetry, in the lowest orders of ϵ, the consideration given in the previous section can be essentially simplified. It is possible to obtain the main results without deliberation over the nonlocal vertexes. Let us retain in Eq. (1.10) only the terms with $r = 0$ and make the following substitution for the local vertexes

$$\bar{g}_k = g_k \left[\int_q h(q) \right]^{k-1} h(0). \tag{6.38}$$

Below, for the sake of simplicity, we will use notation g_k for the values defined in Eq. (6.38). Taking into account that $B^{(k)}_{\kappa_0 \kappa_0} = n/2 + k$ and $A^{(k,m)}_{\kappa_0 \kappa_0 \kappa_0} = 1$, we find

$$\begin{aligned} \frac{dg_1}{dl} &= 2g_1 - 2g_1^2 + (n+2)g_2, \\ \frac{dg_2}{dl} &= \epsilon g_2 - 8g_1 g_2 + \tfrac{3}{2} g_3, \\ \frac{dg_3}{dl} &= -2g_3 - 8g_2^2. \end{aligned} \tag{6.39}$$

In this set we have to keep the equation for the vertex g_3 because its contribution to the second equation has the same order as from the two other terms. The set Eq. (6.39) also has two fixed-point solutions. The nontrivial fixed point is defined by

$$g_1^* = -\frac{n+2}{4(n+8)}\epsilon, \quad g_2^* = \frac{\epsilon}{2(n+8)}, \quad g_3^* = o(\epsilon^2). \tag{6.40}$$

Eigenvalues for the transformation diagonalizing the linearized version of the set Eq. (6.39) can be found from the determinant

$$\begin{vmatrix} 2 - 4g_1^* - \lambda & n+2 & 0 \\ -8g_2^* & \epsilon - 8g_1^* - \lambda & 3(n+4)/2 \\ 0 & -16g_2^* & -2-\lambda \end{vmatrix} = 0. \tag{6.41}$$

This equation, for the fixed point defined in Eq. (6.40), gives the eigenvalues

$$\begin{aligned} \lambda_1 &= 2 - \tfrac{n+2}{n+8}\epsilon, \\ \lambda_2 &= -\epsilon, \\ \lambda_3 &= -2 + \tfrac{6(n+4)}{n+8}\epsilon. \end{aligned} \tag{6.42}$$

The only positive eigenvalue, λ_1, is the same as in Eq. (6.36) and therefore we have the same critical exponents as before. This method can also be generalized for more complicate cases with arbitrary symmetry.

6.3.3 Cubic Symmetry

The model with cubic symmetry is an interesting generalization because it represents the simplest situation when fluctuation effects lead not only to the change of critical exponents but also to appearance of a qualitatively new critical behavior in comparison with the Landau theory. This behavior was first studied by Wallace in 1973 (see also Aharony 1976). The Ginzburg-Landau functional for the system with cubic anisotropy can be written as

$$H = \int \frac{d^d \boldsymbol{r}}{2} \left\{ \tau \phi^2(\boldsymbol{r}) + (\nabla \phi(\boldsymbol{r}))^2 + \frac{u}{4}(\phi^2(\boldsymbol{r}))^2 + \frac{v}{4} \sum_{\alpha=1}^{n} (\phi^\alpha(\boldsymbol{r}))^4 \right\}, \tag{6.43}$$

i.e. in this case we have

$$\begin{aligned} g_{10}^{\alpha_1 \alpha_2} &= \tau \delta_{\alpha_1 \alpha_2}, \\ g_{20}^{\alpha_1 \ldots \alpha_4} &= u R_2^{\alpha_1 \ldots \alpha_4} + v \delta_{\alpha_1 \alpha_2} \delta_{\alpha_1 \alpha_3} \delta_{\alpha_1 \alpha_4}, \end{aligned} \tag{6.44}$$

where the tensor $\hat{R}_k$ is defined in Eq. (1.2) and there is no summation over α_1 in the second term of Eq. (6.44). Using Eq. (1.28) one can easily find

$$\begin{aligned} \frac{du}{dl} &= \epsilon u - \frac{n+8}{2} u^2 - 3uv, \\ \frac{dv}{dl} &= \epsilon v - \frac{9}{2} v^2 - 6uv. \end{aligned} \tag{6.45}$$

This system has four fixed points. Two of them are: the Gaussian μ_0^* ($u = v = 0$) and the Heisenberg point μ_1^* ($u = 2\epsilon/(n+8)$, $v = 0$). They coincide with the analogous points of the isotropic case. In addition to these we also have two more points: the Ising point μ_2^* ($u = 0$, $v = 2\epsilon/9$) and the cubic point μ_3^* ($u = 2\epsilon/3n$, $v = 2\epsilon(n-4)/9n$). At the Ising point, we have the Hamiltonian Eq. (6.43) composed from the sum of n independent Ising Hamiltonians. The cubic fixed point conserves the general symmetry of the initial functional Eq. (6.43). As usual, the critical behavior of the set Eq. (6.45) will be defined by a stable fixed point with the only one positive eigenvalue λ_1. Of course, in order to find this positive eigenvalue one has to add to the set Eq. (6.45) the equation describing evolution of τ. The eignvalues λ_i for each of these points can be found with the help of the method used above (see for instance Sect. 6.3.2). The results of such calculations are shown in the table below.

Point	λ_2	λ_3
μ_0^*	ϵ	ϵ
μ_1^*	$-\epsilon$	$\epsilon(n-4)/(n+8)$
μ_2^*	$\epsilon/2$	$-\epsilon$
μ_3^*	$-\epsilon$	$\epsilon(4-n)/3n$

From this table, we see that the Gaussian and Ising points are unstable, the Heisenberg point is stable when $n < 4$, while the cubic point is stable when $n > 4$. At $n = 4$ the points μ_1^* and μ_3^* have one negative and one zero eigenvalues. Thus, in order to understand which point is stable, one has to analyze higher order in ϵ contributions.

It is interesting to note, that the above phenomenon has a quite general significance. In the case of anisotropic systems the first ϵ-approximation leads to the stable isotropic fixed point when $n < 4$. This point becomes unstable when $n > 4$. At $n = 4$ we always have zero in this approximation. In the second ϵ-order this point at $n = 4$ is split into a number of different points. The number and stability of these points is usually defined by the symmetry of the system under consideration. A more detailed analysis of the above situation was given by Toledano *et al* in 1985. It is important that in this situation we deal with new phenomenon not allowed in the Landau theory, i.e. the symmetry of the system at the critical point is higher than the symmetry of the initial phase. This phenomenon is sometimes called "asymptotic symmetry."

The set Eq. (6.45), having been solved with respect to the functions $u(l)$ and $v(l)$, defines a family of phase flow lines $v = v(u)$ on the (u, v)-plane. In this case the flow lines are determined by the equation

$$\frac{du}{dv} = \frac{2\epsilon u - (n+8)u^2 - 6uv}{2\epsilon v - 9v^2 - 12uv}, \tag{6.46}$$

which can be integrated in an evident form. To do this, let us define parameter $w = u/v$ as a new independent variable. Then, Eq. (6.46) can be reduced to the linear, with respect to v, equation

$$\frac{dv}{dw} = \frac{2\epsilon - v(9+12w)}{w[3+(4-n)w]}.$$

This equation can be integrated in elementary functions as

$$u = cw^{-2}\left|w + \frac{3}{4-n}\right|^{\frac{3n}{4-n}} \pm \frac{2\epsilon w^2}{n+8}\left[w^2 - \frac{3w}{n+2} + \frac{3}{n(n+2)}\right],$$

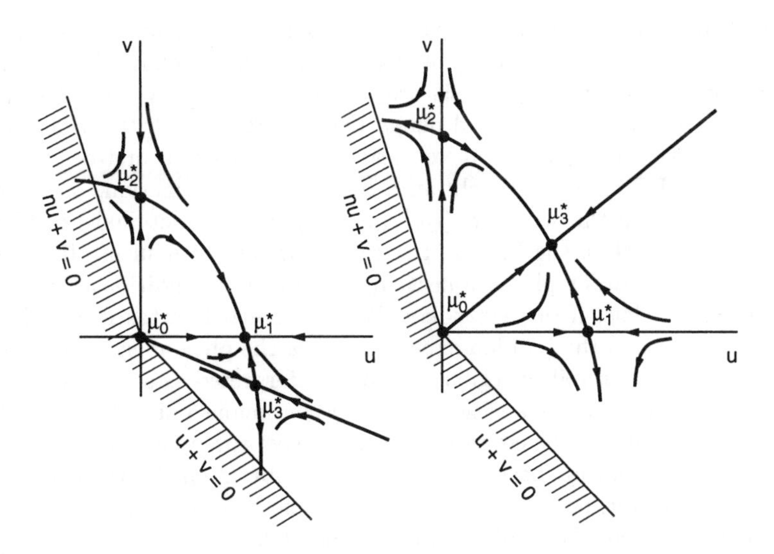

Figure 6.1: The phase portrait of RG equations for the cubic symmetry Hamiltonian at $n < 4$ (a) and $n > 4$ (b).

where $n \neq 4$. For the case when $n = 4$ one can obtain

$$u = cw^{-2}e^{-4w} \pm \frac{2\epsilon w^{-2}}{12}\left[w^2 - \frac{w}{2} + \frac{1}{8}\right].$$

These equations, along with the relation $v = u/w$, give a parametric (parameter w) representation for the family of flow lines controlled by an arbitrary integration constant c. The family of curves, defined by the lines $u = u(v, c)$, constitutes the so-called phase portrait of the set Eq. (6.45). As is seen from Fig. 6.1, there are regions of different behavior. The separatrixes $u = 0$, $u = 3v/(n-4)$ at $n < 4$ and $u = 0$, $v = 0$ at $n > 4$ divide regions of attraction to the stable fixed points. There are also regions of initial parameters (u_0, v_0) from where flow lines never come to a stable point but sooner or later leave the region of the positive definition of the quatric form in the functional Eq. (6.43). This region is defined by the following conditions

$$u + v \geq 0, \qquad nu + v \geq 0. \tag{6.47}$$

The phenomenon, that flow lines cannot reach the stable fixed point, shows that the second-order phase transition is impossible. Really, if conclusions based on the scaling hypothesis are correct, then the system should inevitably come to a stable fixed point. On the other hand, the violation of conditions Eq. (6.47) in the mean field theory leads to a first-order phase transition. In this case, to obtain a finite partition function, one has to add a term with the structure $g_3\phi^6$ ($g_3 > 0$) to the functional Eq. (6.43). In the framework of the Landau theory, the system with such a functional exhibits a first-order phase transition if one of the conditions Eq. (6.47) is violated. Thus, if one considers the renormalized vertexes u and v as effective values in the Landau expansion, then the phenomenon of the flow lines leaving the stability region should be interpreted as a transformation of a continuous phase transition into a discontinuous one. Naturally, this statement should not be considered as a strictly proven fact but rather as a reasonable assumption based on qualitative deliberation. In order to obtain a rigorous answer, one has to calculate free energy of such a system in the fluctuation region. This will be done in Sect. 6.5.

Now let us consider one of the possibilities averting flow lines, "runaway," from the stability region. If the assumption concerning the physical sense of this runaway is correct, then the phenomenon considered below represents a real mechanism for a transformation of first-order phase transitions into continuous ones.

In order to obtain runaway regions, on the phase plane of RG equations, we should satisfy the following conditions: the field $\boldsymbol{\phi}$ is a vector in isotopic space ($n \neq 1$) and the Ginzburg-Landau functional has a particular symmetry which is different from $\mathcal{O}(n)$. The components ϕ_α of the vector field are fluctuating in the critical region under the condition that all of them have the same factor τ at the ϕ^2-term. However, one can imagine a situation (for instance uniaxial compression), when the equivalency of the components is destroyed. The latter will lower the symmetry of the Hamiltonian $H[\boldsymbol{\phi}]$ and in the general case, one may expect not only the appearance of new stable fixed points, absent in a more symmetrical situation, but also the suppression of the initial vertexes into regions of attraction of other, already existing stable points. Naturally, this will lead to the transformation of a first-order transition to a second-order one.

The appearance of the symmetry-breaking field can be taken into account with the help of the addition

$$\Delta H = \frac{\theta}{n}\left[(n-m)\sum_{\alpha=1}^{m}\phi_\alpha^2 - m\sum_{\alpha=m+1}^{n}\phi_\alpha^2\right],$$

which evidently makes m and $n-m$ components of ϕ_α nonequivalent. Let us assume for definiteness that $\theta > 0$. We also define new vertexes r_1 at $\boldsymbol{\phi}_1^2 = \sum_{\alpha=1}^{m} \phi_\alpha^2$ and r_2 at $\boldsymbol{\phi}_2^2 = \sum_{\alpha=m+1}^{n} \phi_\alpha^2$ in the full Hamiltonian as

$$r_1 = \tau + (1 - \frac{m}{n})\theta \quad \text{and} \quad r_2 = \tau - \frac{m}{n}\theta.$$

When the value θ is quite large (in the framework of ϵ-expansion it should be of the order of unity), the field $\boldsymbol{\phi}_2$ practically does not fluctuate. After carrying out the integration in the partition function over this field, the functional $H[\boldsymbol{\phi}]$ will be reduced to $H[\boldsymbol{\phi}_1]$, where in the lowest order the effective vertex is

$$\tau_1 = r_1 + u \int_q \frac{1}{r_2 + q^2}.$$

The new functional $H[\boldsymbol{\phi}_1]$ describes the phase transition in the m-component system with the cubic anisotropy. The stability regions Eq. (6.47) are now reduced to $u + v \geq 0$ and $mu + v \geq 0$ and the separatrix of the attraction region to the stable fixed point (at $m < 4$) is reduced to the line $w(m) = u(m)/v(m) = 3/(m-4)$, which evidently does not coincide with the line $w(n) = 3/(n-4)$. Now, the initial point (u_0, v_0) may be found inside the attraction region.[2] Intuitively, it is quite clear that if the initial value θ is small enough, the critical process in the system should develop approximately in the same manner as at $\theta = 0$. Even if one takes into account the renormalization increase of $\theta(l)$ at $l \to \infty$ it is possible to imagine the situation, when the flow lines $(u(l), v(l))$ will cross the stability boundary $(w(m) = 3/(m-4))$ before $\theta(l)$ reaches the value of the order of unity. The limiting cases of large and small values of θ corresponding to the second and first-order transitions are demarcated by a value θ_t (at $\tau = 0$) defining a tricritical point of the considered system. In other words the points with coordinates

$$r_2(l_m) = 1, \qquad \tau = 0, \qquad w(m) = \frac{3}{m-4} \tag{6.48}$$

is a tricritical point. In the vicinity of this point the phase diagram changes as is shown in Fig. 6.2. In order to find this point in the framework of RG analysis one has to add a couple of equations for θ and r to the equations describing (u, v)-plane. This case was first considered by Domany *et al* in 1977 (see also Kerszberg *et al* 1978, 1979, and 1981). The tricritical point may appear if an initial point (u_0, v_0) is located near the separatrix of attraction to the stable $\mathcal{O}(n)$ fixed point. One can assume that this point is quite close to the cubic fixed point, $u^* = 2\epsilon/3n$ and $v^* = 2\epsilon(n-4)/9n$. Therefore, all equations in this region can be linearized with respect to

[2] As we have $m < n$, the separatrix $v = (m-4)u/3$ is now below the line $v = (n-4)u/3$.

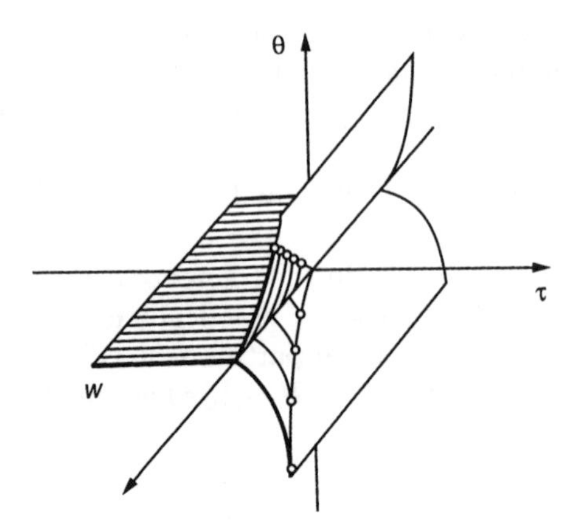

Figure 6.2: The phase diagram in the variables (τ, θ, w) in the vicinity of tricritical points. The dashed regions correspond to the first-order phase transition, white regions correspond to the critical behavior. The circled lines are lines of tricritical points.

$\Delta u = u - u^*$, $\Delta v = v - v^*$ in the lowest approximation in ϵ. This leads to the following equations describing evolution of θ and r

$$\begin{aligned}\frac{d\theta}{dl} &= \left[\lambda_\theta - \Delta u - \tfrac{3}{2}\Delta v\right]\theta, \\ \frac{dr}{dl} &= \lambda_r r + \frac{2\epsilon(n-1)}{3n} + \left[\tfrac{n+2}{2}\Delta u + \tfrac{3}{2}\Delta v\right](1-r),\end{aligned} \tag{6.49}$$

where the eigenvalues λ_θ and λ_r are defined by

$$\lambda_\theta = 2 - \frac{n-2}{3n}\epsilon, \qquad \text{and} \qquad \lambda_r = 2 - \frac{2(n-1)}{3n}\epsilon.$$

The general solution of the set Eq. (6.49) can be written as

$$\theta(l) = \bar{\theta}(l)e^{\lambda_\theta l}, \quad r(l) = -\frac{n+2}{2}\Delta u(l) - \frac{3}{2}\Delta v(l) + t(l),$$

where the functions $\bar{\theta}(l)$ and $t(l)$ are defined by

$$\bar{\theta}(l) = \theta_0 \exp\left(-\int_0^l dl\left[\Delta u(l) + \frac{3}{2}\Delta v(l)\right]\right) \tag{6.50}$$

and

$$t(l) = t_0 \exp\left\{\left[-\int_0^l dl \left[\frac{n+2}{2}\Delta u(l) + \frac{3}{2}\Delta v(l)\right]\right] + \lambda_r l\right\} \equiv \bar{t}(l)e^{\lambda_r l} \tag{6.51}$$

The first two conditions in Eq. (6.48) can be rewritten in terms of the functions $\theta(l)$ and $t(l)$ as

$$t + \frac{m}{n}\theta = 1, \qquad t - \left(1 - \frac{m}{n}\right)\theta = 0. \tag{6.52}$$

Let us define the value $z = \bar{\theta}/\bar{t}^{\phi}$, where the exponent ϕ is equal to the ratio λ_θ/λ_r. In the lowest order in ϵ, one can get from Eq. (6.52) that $z = n/(n-m)$. Also, taking into account the last of the tricriticality conditions Eq. (6.48) that $w(m) = 3/(m-4)$, let us choose (u_0, v_0) so that the flow line will go through the point $(u(l_m), v(l_m))$ satisfying the condition $u(l_m)/v(l_m) = w(m)$. As the value of z is already found, using Eqs. (6.50) and (6.51) one can see that at the tricritical point, $\theta_t = A_m t_t \phi$, where A_m is a nonuniversal value defined by the equation

$$A_m = \frac{n}{n-m} \exp\left[-\frac{n}{2}\int_0^{l_m} dl \Delta u(l)\right].$$

The change of the θ sign will lead to the same picture with the replacement m on $n-m$ and *vice versa.* The ratio

$$\frac{A_m}{A_{n-m}} = \frac{m}{n-m} \exp\left[-\frac{n}{2}\int_{l_{n-m}}^{l_m} dl \Delta u(l)\right] = \frac{n}{n-m}\exp(I)$$

is a universal value usually called "universal ratio of amplitudes." In order to prove the universality one has to change the variable of integration from l to w, then I takes the form

$$I = n\int_{w(n-m)}^{w(m)} dw \left\{-\frac{1}{v}\frac{dv}{dw} - \frac{9+11w}{w[(4-n)w+3]}\right\}.$$

Now the integration can be easily done. However, an explicit expression for I is not that important as the fact that its value is completely defined by the numbers n and $n-m$ and does not depend on a particular path of the flow line trajectory between the points l_m and l_{n-m}. For the most interesting case $n = 3$, $m = 1$ Domany *et al* (1977) obtained

$$I = \ln\left[\frac{27}{64}(2^{20}/3^7 \times 67)^{1/3}\right],$$

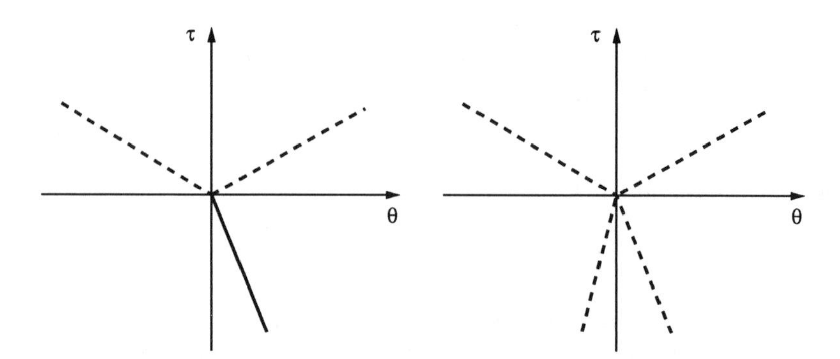

Figure 6.3: The phase diagram with a bicritical (a) and tetracritical (b) points obtained in the mean field approximation. The continuous lines are the first-order transition lines, dashed are the second-order lines.

so that the amplitude ratio is given

$$\frac{A_1}{A_2} = \left(\frac{9}{134}\right)^{1/3} \approx 0.406$$

6.3.4 Interacting Fields

In the simplest case of two isotropic interacting fields the Ginzburg-Landau functional has the form

$$H = \int \frac{d^d \boldsymbol{r}}{2} \left\{ \sum_{s=1}^{2} \left[\tau_s \boldsymbol{\phi}_s^2 + (\boldsymbol{\nabla}\boldsymbol{\phi}_s)^2 + \frac{1}{4} u_s (\boldsymbol{\phi}_s^2)^2 \right] + \frac{1}{4} v \boldsymbol{\phi}_1^2 \boldsymbol{\phi}_2^2 \right\}, \tag{6.53}$$

where $\boldsymbol{\phi}_s$ is an n-component vector. Before starting to consider the fluctuational region, let us study the phase diagram of this system in the framework of the Landau theory. The conventional analysis of this system leads to the following two types of diagrams shown in Fig. 6.3 in coordinates $\theta = (\tau_1 - \tau_2)/2$ and $\tau = (\tau_1 + \tau_2)/2$. In this figure the phase transition is continuous along the lines $\tau = \pm\theta$ in the region $\tau \geq 0$. Below these lines the type of the phase transition is essentially dependent on relations between values of interaction constants in the Hamiltonian Eq. (6.53). In the case of a strong coupling between fields ($v^2 \geq u_1 u_2$), there are possible only phases with $\boldsymbol{\phi}_1 \neq 0$, $\boldsymbol{\phi}_2 = 0$ and $\boldsymbol{\phi}_1 = 0$, $\boldsymbol{\phi}_2 \neq 0$. The transition between these phases is a first-order transition happening on the line

$$(\sqrt{u_1} - \sqrt{u_2})\tau + (\sqrt{u_1} + \sqrt{u_2})\theta = 0.$$

This line is shown as a solid line in Fig. 6.3. When we are in the region with a weak coupling $v^2 < u_1 u_2$, a mixed state with $\boldsymbol{\phi}_1 \neq 0$, $\boldsymbol{\phi}_2 \neq 0$ is possible with continuous transitions defined by the lines

$$(u_1 - v)\tau - (u_1 + v)\theta = 0, \qquad (u_2 - v)\tau - (u_2 + v)\theta = 0.$$

The points corresponding to the origin of reference systems in Fig 6.3 (a) and (b) are called bicritical and tetracritical points according to the number of continuous transitions lines.

Now, we are going to find out how these phase diagrams change under the influence of fluctuations. Let us assume that the system is quite close to the vicinity of the lines crossing and both variables τ_1 and τ_2 are close to zero. Hence, both fields $\boldsymbol{\phi}_1$ and $\boldsymbol{\phi}_2$ fluctuate substantially in this region. In order to study this region we have to obtain RG equations for the vertexes τ_s, u_s and v. In the first ϵ-approximation these are equations of the set Eq. (1.31). Substituting into this set the tensors $h^{\alpha_1\alpha_2}(\boldsymbol{q})$, $g^{\alpha_1\alpha_2}_{1,s_1s_2}$ and $g^{\alpha_1\ldots\alpha_4}_{2,s_1\ldots s_4}$ in the form

$$\begin{aligned}
h^{\alpha_1\alpha_2}(\boldsymbol{q}) &= \delta_{\alpha_1\alpha_2}\exp(-\boldsymbol{q}^2), \qquad g^{\alpha_1\alpha_2}_{1,s_1s_2} = \delta_{\alpha_1\alpha_2}\delta_{s_1s_2}\tau_{s_1}, \\
g^{\alpha_1\ldots\alpha_4}_{2,s_1\ldots s_4} &= u_{s_1}\delta_{s_1s_2}\delta_{s_1s_3}\delta_{s_1s_4}R_2^{\alpha_1\ldots\alpha_4} \\
&+ \tfrac{v}{3}\left\{\delta_{\alpha_1\alpha_2}\delta_{\alpha_3\alpha_4}\delta_{s_1s_2}\delta_{s_3s_4} + 2\,permutations\right\}
\end{aligned}$$

one can obtain the set

$$\begin{aligned}
\tfrac{d\tau_s}{dl} &= 2\tau_s + \tfrac{n_s+2}{2}u_s + \tfrac{n_{t\neq s}}{2}v - 2\tau_s^2, \\
\tfrac{du_s}{dl} &= \epsilon u_s - \tfrac{n_s+2}{2}u_s^2 - \tfrac{n_{t\neq s}}{2}v^2, \\
\tfrac{dv}{dl} &= v[\epsilon - \tfrac{n_1+2}{2}u_1 - \tfrac{n_2+2}{2}u_2 - 2v].
\end{aligned} \tag{6.54}$$

The first equation of this set defines renormalization of initial critical temperatures

$$\delta\tau_s = -\frac{n_s+2}{4}u_s - \frac{n_{t\neq s}}{4}v_s. \tag{6.55}$$

The set Eq. (6.54) has the fixed points shown in the table below (see Kosterlitz *et al* 1976)

Point	u_1^*	u_2^*	v^*
μ_0^*	0	0	0
μ_1^*	$2\epsilon/(n_1+8)$	0	0
μ_2^*	0	$2\epsilon/(n_2+8)$	0
μ_3^*	$2\epsilon/(n_1+8)$	$2\epsilon/(n_2+8)$	0
μ_4^*	$2\epsilon/(n_1+n_2+8)$	$2\epsilon/(n_1+n_2+8)$	$2\epsilon/(n_1+n_2+8)$
μ_5^*	$\epsilon a_1/9$	$\epsilon a_2/(n_1+n_2+7)$	ϵx

(6.56)

where a_1 and a_2 are defined by

$$a_1 = 1 + \sqrt{1 - 9(n_1+n_2-1)x^2}, \qquad a_2 = 1 + \sqrt{1-(n_1+n_2+7)x^2},$$

with the value $x(n)$ taken as a real root of the cubic equation

$$9(4n^2+29n+88)x^3 - 6(2n^2+28n+179)x^2$$

$$+(n^2+5n+472)x + 6(n-11) = 0.$$

This equation gives rational values of roots at some particular values of n: $x=0$ at $n=11$, $x=1/16$ at $n=4$, $x=1/3$ at $n=2$, $x=10/33$ at $n=1$, and $x=2/3$ at $n=-1$. However, in the general case $x(n)$ is not a rational value.

In the table (Eq. (6.56)) the first fixed point μ_0^* is the trivial point unstable at $d \le 4$. The points μ_1^* and μ_2^* are also always unstable. From the three points μ_3^*, μ_4^*, and μ_5^* one is always stable ($d \le 4$) depending on the values n_1 and n_2. The attraction regions for these points are shown on the plane (n_1, n_2) in Fig. 6.4. When $n_1+n_2<4$, the isotropic point μ_4^* with the $\mathcal{O}(n_1+n_2)$ symmetry is stable, in full agreement with the deliberations of the previous section. Here we again deal with the phenomenon of asymptotic symmetry. When $n_1+n_2>4$ and $2(n_1+n_2)+n_1n_2>32$, the point μ_3^* with the symmetry $\mathcal{O}(n_1)\oplus\mathcal{O}(n_2)$ is stable. This point corresponds to splitting of the fields $\boldsymbol{\phi}_1$ and $\boldsymbol{\phi}_2$. At last, when $n_1+n_2>4$ but $2(n_1+n_2)+n_1n_2<32$, the point μ_3^* with the lowest possible symmetry is stable. The symmetry of this point is $\mathcal{O}(n_1)\otimes\mathcal{O}(n_2)$.

It is necessary to mention that one can speak about stability of the points μ_3^*, μ_4^*, μ_5^* only when $\tau_1=\tau_2$. In the other case (for instance $\tau_1<\tau_2$) when a flow line will tend to one of these points (for example μ_3^*) the renormalized value τ_1 will tend to zero, while τ_2 will stay as a positive finite value. As a result the point μ_3^* happens to be unstable with respect to the parameter τ_2 and phase trajectories will flow by it and end up at the

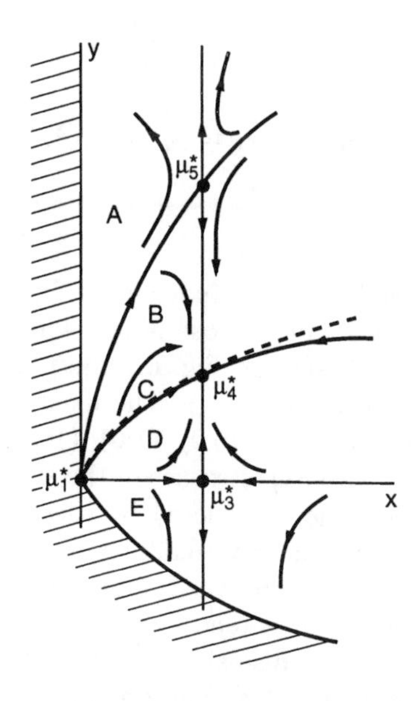

Figure 6.4: The phase portrait of the RG equations in x, y variables.

point μ_1^*. Such behavior, when the critical regime induced by the point μ_3^* eventually changes into the regime with the $\mathcal{O}(n_1)$ symmetry is called "crossover."

It is interesting to notice that the fluctuation renormalization of the trial critical temperatures τ_s is essentially dependent on n_s the number of field components (see Eq. (6.55)). The latter may lead to the following effect, impossible in the Landau approximation. Let us assume that $\tau_1 < \tau_2$ but $\delta\tau_1$ is less than $\delta\tau_2$ to the extent that $\tau_1^* > \tau_2^*$. As a result one has a second-order transition with the ordering of the field ϕ_2 instead of ϕ_1 which must be ordered in the mean-field approximation.

Let us now study the flow lines portrait of the set Eq. (6.54). In what follows we will only be interested in the phase transition type and the structure of the ordered phase. This knowledge can be gained from the study of vertex ratios like $x = u_2/u_1$ and $y = v/u_1$. The introduction of these ratios makes the phase space two-dimensional and essentially simplifies consideration. Using the set Eq. (6.54) one can obtain the equations for x, y functions

$$\begin{aligned}\frac{1}{u_1}\frac{dx}{dl} &= (n+8)x + n_2xy^2 - (n_2+8)x^2 - n_1y^2, \\ \frac{1}{u_1}\frac{dy}{dl} &= 6y - (n_2+2)xy - 4y^2 + n_2y^3.\end{aligned} \tag{6.57}$$

This system of equations does not include all the fixed points of Table 6.56. Owing to the division by u_1, only those points which do not have $u_1^* = 0$ can be considered with this set. In Fig. 6.5 the phase portrait of the set Eq. (6.57) is shown when $n_1 = n_2 = 1$. The fixed points μ_1^*, μ_3^*, μ_4^*, and μ_5^* have the following coordinates: (0;0), (1;0), (1;1), (1;3). The flow lines crossing μ_1^*, μ_5^* and μ_1^*, μ_3^* are separatrixes of the stable fixed point μ_4^*. All flow lines outside of the region restricted by these separatrices leave the stability region of the functional Eq. (6.53). The boundaries of this region are constituted by the lines $x = 0$ at $y > 0$ and $y^2 = x$ at $y < 0$, and also the infinite point, as the flow lines tend to it at $u_1 \to 0$. In Fig. 6.5, the line $y^2 = x$ ($v^2 = u_1u_2$) at $y > 0$ is also shown as a dash-dotted line as well as the flow line leaving the point μ_1^* and entering to μ_4^* at the same angle as the dash-dotted line.

Let us consider the case when the trial values of the vertexes u_1 and u_2 satisfy the inequality $u_2^0/u_1^0 = x_0 < 1$, i.e. phase trajectories are originated at the left-hand side of the line joining the fixed points μ_3^*, μ_4^*, and μ_5^*. If the trial point x_0, y_0 is above the separatrix $\mu_1^*\mu_5^*$, then the flow will leave the stability region and the first-order phase transition is to happen in the system. It is important that this transition corresponds to the $\boldsymbol{\phi}_2$ field ordering. The latter can be seen from the fact that along these flow lines the value x changes sign, i.e. u_2 becomes negative, which means that the field $\boldsymbol{\phi}_2$ is loosing its stability. Such a kind of transition happens at positive values of τ_1 and τ_2 and the structure of the ordered phase does not depend on the sign of the difference $\tau_1 - \tau_2$. In order that this phenomenon may happen it is important to have both of the fields in the critical regime, i.e. $|\tau_1 - \tau_2| \ll 1$. If this equality is fulfilled for the trial values τ_s, then, as we have from Eq. (6.55) $\delta\tau_1 = \delta\tau_2$, it will be also preserved for the renormalized values τ_s. Thus, if $\tau_2 < \tau_1$, the phase transition will occur in the phase given by the Landau theory, however, the interaction of critical fluctuations changes the kind of transition. In contrast to the case with $\tau_2 > \tau_1$, instead of the phase with $\boldsymbol{\phi}_1 \neq 0$, $\boldsymbol{\phi}_2 = 0$, the fluctuation theory leads to the first-order phase transition into an anomalous, from the point of view of the Landau theory, phase with $\boldsymbol{\phi}_1 = 0$, $\boldsymbol{\phi}_2 \neq 0$. This phase is stabilized by the interaction of critical fluctuations and hence, may only exist in the critical region. Under the decrease of temperature, it must turn

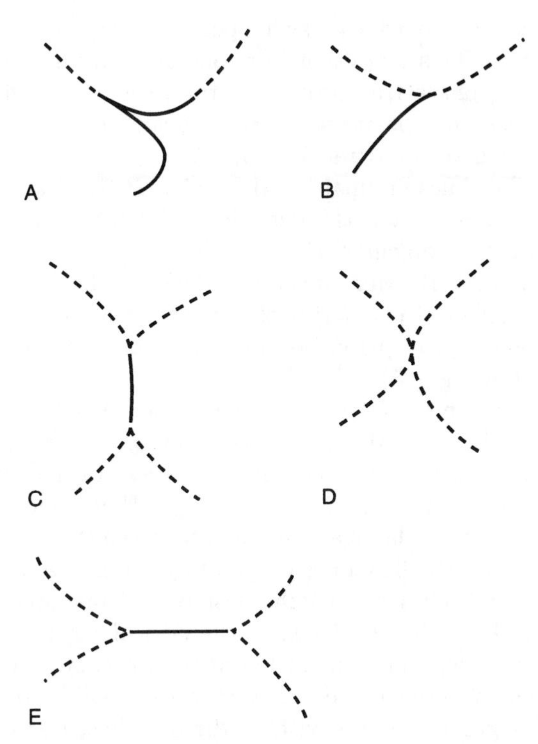

Figure 6.5: Possible types of phase diagrams. The letter notations correspond to the localization of the trial values x, y on the phase portrait in Fig. 6.4.

into the "normal" phase, which may happen in the region where fluctuations are still essential. This transition from one ordered phase to the other is also a first-order phase transition. On the phase diagram plane such a situation leads to the appearance of a bird "beak" (in Fig. 6.5, the diagram A) drawn by the first-order transition lines.

When the flow lines originate in the region B (see Fig. 6.5), restricted by the separatrix $\mu_1^*\mu_5^*$ and the dash-dotted line, or from the region D, restricted by the integral curve $\mu_1^*\mu_4^*$ and the separatrix $\mu_1^*\mu_3^*$, they will end up at the fixed point μ_4^*. More strictly, in the case when $\tau_1 \neq \tau_2$, this is the point corresponding to the ordering of one of the fields. The phase diagram for this situation is principally the same as in the Landau theory (see the diagrams B, D in Fig. 6.5).

An essentially different situation occurs when flow lines are originated in the region C between the dash-dotted line $y = \sqrt{x}$ and the integral curve $\mu_1^*\mu_4^*$. Here we have, for the initial vertexes of the Ginzburg-Landau functional, inequality $y_0^2 < x_0$ (or $v^2 < u_1 u_2$). Under such inequality the mean field theory gives the phase diagram with the tetracritical point (see Fig. 6.3). However, the flow lines originating in this region on the way to the fixed point will always cross the dash-dotted line and the inequality $v^2 < u_1 u_2$ will be violated. Thus, in the critical region we will have in this case only the diagram with the bi-critical point, but under the further decrease of temperature the inequality $v^2 < u_1 u_2$ will be restored. Hence, on the phase diagram, the lines with continuous phase transitions from the state with $\phi_1 \neq 0$, $\phi_2 = 0$ and $\phi_1 = 0$, $\phi_2 \neq 0$ into a mixed phase with $\phi_1 \neq 0$, $\phi_2 \neq 0$ appear. The final phase diagram for this case is shown in Fig. 6.6, diagram C.

At last, all flow lines originating below the separatrix $\mu_1^*\mu_3^*$ (the region E) leave the stability region of the functional. Now, independently of the sign $\tau_1 - \tau_2$ both Landau fields are ordered. The consequent phase diagram (E) is also shown in Fig. 6.6.

Thus, the case of interacting fields in the critical region, has a number of possibilities. This case is interesting not only from the theoretical point of view, but also because it can be realized in experiment. First of all this has a direct relation to systems with two close phase transitions (for instance, structural and magnetic phase transitions), though it can also be realized in a number of other systems.

In conclusion it is necessary to note that the considered qualitative deliberations concerning the change of a kind of a phase transition should always be confirmed by direct calculations of the free energy in the critical region. The latter is especially important when one deals with the situation of anomalous phases.

6.3.5 Logarithmical Corrections

Logarithmical corrections to critical behavior were found by Larkin and Khmelnitsky in 1969. They arise in the ϕ^4-model at $d = 4$. In order to obtain them in the framework of the SE method it is quite sufficient to use the local form of the Ginzburg-Landau functional. The equation leading to such corrections can be obtained from Eq. (1.8). In this case one has to retain only two vertexes $g_1 = \tau$ and g_2. As a result in the critical region we have equation

$$\left\{1 + \tau[1 - (n+2)g_2]\frac{\partial}{\partial \tau} - (n+8)g_2^2\frac{\partial}{\partial g_2}\right\} G(\tau, g_2) = 0. \tag{6.58}$$

Let us define the functions $\tau(\lambda)$ and $g_2(\lambda)$ with the help of equations

$$\begin{aligned} \lambda\tfrac{d\tau(\lambda)}{d\lambda} &= \tau(\lambda)[1-(n+2)g_\lambda], & \tau(1) &= \tau, \\ \lambda\tfrac{dg_2(\lambda)}{d\lambda} &= -(n+8)g_2^2(\lambda), & g_2(1) &= g_2. \end{aligned} \tag{6.59}$$

Now, Eq. (6.58) can be rewritten as

$$\left(\lambda\frac{d}{d\lambda} + 1\right) G[\tau(\lambda), g_2(\lambda)] = 0,$$

from which one gets

$$G[\tau(\lambda), g_2(\lambda)] = G(\tau, g_2)/\lambda. \tag{6.60}$$

The set Eq. (6.59) has solutions

$$\begin{aligned} \tau(\lambda) &= \tau\lambda[1 - g_2 \ln|\lambda|]^{-(n+2)/(n+8)}, \\ g_2(\lambda) &= g_2[1 - g_2(n+8)\ln|\lambda|]^{-1}. \end{aligned} \tag{6.61}$$

Let us choose λ_0 so that $\tau(\lambda_0) = 1$. Then in the limit $\tau \to 0$ from the set Eq. (6.61) one can get

$$\lambda_0 \approx \frac{1}{\tau}[g_2(n+8)\ln|\tau|]^{(n+2)/(n+8)}.$$

Substituting this value into Eq. (6.59) one finally arrives at

$$G(\tau, g_2) \sim \frac{1}{|\tau|}[\ln|\tau|]^{(n+2)/(n+8)} G(1, 0).$$

6.4 The $\phi^4 + \phi^6$-Model

As it has been shown above, the problem of critical behavior in the lowest ϵ approximation can be reduced to the analysis of equations for vertexes g_1 and g_2 for the $\mathcal{O}(n)$-symmetrical model. The same results, in principle, can be obtained without consideration of the dimension $d = 4 - \epsilon$ but by doing the same calculations at $d = 3$ with the assumption that vertexes g_k are small and satisfy inequality $g_{k>2}/g_{1,2} \ll 1$. In this case, the method is equivalent to the technique developed by Sokolov (1979). In this section we will demonstrate this approach along with the elimination of generating terms on the $\phi^4 + \phi^6$-model.

Usually, in the theory of phase transitions one has to retain the powers ϕ^4 and ϕ^6 in the Ginzburg-Landau functional in the vicinity of a tricritical point where the vertex g_2 tends to zero. Therefore, we assume that the three vertexes g_1, g_2, and g_3 are small and have the same order of value, while for the other vertexes we have the inequality $g_{k>3}/g_{1,2,3} \ll 1$. At $d = 3$ we have the trivial fixed point eigenvalues equal to $\epsilon_k = 3 - k$ and when $k > 3$ all eigenvalues are negative. In this sense the vertexes with $k > 3$ are unessential and the assumption $g_k/g_{1,2,3} \ll 1$ seems to be justified. However, this does not mean that consideration can be restricted only to the three equations for the vertexes $g_{1,2,3}$. The generating terms in the left-hand sides of the equations describing evolution of $g_{k>3}$ create terms like $g_2 g_3$, g_2^2, and g_3. Thus, in order to obtain the set of equations for $g_{1,2,3}$, one has also to retain $g_{k>3}$ and to eliminate the generating terms giving contributions of the same order. In the lowest order, with respect to the vertexes g_2 and g_3, we have to consider also equations for g_4 and g_5. As a result the set of equations can be written in the form

$$
\begin{aligned}
\frac{dg_1}{dl} &= 2g_1 + (n+2)g_2 - 2g_1^2, \\
\frac{dg_2}{dl} &= g_2 + \frac{3(n+4)}{2} g_3 - 8g_1 g_2, \\
\frac{dg_3}{dl} &= 2(n+6)g_4 - 8g_2^2 - 12g_1 g_3, \\
\frac{dg_4}{dl} &= -g_4 + \frac{5(n+8)}{2} g_5 - 24 g_2 g_3, \\
\frac{dg_5}{dl} &= -2g_4 - 18g_3^2.
\end{aligned}
$$

Now, following the scheme developed for the ϕ^4-model (see Sect. 6.3.1), let us shift the variable g_5: $g_5 = u_5 - 9g_3^2$. Taking into account that $g_3 dg_3/dl$

is of the order of g_3^3, one can obtain equations for g_4 and u_5 as

$$\begin{aligned}\frac{dg_4}{dl} &= -g_4 + \frac{5(n+8)}{2}u_5 - 24g_2g_3 - \frac{45(n+8)}{2}u_5,\\ \frac{du_5}{dl} &= -2u_5 + o(g_5g_{1,2,3}).\end{aligned}$$

The last equation does not contain generating terms and has the eigenvalue $\lambda_5 = -2 + o(g_{1,2,3})$. Taking into account that $u_5^* = 0$ the value u_5 can be omitted in all other equations. At the first glance it seems that for the elimination of g_4 one has to substitute $u_4 = g_4 - 45(n+8)g_3^2/2 - 24g_2g_3$. However, the situation here a little more complicated because $dg_2/dl \approx g_2 + 3(n+4)g_3/4$ has the same order of value as g_2. The latter means that the value u_4 should be chosen in the form $u_4 = g_4 - (ag_3^2 + bg_2g_3)$ where the constants a and b are to be found from the condition

$$\frac{du_4}{dl} = -2u_4 + o(g_4g_{1,2,3}).$$

As a result one can find that the substitution

$$g_4 = u_5 - 9(n+24)g_3^2 - 24g_2g_3$$

will do the job. With the assumption that the system is already on the critical surface (i.e. $g_1 \approx (n+2)g_2/2$) one can find

$$\begin{aligned}\frac{dg_2}{dl} &= g_2 + 3(n+4)g_3 + 4g_2^2,\\ \frac{dg_3}{dl} &= -24(n+6)g_2g_3 - 8g_2^2 - 9(n+6)(n+24)g_3^2 + 6(n+2)g_2g_3.\end{aligned}$$

This system of equations constitutes a closed set. However, this set cannot be considered a final version as it is still impossible to use it for the study of a critical (at $g_3 = 0$), as well as a tricritical behavior (at $g_2 = 0$). In order to include these two physically important cases in consideration, one has to eliminate in the first equation the term proportional to g_3 and in the second, the term proportional to g_2. This can be done with the help of a linear transformation of the vertexes $g_{2,3} \to \tilde{g}_{2,3}$ and leads to the set

$$\begin{aligned}\frac{d\tilde{g}_2}{dl} &= \tilde{g}_2 - 2(n+2)\tilde{g}_2^2 - 3(n+26)(n+4)\tilde{g}_2\tilde{g}_3,\\ \frac{d\tilde{g}_3}{dl} &= -6(n+14)\tilde{g}_2\tilde{g}_3 - 36(3n+22)\tilde{g}_3^2.\end{aligned} \tag{6.62}$$

For the following it is convenient to change normalization of the vertexes with the help of substitutions $u_2 = 2(n+8)\tilde{g}_2$ and $u_3 = 36(22+3n)\tilde{g}_3$. As a result, the set Eq. (6.62) can be written in a compact form as

$$\begin{aligned} \frac{du_2}{dl} &= u_2 - u_2^2 - a(n)u_2u_3, \\ \frac{du_3}{dl} &= -u_3^2 - b(n)u_2u_3, \end{aligned} \tag{6.63}$$

where the factors $a(n)$ and $b(n)$ are defined by

$$a(n) = \frac{(n+4)(n+26)}{12(3n+22)} \quad \text{and} \quad b(n) = \frac{3(n+14)}{n+8} \tag{6.64}$$

The set Eq. (6.63) can be integrated (see Sokolov 1979) in the form

$$l = \int dy \left\{ 1 - \frac{b}{b-1} \frac{y^{(1-a)}e^y}{c_1 - \gamma(1-a,-y)\exp[-i\pi a]} \right\} + c_2, \tag{6.65}$$

where $y = 1/u_3$; c_1 and c_2 are integration constants, $\gamma(\alpha, x)$ is the Euler's gamma function.

However, though the integral Eq. (6.65), formally solves the problem, it is inconvenient for practical uses. In order to construct the line flow portrait it is better to use one of the following methods. In the first, one has to analyze the equations for u_2 and u_3 only for in the most interesting regions for physical applications ($g_2 \ll g_3$, $g_2 \sim g_3 \ll 1$ and $g_3 \ll g_2$). This method was originally considered by Sokolov 1979. The other method consists of the use of isoclines. In this approach, for one of the vertexes one has to find lines along which the derivative with respect to l is equal to zero ($dg_i/dl = 0$). Then one has to draw flow lines perpendicular to the axis corresponding to the consequent vertex (g_i). In all cases when a phase portrait can be reduced to a two dimensional picture, the topology of phase flow lines in the framework of the isoclines method is uniquely defined. Quite frequently, qualitative estimates are sufficient for understanding the behavior of a system in the critical region. In the case under consideration, isoclines for the horizontal axis ($du_2/dl = 0$ see Fig. 6.6) on the plane (u_2, u_3) can be reduced to the two straight lines $u_2 = 0$ and $1 - u_2 - a(n)u_3 = 0$, while the vertical isoclines ($du_3/dl = 0$) are defined by the lines $u_3 = 0$ and $u_3 + b(n)u_2 = 0$ (see Eq. (6.63)). Evidently, the crossing points for the horizontal and vertical isoclines are the fixed point of the problem. In this

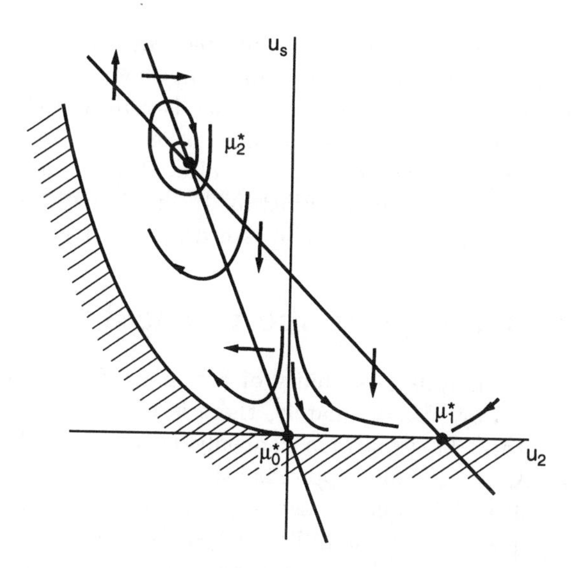

Figure 6.6: The phase portrait of RG equations in the tricritical region. The shaded region follows the stability boundary of the Ginzburg-Landau functional ($u_3 = 0$ when $u_2 > 0$ and $u_3 \approx 4u_2^2$ when $u_2 < 0$).

case we have

Point	u_2^*	u_3^*
μ_0^*	0	0
μ_1^*	1	0
μ_2^*	$-(a(n)b(n)-1)^{-1}$	$b(n)/(a(n)b(n)-1)$

(6.66)

Apparently, the flow lines with $u_3 = 0$ would correspond to conventional critical behavior, while the flow lines with $u_2 = 0$ give a tricritical behavior. If the initial value of u_2 is positive (even when it is very small) the system sooner or later will come to the point μ_1^* (see Eq. (6.66)). Such behavior gives a crossover from the tricritical regime to the critical one. On the contrary, when $u_2 < 0$, flow lines move into the direction of the attractive focus μ_2^* which has complex eigenvalues λ_i. If this behavior were realized it would correspond to oscillatory dependence of critical exponents on temperature. Such behavior has never been confirmed by experiment. Therefore, one may consider the point μ_2^* as "unphysical". Another fact pointing to the unphysical nature of this point is: physically reasonable values for n ($n \leq 20$) give a value for u_3 larger than one. Hence, the perturbation theory in powers of small values of u_2 and u_3 is not justified in this region. However, if n is very large, the values u_2 and u_3 in the point μ_2^* are less than one. This should not be considered as evidence that with the increase of the number of components behavior of the isotropic system in the vicinity of the tricritical point is radically changed. Most likely, this is evidence that in this region all other terms in the Ginzburg-Landau functional play equally important roles. Really, one can show that for an arbitrary Ginzburg-Landau functional, containing all local vertexes, eigenvalues for any fixed point are real (see Appendix B).

6.5 Free Energy in the Critical Region

As is known, the analysis of stability of a bilinear form can be made in the most general way. Unfortunately, there is no such general approach for a quatric form. This question can only be answered by solution of RG equations. As a particular approximation and Hamiltonian essentially affect the form of the RG equation set, any general statement or theorem is of extreme importance in the stability analysis. Below, following Jacobson and Amit (1981), we consider some general properties of the ϕ^4-model Ginzburg-Landau functional in the vicinity of the stability boundary.

6.5.1 Quatric Form in the Instability Region

Let us consider the functional of the form

$$H = \int \frac{d^d\boldsymbol{r}}{2} \left\{ g_1\boldsymbol{\phi}^2 + (\boldsymbol{\nabla}\boldsymbol{\phi})^2 + \frac{1}{4} g_2^{\alpha\beta\gamma\delta}\phi^\alpha\phi^\beta\phi^\gamma\phi^\delta \right\}. \tag{6.67}$$

In the critical region the vertex $\hat{g}_2$ is being renormalized according to Eq. (1.28) (as before we restrict consideration to the first-order of ϵ). Now we are going to show that there is an important property of the functional Eq. (6.67) relating to stability of the system. Namely, if the vertexes ever leave the stability region they will never return to it again. In order to prove this statement, let us introduce the fourth order form by

$$g_2(\boldsymbol{x}) = g_2^{\alpha\beta\gamma\delta} x_\alpha x_\beta x_\gamma x_\delta.$$

With the help of this form one can reduce Eq. (1.28) to

$$\frac{dg_2(\boldsymbol{x})}{dl} = \epsilon g_2(\boldsymbol{x}) - 9\hat{S} g_2^{\alpha\beta\rho\nu} g_2^{\rho\nu\gamma\delta} x_\alpha x_\beta x_\gamma x_\delta.$$

We are interested in the formal analysis of the $dg_2(\boldsymbol{x})/dl$ behavior beyond the stability region. In this region it is informative to use the Schwarz inequality in the form

$$g_2^{\alpha\beta\rho\nu} g_2^{\rho\nu\gamma\delta} x_\alpha x_\beta x_\gamma x_\delta \geq (g_2^{\alpha\beta\gamma\delta} x_\alpha x_\beta x_\gamma x_\delta)^2 \equiv g_2^2(\boldsymbol{x}).$$

From this inequality follows that

$$\frac{dg_2(\boldsymbol{x})}{dl} \leq \epsilon g_2(\boldsymbol{x}) - 9g_2^2(\boldsymbol{x}).$$

From this inequality, one can see that if the form $g_2(\boldsymbol{x})$ ever loses its positiveness, then when $l \to \infty$ its negativeness is infinitely increasing. At the stability boundary ($g_2(\boldsymbol{x}) = 0$) flow lines can only be directed to the region $g_2(\boldsymbol{x}) < 0$ and their return to the stability region is impossible. However, one should also remember that RG equations do not have much sense beyond the stability boundary. In this domain, magnitudes of vertexes are quickly increasing functions of l and we cannot restrict consideration only with the vertex $\hat{g}_2$ due to the indefinitness of the partition function. Therefore, in order to prove that a first-order phase transition corresponds to such a situation, one has to analyze the free energy of the system when the value of the form $g_2(\boldsymbol{x})$ tends to zero ($g_2(\boldsymbol{x}) \to +0$).

6.5.2 Cubic Symmetry

As usual in order to find the free energy of the system one has to carry out a functional integration in

$$VF = -\ln \int D\boldsymbol{\phi} \exp(-H\{\boldsymbol{\phi}\}). \tag{6.68}$$

In the Landau approximation (sometimes called "tree approximation") one can find that $VF = H(\boldsymbol{M})$, where $\boldsymbol{M}$ is defined by the equation

$$\left.\frac{\delta H}{\delta \boldsymbol{\phi}}\right|_{\phi=M} = 0. \tag{6.69}$$

In the vicinity of the stability boundary this approximation is not applicable since $H(\boldsymbol{M})$, as well as $g_2(\boldsymbol{x})$, is no longer positively defined. Let us expand the functional $H\{\boldsymbol{\phi}\}$ in the vicinity $\boldsymbol{\phi} = \boldsymbol{M}$. Then, we have

$$H\{\boldsymbol{\phi}\} \approx H\{\boldsymbol{M}\} + \int \frac{d\boldsymbol{x}d\boldsymbol{y}}{2} \left.\frac{\delta^2 H\{\boldsymbol{\phi}\}}{\delta\phi^\alpha(\boldsymbol{x})\delta\phi^\beta(\boldsymbol{y})}\right|_{\phi=M} \phi^\alpha(\boldsymbol{x})\phi^\beta(\boldsymbol{y}) \tag{6.70}$$

Substituting this expression into Eq. (6.68) one arrives at

$$F = H\{\boldsymbol{M}\} + \frac{1}{2}\ln\det \left.\frac{\delta^2 H\{\boldsymbol{\phi}\}}{\delta\phi^\alpha(\boldsymbol{x})\delta\phi^\beta(\boldsymbol{y})}\right|_{\phi=M} + const. \tag{6.71}$$

Using the Fourier transformation technique and identity $\ln\det\hat{A} = \mathrm{Tr}\ln\hat{A}$, one can find the following expression for the $\boldsymbol{M}$-dependent part of F

$$F = \frac{1}{2}g_1\boldsymbol{M}^2 + \frac{1}{8}g_2^{\alpha\beta\gamma\delta}M^\alpha M^\beta M^\gamma M^\delta$$

$$+\frac{1}{2}\mathrm{Tr}_{\alpha\beta}\int_q \ln\left[(g_1+\boldsymbol{q}^2)\delta_{\alpha\beta} + \frac{3}{2}g_2^{\alpha\beta\gamma\delta}M^\gamma M^\delta\right]. \tag{6.72}$$

Carrying out the integration in Eq. (6.72) one has to cut the integral off at some value Λ. Let us extinguish the divergence in Eq. (6.72) with the help of the regularization procedure used in the field theoretical RG (see Chapter 4). Here we realize this procedure with the help of subtractions. After the subtraction of the most divergent ($\Lambda^d\ln\Lambda$) part, we still have a Λ^{d-2} divergence. The second subtraction eliminating this divergence is not sufficient as at $d=4$ we have a logarithmic contribution $\ln\Lambda$, which should also be eliminated. As a result of the third subtraction we obtain

$$F = \frac{1}{2}g_1\boldsymbol{M}^2 + \frac{1}{8}g_2^{\alpha\beta\gamma\delta}M^\alpha M^\beta M^\gamma M^\delta$$

$$+\frac{1}{2}\mathrm{Tr}_{\alpha\beta}\int_q\left[\ln(\boldsymbol{q}^2\delta_{\alpha\beta}+a_{\alpha\beta}) - \ln\boldsymbol{q}^2\delta_{\alpha\beta} - \frac{a_{\alpha\beta}}{\boldsymbol{q}^2} + \frac{a_{\alpha\phi}a_{\gamma\beta}}{2\boldsymbol{q}^2(\boldsymbol{q}^2+1)}\right], \tag{6.73}$$

where

$$a_{\alpha\beta} = g_1\delta_{\alpha\beta} + \frac{3}{2}g_2^{\alpha\beta\gamma\delta}M^\gamma M^\delta \equiv g_1\delta_{\alpha\beta} + B_{\alpha\beta}. \tag{6.74}$$

Now in Eq. (6.73) the momentum integration is being carried out in the infinite region and we also have to use the renormalized vertexes defined by Eq. (1.28) and Eq. (6.30) instead of the trial vertexes. The expression for F defined by Eq. (6.73) is usually called a "loop approximation". This accounts for the following. The free energy considered as a function of temperature and order parameter can be represented as

$$F = \sum_{N=1}^{\infty}\frac{1}{N!}\Gamma^{(N)}_{\alpha_1\ldots\alpha_N}M^{\alpha_1}\ldots M^{\alpha_N}\Big|_{q_i=0}, \tag{6.75}$$

where $\hat{\Gamma}^{(N)}$ are irreducible vertexes (see Chapter 4). The series Eq. (6.75) can be summed up if one restricts consideration only to the graphs shown in Fig. 6.7. Such a summation was first done in quantum electrodynamics by Coleman and Weinberg in 1973. In the theory of phase transitions this method was applied to the system with cubic anisotropy by Luksutov and Pokrovskii in 1975 (see also Rudnick 1978).

Let us now return to the value of the free energy, Eq. (6.73). After

uM^2, $\mathcal{B}\epsilon^2 M^4$ and the term $\Phi(g_1 + uM^2/2)$ are of the same order and they compensate each other. For large values of g_1 this minimum gives a higher value of the free energy than the minimum $M = 0$. Hence, the state with $M \neq 0$ is metastable. With the decrease of g_1 down to some critical value $g_1 = \tau_c$, the second minimum reaches the energy level of the minimum with $M = 0$. The parameters τ_c and M_1 can be defined from the equations

$$\left.\frac{dF(M)}{dM}\right|_{M=M_1} = 0, \qquad F(M_1) = F(0). \tag{6.82}$$

Equal values of the free energy at $M = 0$ and $M = M_1$ is a condition for the phase equilibrium. The equilibrium curve, resulting from the solution of Eq. (6.82) is usually called binodal and the value M_1 is a value of a discontinuous change of the order parameter on this curve. Substituting into Eq. (6.82) the apparent representation for F (see Eq. (6.81)) one can obtain in the main order in ϵ the following couple of equations

$$\begin{aligned} b + \tfrac{1}{2}\mathcal{B}c + \tfrac{1}{2}(n-1)A^2 c(\ln Ac - 2) &= 0, \\ bc + \tfrac{1}{4}\mathcal{B}c^2 + \tfrac{1}{4}(n-1)A^2 c^2 \left(\ln Ac - \tfrac{5}{2}\right) &= 0, \end{aligned} \tag{6.83}$$

where the values A, b, and c are introduced to evidently extract the value ϵ from g_1, u, and M_1^2:

$$g_1 = b\epsilon, \qquad u = 2A\epsilon, \qquad M_1^2 = c/\epsilon.$$

The solution of the set Eq. (6.83) can be represented as

$$\begin{aligned} b_b &= \tfrac{5A^2}{8}(n-1)c_b, \\ c_b &= A^{-1}\exp\left[\tfrac{-\mathcal{B}+3(n-1)A^2/4}{(n-1)A^2}\right]. \end{aligned}$$

If the phase with $M \neq 0$ has been created (in the stable or metastable region), it cannot vanish unless the minimum in $F(M)$ vanishes.[4] The value of the discontinuous change of the order parameter M_s and the spinodal curve are defined by the equations

$$\left.\frac{d^2F(M)}{dM^2}\right|_{M=M_s} = 0, \qquad \left.\frac{dF(M)}{dM}\right|_{M=M_s} = 0.$$

[4]This is the so-called overheating spinodal or lability boundary for the phase with $M \neq 0$.

In the apparent form, this set can be written as

$$\begin{aligned} b+\tfrac{3}{2}\mathcal{B}c+\tfrac{3}{2}(n-1)A^2c\left(\ln Ac-\tfrac{5}{2}\right)+7\tfrac{(n-1)}{4}A^2c &= 0, \\ b+\tfrac{1}{2}\mathcal{B}c+\tfrac{1}{2}(n-1)A^2c\left(\ln Ac-\tfrac{5}{2}\right)+\tfrac{(n-1)}{4}A^2c &= 0. \end{aligned}$$

The solution of the spinodal set can be represented as

$$\begin{aligned} b_s &= \tfrac{A^2}{2}(n-1)c_s, \\ c_s &= A^{-1}\exp\left[\tfrac{-\mathcal{B}+(n-1)A^2}{(n-1)A^2}\right]. \end{aligned}$$

In conclusion, though we considered only the simplest form of the Ginzburg-Landau functional with cubic symmetry, these calculations are applicable to an essentially more general case. Jacobson and Amit (1981) showed that the general form Eq. (6.67) has positive eigenvalues λ_i. Therefore, if flow lines approach to the stability boundary, a nontrivial solution $M^2 \sim \epsilon^{-1}$ is always present and the free energy has a form typical for the fist order phase transitions.

6.5.3 Interacting Fields

In this case we have to apply the loop approximation to the functional Eq. (6.53) (see Lisyansky and Filippov (1986a)). Now, instead of Eq. (6.71) we arrive at

$$F = H\{\boldsymbol{M}_i\} + \frac{1}{2}\ln\det \left.\frac{\delta^2 H\{\boldsymbol{\phi}_i\}}{\delta\phi_k^{\alpha}(\boldsymbol{x})\delta\phi_j^{\beta}(\boldsymbol{y})}\right|_{\boldsymbol{\phi}_i=\boldsymbol{M}_i} + const. \tag{6.84}$$

Instead of the matrix $a_{\alpha\beta}$ now we have a matrix which also has indexes related to different fields so that the generalization of Eq. (6.74) takes the form

$$\begin{aligned} a_{kj}^{\alpha\beta} &= \tau_k\delta_{kj}\delta_{\alpha\beta} + u_k\delta_{kj}\left(M_k^{\alpha}M_j^{\beta} + \frac{1}{2}\delta_{\alpha\beta M_k^2}\right) \\ &+v\left[M_k^{\alpha}M_j^{\beta}(1-\delta_{kj}) + \frac{1}{2}\delta_{\alpha\beta}\delta_{kj}\sum_{m=1}^{2}M_m^2(1-\delta_{km})\right] \\ w&\left[M_k^{\alpha}M_j^{\beta}(1-\delta_{kj})(1+\delta_{\alpha\beta}) + \delta_{kj}\sum_{m=1}^{2}M_m^{\alpha}M_m^{\beta}(1-\delta_{km})\right]. \end{aligned} \tag{6.85}$$

After diagonalization of the matrix $\hat{a}$ with respect to the indexes α, β and k, j and carrying out the integration using the same subtracting procedure as in Sect. 6.5.2 we arrive at

$$F = \frac{1}{V} H(\boldsymbol{M}_1, \boldsymbol{M}_2) + \sum_{i,\alpha} \Phi(\lambda_i^\alpha),$$

where $\boldsymbol{M}_{1,2}$ stands in Hamiltonian Eq. (6.53) instead of the field operators $\boldsymbol{\phi}_{1,2}$ and the values λ_i^α are eigenvalues of the matrix $\hat{a}$. In the general case the diagonalization procedure is quite cumbersome. Here we consider only three cases at $\tau_1 < \tau_2$:

1. The normal phase of the Landau theory $\boldsymbol{M}_1 \neq 0$, $\boldsymbol{M}_2 = 0$.
2. The anomalous phase at $u_2 \to 0$ with $\boldsymbol{M}_1 = 0$, $\boldsymbol{M}_2 \neq 0$.
3. The anomalous phase at $v \to -2w - \sqrt{u_1 u_2}, w > 0$ with $\boldsymbol{M}_1 \uparrow\uparrow \boldsymbol{M}_2 \neq 0$.

Let us consider all three variants successively.

Normal Phase

In this case, if one chooses one of the coordinate axes along the vector $\boldsymbol{M}_1$, the matrix $\hat{a}$ will take a diagonal form with the eigenvalues

$$\lambda_i^1 = \tau_i + \frac{3}{2} u_i \boldsymbol{M}_i^2 \quad \text{and} \quad \lambda_i^\beta = \tau_i + \frac{1}{2} u_i \boldsymbol{M}_i^2,$$

where the exponent β accepts $n-1$ numbers (from 2 to n). The free energy has the form

$$\begin{aligned} F_N = &\frac{\tau_1 \boldsymbol{M}_1^2}{2} + \frac{1}{8} u_1 (\boldsymbol{M}_1^2)^2 \\ &+\Phi\left(\tau_1 + \frac{3}{2} u_1 \boldsymbol{M}_1^2\right) + (n-1)\Phi\left(\tau_1 + \frac{1}{2} u_1 \boldsymbol{M}_1^2\right) \\ &+ \Phi\left(\tau_2 + \left(w + \frac{1}{2} v\right) \boldsymbol{M}_1^2\right) + (n-1)\Phi\left(\tau_2 + \frac{1}{2} v \boldsymbol{M}_1^2\right). \end{aligned} \tag{6.86}$$

The energy for the paramagnetic and ordered phases are equal when $F_I = 0$. Let us prove that the energy defined by Eq. (6.86) is equal to zero when $M_1^2 = -\tau_1/2u_1$. Such a situation corresponds to the second-order phase transition at $\tau_1 = 0$. In the framework of ϵ-expansion the parameters τ_1 and u_1 (just as well as v, w) are small values of the order of ϵ. Taking into account the explicit expression for the function $\Phi(x)$ (see Eq. (6.77)) one can conclude that the presence of the last four terms in Eq. (6.86) are unessential because they are small as $\epsilon^2 \ln \epsilon$, while the first two terms are of the order of ϵ. As a result, the free energy has the same expression as given by the steepest descent method

$$F_I \approx \frac{\tau_1}{2} \boldsymbol{M}_1^2 + \frac{u_1}{8} (\boldsymbol{M}_1^2)^2,$$

from which the statement about the continuous phase transition becomes evident.

Anomalous Phase I

In this case the free energy can be obtained from Eq. (6.86) by the exchange of indexes $1 \leftrightarrow 2$. However, the fact that $u_2 \to 0$ at the stability boundary leads to a cardinal change in the behavior of the system in comparison with the previously considered case. In the absence of the term $u_2 \boldsymbol{M}_2^2$ the condition $F_2 = 0$ is satisfied only if the functions $\Phi(x)$ give contributions of the same order as $\tau_2 \boldsymbol{M}_2^2$. As the vertexes τ_2, v, and w are of the order of ϵ as before, one can omit in the expression for the free energy terms smaller than $\tau_2 \boldsymbol{M}_2^2$ to arrive at

$$F_{AI} = \frac{\tau_2 \boldsymbol{M}_2^2}{2} + \Phi\left(\left(w + \frac{1}{2}v\right) \boldsymbol{M}_2^2\right) + (n-1)\Phi\left(\frac{1}{2}v \boldsymbol{M}_2^2\right).$$

Now, taking into account evident expression for the function $\Phi(x)$, one can see that the condition $F_{AI} = 0$ can be satisfied with a nontrivial solution $M_2^2 \approx 1/\epsilon$. This solution, as before, corresponds to a discontinuous appearance of the phase with $\boldsymbol{M}_2 \neq 0$. Considering again conditions $F_{AI} = 0$ and $dF_{AI}/dM_2 = 0$ one can obtain the coexistence surface for these two phases (binodal) $\tau_2^b = \tau(u, w)$ and the value of the order parameter M_2^b on this surface. The results of such calculations are written below

$$\tau_2^b = \frac{1}{4}\left[\left(w + \frac{v}{2}\right)^2 + (n-1)\left(\frac{v}{2}\right)^2\right] \exp\left\{\frac{3}{2} - \left[\left(w + \frac{v}{2}\right)^2\right.\right.$$

$$\left.\left. \times \ln\left(w + \frac{v}{2}\right) + (n-1)\left(\frac{v}{2}\right)^2 \ln\frac{v}{2}\right]\left[\left(w + \frac{v}{2}\right)^2 + (n-1)\left(\frac{v}{2}\right)^2\right]^{-1}\right\},$$

$$(M_2^b)^2 = 4\tau_2^b\left[\left(w + \frac{v}{2}\right)^2 + (n-1)\left(\frac{v}{2}\right)^2\right]^{-1}.$$

When $n = 1$ the spinodal equation can be essentially simplified,[5] and can be reduced to

$$\tau_2^b = \frac{1}{8}e^{3/2}v,$$

[5] In this case we also have $w = 0$.

with a jump of the order parameter

$$(M_2^{(b)})^2 = \frac{1}{v} e^{3/2}.$$

This equation gives justification of the initial assumption $(M_2^{(b)})^2 \approx 1/\epsilon$.

In order to find the surface of the absolute instability of the low temperature phase (overheating spinodal) $\tau_2^{(s)} = \tau(v, w)$ and discontinuous change of the order parameter on this surface, one needs to find an inflection point for the function F_{AI}; i.e. to find the second derivative d^2F_{AI}/dM_2^2 and equalize it to zero along with the first derivative. The result of this procedure can be written as

$$\tau_2^{(s)} = 2\tau_2^{(b)} e^{-1/2} \quad \text{and} \quad (M_2^{(s)})^2 = (M_2^{(b)})^2 e^{-1/2}.$$

The phase with $\boldsymbol{M}_2 \neq 0$ occurs at $\tau > 0$. The latter means that even at $\tau_2 > \tau_1$ the first-order phase transition occurs before the second-order transition into the normal Landau phase.[6] Therefore, the flow of phase trajectories with the change of the u_2 sign really corresponds to a discontinuous creation of the anomalous phase.

Anomalous Phase II

In this case ($\boldsymbol{M}_1 \neq 0$, $\boldsymbol{M}_2 \neq 0$,) the crossing of the stability boundary $v = -2w - \sqrt{u_1 u_2}$ occurs when $w > 0$. As a result, the lowest free energy corresponds to collinear vectors $\boldsymbol{M}_1$ and $\boldsymbol{M}_2$. The latter essentially simplifies the procedure of the matrix $\hat{a}$ diagonalization at $u = -v - 2w$

$$\Delta F_{AII} \equiv \sum_{i,\alpha} \Phi(\lambda_i^\alpha) = \Phi\left(\tau + \frac{u}{2}(\boldsymbol{M}_1 + \boldsymbol{M}_2)^2\right) \Phi(\tau) + (n-1)$$

$$\times \left\{ \Phi\left(\tau + \frac{u}{2}\boldsymbol{M}_1^2 + \frac{v}{2}\boldsymbol{M}_2^2\right) + \Phi\left(\tau + \frac{u}{2}\boldsymbol{M}_2^2 + \frac{v}{2}\boldsymbol{M}_1^2\right) \right\},$$

where $\tau = (\tau_1 + \tau_2)/2$. Here again one can neglect τ in the function Φ and reduce ΔF_{AII} to

$$\Delta F_{AII} \approx \Phi(2uM^2) + 2(n-1)\Phi((u+v)M^2/2).$$

In obtaining this expression the following considerations have been made: When $v \to -2w - \sqrt{u_1 u_2}$ the value $x = u_2/u_1$ tends to unity (i.e. $u_{1,2} \approx u$) and also that $|\tau_1 - \tau_2| \ll \tau_{1,2} \approx \tau$ along with $|M_1^2 - M_2^2| \ll M_{1,2}^2 \approx M^2$. The

[6]This consideration is completely applicable only when τ_1 and τ_2 are quite close to each other so that both fields, M_1 and M_2, fluctuate.

binodal equation and discontinuous change of the order parameter $M^{(b)}$ on this curve can be written as

$$\tau^{(b)} = \frac{1}{8} A \exp\left\{\frac{3}{2} - \frac{1}{A}\left[4u^2 \ln 2u + \frac{n-1}{2}(u+v)^2 \ln \frac{u+v}{2}\right]\right\},$$

$$(M^{(b)})^2 = 8\tau^{(b)} A^*,$$

where $A = 4u^2 + (n-1)(u+v)^2/2$. For the spinodal one can obtain

$$\tau^{(s)} = 2e^{-1/2}\tau^{(b)} \quad \text{and} \quad (M^{(s)})^2 = (M^{(b)})^2 e^{-1/2}.$$

As the transition to the phase with $\boldsymbol{M}_1 \neq 0$, $\boldsymbol{M}_2 \neq 0$ occurs when $\tau > 0$ the discussion of the previous case is completely justified for this situation too. Thus, the fluctuation instability also leads in this case to the appearance of the anomalous intermediate phase with $\boldsymbol{M}_1 \neq 0$, $\boldsymbol{M}_2 \neq 0$. This phase is stabilized only in the critical region by strong fluctuations.

Chapter 7

Competing Interactions

In practically every physical system undergoing a phase transition we encounter a situation where the equilibrium can only be reached with the balance of competing forces. For instance, in order to describe the liquid state one has to take into account at least two kinds of interactions: repulsive (usually short range) and attractive forces. Namely, the competition between these two forces defines all properties of a liquid state. The absence of one of these constituents leads to either instability (in the case of only attractive forces) or to the absence of the phase transition in the system.[1] Even when one has a simple situation with an antiferromagnetic ordering, competition in the interactions of magnetizations of different sublattices is the main driving force of the transition. In a quite general case, competitive interactions can be formally treated as a system of at least two interacting fields. Therefore, study of particular systems, in the critical region, with more than one essential interacting field should reveal a number of subtleties specific to competing interactions.

7.1 Heisenberg Magnets

In this section, following the work by Ivanchenko *et al* (1984), we consider Heisenberg magnets with two exchange mechanisms. The Ginzburg-Landau functional for such a system is derived in Chapter 2 (see Sect. 2.2.2). Here we will slightly change notations by multiplicative renormalization of interacting fields. Such renormalization transforms the Hamiltonian to a form convenient for RG analysis. In these notations the Hamiltonian can be

[1]The short range repulsive forces cannot lead to a phase transition in the system.

written as

$$H = \int dr \left\{ \frac{\tau_a}{2}\phi_a^2(r) + \frac{\tau_f}{2}\phi_f^2(r) + \frac{c_a}{2}\left(\nabla\phi_a(r)\right)^2 \right.$$
$$+\frac{c_f}{2}\left(\nabla\phi_f(r)\right)^2 + \frac{\tilde{u}_a}{8}\left(\phi_a^2(r)\right)^2 + \frac{\tilde{u}_f}{8}\left(\phi_f^2(r)\right)^2$$
$$\left. +\frac{\tilde{v}}{4}\phi_a^2(r)\phi_f^2(r) + \frac{\tilde{w}}{2}\left(\phi_a(r)\cdot\phi_f(r)\right)^2 \right\}, \tag{7.1}$$

where $\phi_{a,f}$ are fields related to antiferromagnetic and ferromagnetic ordering. It is necessary to remind ourselves that the Hamiltonian Eq. (7.1) is derived from the traditional microscopic Heisenberg Hamiltonian Eq. (2.20) for which the Fourier transform of the exchange interaction $J(q)$ has two sets of maxima in the Brillouin zone. These sets are formed by the maximum at $q = 0$, which gives ferromagnetic ordering, the set corresponding to the symmetry related maxima $J(q_j)$ at the corners of the Brillouin zone. This latter set by itself would lead to the antiferromagnetic ordering. When the difference in the values of maxima $\Delta = |J(0) - J(q_j)|$ is less than the value of the critical temperature, competition between antiferro and ferromagnetic fields may lead to a very complex picture of phase transitions in the critical region. In order to reveal this picture one has to consider RG behavior of the system.

7.1.1 RG Equations

As usual, the critical thermodynamics of the system is defined by the behavior of the renormalized vertices u_a, u_f, $\tilde{v}$, and $\tilde{w}$ when one approaches the critical temperature T_c. The process of the renormalization of these vertices can be described with the help of the general equation obtained in the first ϵ-order in Chapter 6 (see Eq. (1.31)). In this case, the tensor $\hat{g}_{20}$ can be represented as

$$g_{20,s_1\ldots s_4}^{\alpha_1\ldots\alpha_4} = \frac{\tilde{u}_s}{3}(\delta_{\alpha_1\alpha_2}\delta_{\alpha_1\alpha_2}\delta_{\alpha_1\alpha_4} + 2\,permutations)$$
$$+\frac{\tilde{v}}{3}(\delta_{\alpha_1\alpha_2}\delta_{\alpha_3\alpha_4}\delta_{s_1s_2}\delta_{s_3s_4}(1-\delta_{s_1s_3}) + 2\,permutations)$$
$$+\frac{\tilde{w}}{3}(\delta_{\alpha_1\alpha_2}\delta_{\alpha_3\alpha_4}(1-\delta_{s_1s_2})(1-\delta_{s_3s_4}) + 2\,permutations), \tag{7.2}$$

where the indexes s_i accept notations a, f. Changing the scale of the field operators $\phi_{a,f} \to c_{a,f}\phi_{a,f}$ one can introduce notations

$$v = \frac{\tilde{v}}{c_a c_f}, \quad w = \frac{\tilde{w}}{c_a c_f}, \quad u_a = \frac{\tilde{u}_a}{c_a^2}, \quad u_f = \frac{\tilde{u}_f}{c_f^2},$$

which after the substitution of Eq. (7.2) into Eq. (1.31) finally leads to the set

$$\begin{aligned}
\frac{du_{a,f}}{dl} &= \epsilon u_{a,f} - \left\{ \frac{n+8}{2} u_{a,f}^2 + \frac{nv^2}{2} + 2vw + 2w^2 \right\}, \\
\frac{dv}{dl} &= \epsilon v - \left\{ (u_a + u_f) \left[\frac{(n+2)v}{2} + w \right] + 2v^2 + 2w^2 \right\}, \qquad (7.3) \\
\frac{dw}{dl} &= w \left\{ \epsilon - \left[u_a + u_f + 4v + (n+2)w \right] \right\}.
\end{aligned}$$

Now we have to find all fixed points for this set in order to describe the evolution of the renormalized vertices, which in turn gives all possible phase diagrams for this system.

7.1.2 Fixed Points

The set Eq. (7.3) has the fixed points shown in the table below. Besides,

	u_a^*	u_f^*	v^*	w^*
μ_0^*	0	0	0	0
μ_1^*	$2\epsilon/(n+8)$	0	0	0
μ_2^*	0	$2\epsilon/(n+8)$	0	0
μ_3^*	$2\epsilon/(n+8)$	$2\epsilon/(n+8)$	0	0
μ_4^*	$\epsilon/(n+4)$	$\epsilon/(n+4)$	$\epsilon/(n+4)$	0
μ_5^*	$\epsilon n/(n^2+8)$	$\epsilon n/(n^2+8)$	$\epsilon(4-n)/(n^2+8)$	0
μ_6^*	$2\epsilon/(n^2+8)$	$2\epsilon/(n^2+8)$	$2\epsilon/(n^2+8)$	$\epsilon(n-2)/(n^2+8)$
μ_7^*	$\epsilon/(n+8)$	$\epsilon/(n+8)$	$\epsilon/(n+8)$	$\epsilon/(n+8)$

Table 7.1: Fixed points for $n \neq 1, 2$.

for $n = 2$ an additional point exists

$$u_f^* = u_a^* = \frac{\epsilon}{10}, \qquad v^* = \frac{3\epsilon}{10}, \qquad w^* = -\frac{\epsilon}{10}.$$

All fixed points given by Table 7.1 are unstable ($n \neq 1$). The case $n = 1$ is a particular case, because the invariant $(\phi_a \phi_f)^2$ is now reduced to $\phi_a^2 \phi_f^2$ and formally $\tilde{w} \to 0$. The points μ_6^* — μ_8^* are absent and the point μ_4^* is transformed into an attractive center.[2]

A simple look over Table 7.1 tells us that all fixed points (except the trivial Heisenberg points μ_1^* and μ_2^*) have equal values of the parameters

[2]In this case the set Eq. (7.3) is reduced to the set Eq. (6.54).

u_a^* and u_f^*. The latter is a consequence of a symmetry of the set Eq. (7.3) with respect to the exchange $u_a \leftrightarrow u_f$. Let us restrict the following deliberation to the hyperplane $u_a = u_f = u$ and consider the quatric form of the Hamiltonian Eq. (7.1)

$$H_I = \frac{1}{4}\int dr \left\{ u \left[\frac{(\phi_a^2)^2 + (\phi_f^2)^2}{2} \right] + v\phi_a^2\phi_f^2 + 2w(\phi_a\phi_f)^2 \right\}. \tag{7.4}$$

If we make the substitutions

$$\phi_a = \frac{1}{\sqrt{2}}(\phi_1 - \phi_2), \qquad \phi_f = \frac{1}{\sqrt{2}}(\phi_1 + \phi_2)$$

in Eq. (7.1), then the Hamiltonian is transformed to

$$H_I = \frac{1}{4}\int dr \left\{ u_1 \left[\frac{(\phi_1^2)^2 + (\phi_2^2)^2}{2} \right] + v_1\phi_1^2\phi_2^2 + 2w_1(\phi_1\phi_2)^2 \right\}, \tag{7.5}$$

where the new interaction parameters u_1, v_1, w_1 are introduced. They are related to the old parameters by

$$u_1 = \frac{u+v}{2} + w, \quad v_1 = \frac{u+v}{2} - w, \quad w_1 = \frac{u-v}{2} \tag{7.6}$$

One can also check that the set Eq. (7.6) is invariant under the exchange transformation $u_1 \leftrightarrow u$, $v_1 \leftrightarrow v$, $w_1 \leftrightarrow w$. The latter is a natural consequence of the existence of the inverse transformation. The set Eq. (7.6) constitutes an additional symmetry present in the system of the fixed points. The point μ_3^* under the transformation Eq. (7.5) turns into the point μ_7^* and in this sense these two points are complementary. Likewise, the point μ_5^* is complementary to the point μ_6^*. The point μ_4^* is an invariant point of this transformation in the sense that it transforms into itself. The above symmetry can be revealed if one changes the variables in Eq. (7.3) according to

$$u + v = y, \qquad u - v = x, \qquad w = z/2.$$

As a result the set Eq. (7.3) has the form explicitly revealing the symmetry $x \leftrightarrow z$, i.e.

$$\begin{aligned} \frac{dx}{dl} &= \epsilon x - x\left[\tfrac{n+2}{2}x + 3y - z\right], \\ \frac{dz}{dl} &= \epsilon z - z\left[\tfrac{n+2}{2}z + 3y - x\right], \\ \frac{dy}{dl} &= \epsilon y - \left[x^2 + z^2 + (z+x)y + \tfrac{n+4}{2}y^2\right]. \end{aligned}$$

7.1.3 Flow Lines and Phase Diagrams

An instructive analysis of the set Eq. (7.3) can be made at $n = 1$, when the Hamiltonian Eq. (7.1) lacks terms of the type $(\phi_a \cdot \phi_f)^2$. Formally this situation corresponds to the case $w = 0$. However, one should be aware that the trial value $\tilde{v}$ is three times as larger than in the case with $n \neq 1$. This accounts for the following transformation $\tilde{v} + 2\tilde{w} \to \tilde{v}$. Indeed, the invariants $(\phi_a \phi_f)^2$ and $\phi_a^2 \phi_f^2$ are now indistinguishable and are summed up as well as the equations for $2d\tilde{w}/dl$ and $d\tilde{v}/dl$. One can easily verify that the sum of these equations leads to the equation for the new vertex $\tilde{v}$. When $\tilde{w}$ is absent, the points μ_6^*, $\mu_{8\pm}^*$ are also absent. The point μ_4^* becomes a stable fixed point. It is also necessary to mention that in the linear ϵ-approximation the point μ_5^*, excepting the negative eigenvalues, has a zero eigenvalue. However, stricter analysis in higher ϵ orders shows that this point is a saddle point.

When $n = 1$, instead of the set Eq. (7.3), owing to the reduction of the number of vertices in the functional Eq. (7.1), we have

$$\begin{aligned} \frac{du_{a,f}}{dl} &= \epsilon u_{a,f} - \tfrac{n+8}{2} u_{a,f}^2 - \tfrac{n}{2} v^2, \\ \frac{dv}{dl} &= v \left[\epsilon - \tfrac{n+2}{2}(u_a + u_f) - 2v \right]. \end{aligned} \tag{7.7}$$

Though this set is simplified in comparison with the set Eq. (7.3), nonetheless it is still quite difficult for analysis. Additional simplification can be made using the fact that the structure of the ordered phase is defined not by the values of the vertices $u_{a,f}$ and v but rather by their ratios. Therefore, one can rewrite the set Eq. (7.6) in terms of $x = u_f/v$ and $y = u_a/v$

$$\begin{aligned} \frac{1}{v}\frac{dx}{dl} &= \tfrac{1}{2}(-6x^2 + 3xy + 4x - 1), \\ \frac{1}{v}\frac{dy}{dl} &= \tfrac{1}{2}(-6y^2 + 3xy + 4y - 1), \end{aligned} \tag{7.8}$$

conserving the symmetry with respect to the replacement $x \leftrightarrow y$. The set Eq. (7.8) has only two fixed points

$$\mu_a^* : \quad x = y = \frac{1}{3} \quad \text{and} \quad \mu_b^* : \quad x = y = 1.$$

One of them, i.e. μ_b^* is a stable fixed point.

The flow line picture for the set Eq. (7.8) can be qualitatively understood even though one does not know the exact solution of this set. This

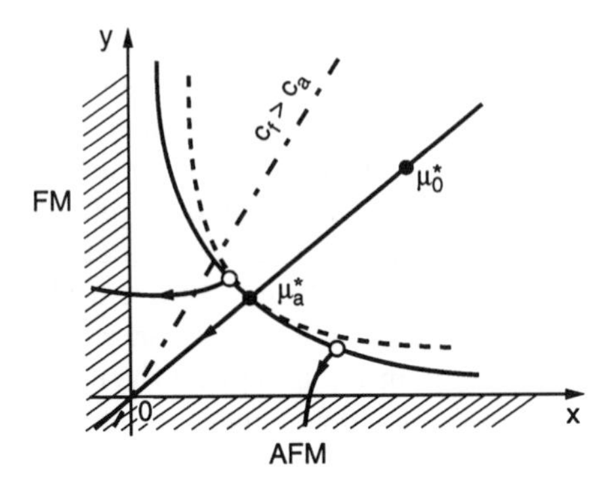

Figure 7.1: The phase portrait of RG equations on the (x, y) plane. The dashed line is the separatrix, the bold line is the hyperbola $xy = 1/9$, the circle is an initial point at $c_a > c_f$.

accounts for the following two facts. First — one of the integrals of the set can be easily found, it is $y = x$. Second — from the uniqueness of the solution the requirement follows that flow lines never cross the line $x = y$ except at the particular points $\mu^*_{a,b}$. At last, one can also find the isoclines for verticals $dx/dl = 0$ and horizontals $dy/dl = 0$, which are parabolas in this case. As a result, one can obtain the picture of flow lines as shown in Fig. 7.1 and in Fig. 7.2. The attraction region for low lines to the stable fixed point is restricted by a line (separatrix) which is shown in this figure as a dashed line. In the case where not all lines can reach the stable fixed point, it is very important to know the region where the trial parameters of the Hamiltonian Eq. (7.1) are located. As the values of these parameters are obtained from microscopic deliberations (see Chapter 2), one sees that on the plane (x, y) these parameters can only be found on the hyperbola $xy = 1/9$. A numerical analysis shows that the separatrix, for the family of critical surfaces (the region of attraction) of the stable fixed point μ^*_b, touches the hyperbola of initial conditions at the single point μ^*_a. Except for this point, all other flow lines, originating at the points of the hyperbola $xy = 1/9$, are leaving the stability region of the Hamiltonian Eq. (7.1) being unstable with respect to one of the parameters u_a or u_f. Thus, for any values of the initial parameters, excepting the point μ^*_a, the effective Hamiltonian is unstable and a second-order phase transition to the ordered

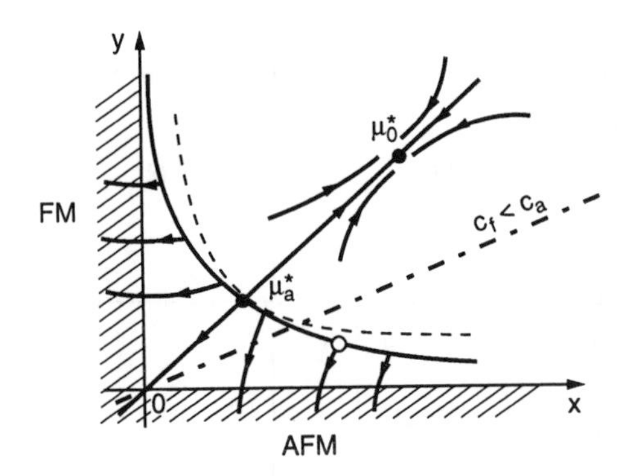

Figure 7.2: The same as in Fig. 7.1 but in the case when $c_a < c_f$. The circles are initial points corresponding to different variants of the initial parameters x_0, y_0.

state is impossible. As we proved in Chapter 6, in this situation one should expect the first-order transition. The structure of the phase for this discontinuous transition is defined by the vertex (u_a, u_f, or both) which leads to the instability of the Hamiltonian. Namely, when u_a changes sign, the field ϕ_a is going to order, while for u_f we will have ferromagnetic ordering. The straight line $u_a = u_f$ below the point μ_a^* is a set of repulsive points for flow lines. Hence, flow lines will never cross this line. The values of the trial parameters u_a and u_f are uniquely defined in terms of the values J_a and J_f (see Chapter 2). If we initially choose $J(0) > J(\boldsymbol{q}_{0i})$, then $u_a < u_f$, so that the initial values x and y along the curve $xy = 1/9$ are positioned on the right hand side of the coexistence phase line $u_a c_a^2 = u_f c_f^2$. As the line $yc_a^2 = xc_f^2$ ($u_a c_a^2 = u_f c_f^2$), depending on the value of the ratio c_a/c_f, may be found either above or below the bisectrice $x = y$, one can consider for a real phase trajectory $(x(l), y(l))$ two qualitatively different situations. In the first case, when we have inequality $u_a c_a^2 > u_f c_f^2$, flow lines begin below the line $x = y$ and leave the stability region changing the y-sign. This case may take place at any value of the ratio c_a/c_f (as we have $u_a < u_f$). The second case is defined by the inequality $u_a c_a^2 < u_f c_f^2$ and can be realized only when $c_a < c_f$.

Let us consider each of the these situations. For the first situation flow lines begin in the region of parameters which in the mean field approxi-

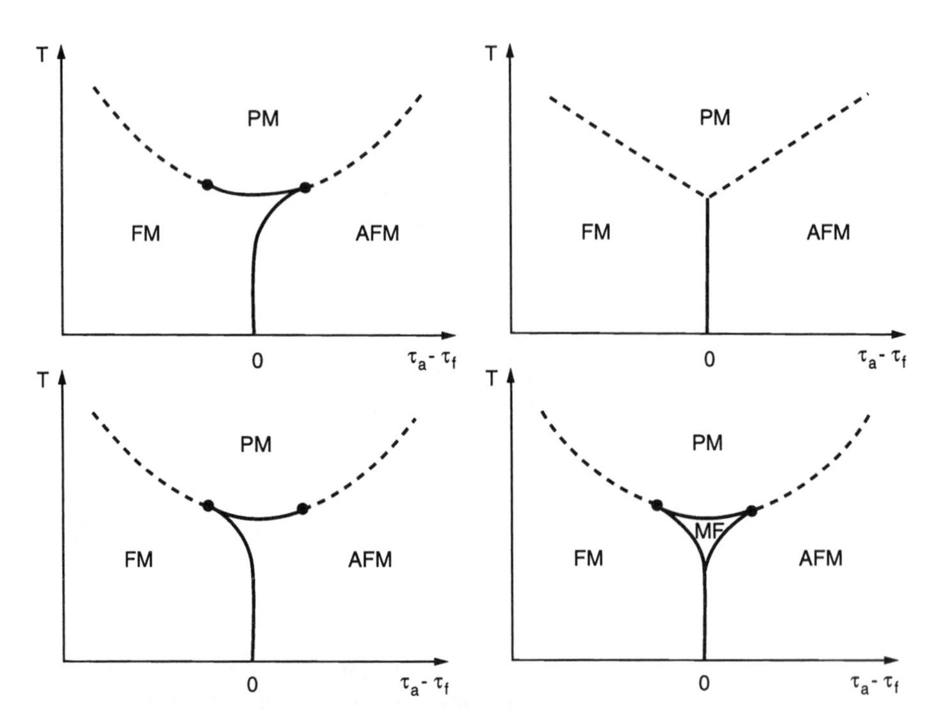

Figure 7.3: Different kinds of phase diagrams (MP — mixed phase).

mation correspond to the ferromagnetic (FM) phase. As the stability is lost for the antiferromagnetic part of the Hamiltonian, RG analysis leads to the antiferromagnetic (AF) phase (see Fig. 7.1 and Fig. 7.2). This case corresponds to the AF state stabilized by strong ferromagnetic fluctuations. The other situation, when flow lines begin in the region of the mean field FM phase and end up with the discontinuous transition to this phase, is not drastically different in the fluctuation region. The influence of antiferromagnetic fluctuations now leads only to the transformation of the continuous transition into the discontinuous one without any sign of the phase transition inversion. Thus, instead of the mean field phase diagram shown in Fig. 7.3 (a) in the variables $T = (\tau_a + \tau_f)/2$, $\tau_a - \tau_f$, we can find three other diagrams as shown in Fig. 7.3 (b,c,d).

If $n \neq 1$ the behavior is somewhat different. As the stable fixed point is absent, all flow lines are leaving the stability region of the Hamiltonian Eq. (7.1). In this case (as for the case $n = 1$) the number of RG equations can be diminished by one by the introduction of ratios $x = u_f/w$,

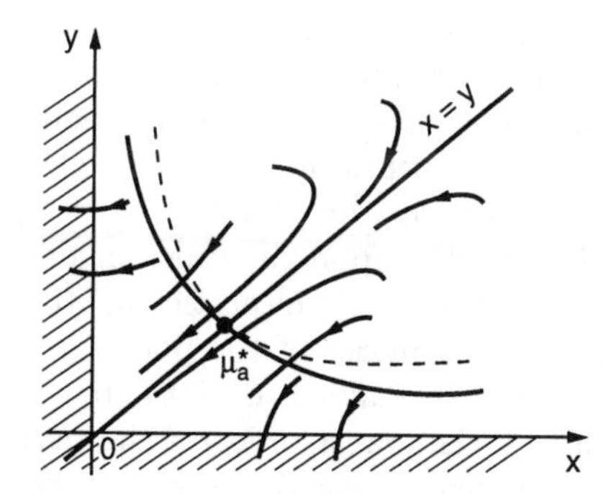

Figure 7.4: The phase portrait on the plane (x, y) at $n \neq 1$. The dashed line is a separatrix surface demarcating different kinds of flow line motion to the stability boundary.

$y = u_a/w$, and $z = v/w$. The RG set for these parameters can be written as

$$\begin{aligned}
\frac{1}{w}\frac{dx}{dl} &= -\frac{n+6}{2}x^2 + xy + 4xz - \frac{nz^2}{2} + (n+2)x - 2z - 2, \\
\frac{1}{w}\frac{dy}{dl} &= -\frac{n+6}{2}y^2 + xy + 4yz - \frac{nz^2}{2} + (n+2)y - 2z - 2, \qquad (7.9) \\
\frac{1}{w}\frac{dz}{dl} &= -\frac{n}{2}z(x+y) + 2z^2 + (n+2)z - (x+y) - 2.
\end{aligned}$$

The fixed points $\mu_0^*, \ldots, \mu_5^*$ are absent in the space (x, y, z). The points μ_6^* and μ_7^* are transformed into $\mu_I^* = (1, 1, 1)$ and $\mu_{II}^* = (2/(n-2), 2/(n-2), 2/(n-2))$. The set of the initial parameters is now defined by the hyperbola $xy = 1$ positioned on the plane $z = 1$. The surface separating two phases is the plane $xc_f^2 = yc_a^2$.

Let us consider a projection of the phase portrait for the set Eq. (7.9) on the plane (x, y). On this plane, the picture of flow lines is similar to the consequent picture for the case $n = 1$. As in the latter case, we have two classes of trajectories demarcated by the separatrix surface (see Fig. 7.4). All flow lines not positioned on the plane $x = y$ repel from this plane and leave the regions of stability for one or the other of the fields (antiferromagnetic or ferromagnetic). The flow lines originating on the plane do not leave this plane until they reach the instability boundary $x = 0,\ y = 0$.

The flow lines all originate on the hyperbola $xy = 1$ on the plane $z = 1$ and they leave it moving down $(dz/dl|_{xy=1,z=1} < 0)$ and off the planes $x = 0,\ y = 0$ (stability boundaries) without crossing the planes $x = y$ and

$z = 0$. Hence, the above described qualitative analysis of different physical situations for $n = 1$ can be completely transferred to the case $n \neq 1$. The flow lines leaving the immediate vicinity of the saddle point μ_I^*, at which the set of trial parameters touches the separatrix surface, are of particular interest. It is necessary to mention that the strict equalities $xy = 1$ and $z = 1$ result partially from the approximations made under the derivation of the Ginzburg-Landau functional. In particular, the integration over the unessential modes, in the vicinities of the points $\boldsymbol{q} = 0$ and $\boldsymbol{q} = \boldsymbol{q}_j$ (see Sect. 2.2.2) corresponding to maxima of $J(\boldsymbol{q})$, renormalizes the trial parameters equally only in the lowest order of the perturbation theory. In the following orders this renormalization is explicitly dependent on $\boldsymbol{q}$, i.e. dependent on which of the maxima $J(\boldsymbol{q} = 0)$ of $J(\boldsymbol{q} = \boldsymbol{q}_j)$ is being renormalized. As a consequence the constraints $xy = 1$ and $z = 1$ change. The latter leads to a deformation of the set of initial parameters and as a result, the overlap region with the separatrix surface changes (may be increased or even vanish at all). In the case when this region is increased we have a set of trial parameters from which flow lines may either $(n = 1)$ end up at the fixed point μ_b^* or $(n > 1)$, moving to the negative values of v, leave the stability region of the Hamiltonian. In the first case we have the second-order phase transition; while in the second the first-order phase transition is into a mixed phase with $\langle \boldsymbol{\phi}_a \rangle \neq 0$ and $\langle \boldsymbol{\phi}_f \rangle \neq 0$. Naturally, the mixed state is only stable in the critical temperature region and out of this region it changes, by means of the continuous phase transition, into a "clean" phase. This leads to the phase diagram shown in Fig. 7.3 (d). In experiments the dependence $J(\boldsymbol{q})$ can be changed with the application of an external pressure or by adding impurities to the material. Therefore, the diagrams shown in Fig. 7.3 (b,c,d) are very sensitive to external parameters of the experiments.

In conclusion, with the decrease of temperature, when the inequality $J(0)c_a > J(\boldsymbol{q}_j)c_f$ holds, the transition from the paramagnetic to the antiferromagnetic phase is a first-order transition. This latter occurs at positive values of "masses" τ_a and τ_f where both fields are fluctuating. The AF phase, unstable from the point of view of the mean-field theory, is stabilized only by the critical fluctuations. Hence, out of the fluctuation region the FM phase is to be established (in accordance with the Landau theory). The transition from the AF to FM state is also a first-order phase transition. As a result, the state diagram has specific "bird beaks" composed by the lines of discontinuous phase transitions. In the general case the function $J(\boldsymbol{q})$ may have a number of different local maxima. A reciprocal inversion of these maxima, resulting from the fluctuation renormalization, may lead to a series of first-order transitions from one ordered phase to another. The above mechanism has a direct relation to the "order-to-order" transitions

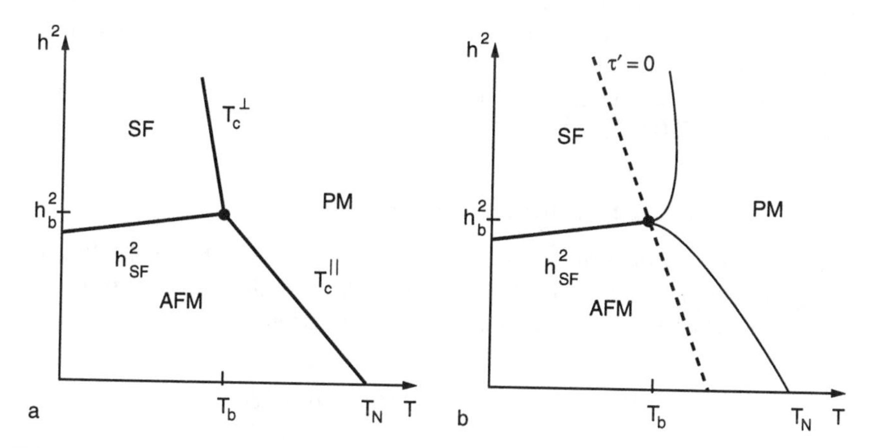

Figure 7.5: The phase diagram of an easy-axis antiferromagnet: (a) mean field approximation; (b) fluctuation theory.

in materials where the exchange inversion is not caused by the variation of lattice parameters and an applied hydrostatic pressure does not change the sequence of transitions.

7.2 Bicritical Points in Antiferromagnets

In this section we consider the critical behavior of anisotropic antiferromagnets (AFM). For the sake of simplicity the consideration is restricted to easy-axis AFM's subjected to the influence of an external magnetic field $\boldsymbol{h}$ in the direction of the easy-axis. Such systems were quite thoroughly considered theoretically and experimentally (see, for instance, Shapira (1984)). A typical phase diagram for these systems, obtained in the mean field approximation, is shown in Fig. 7.5 (a). As one can see from this figure there are three different phases: the PM phase, which is a paramagnetic phase at $\boldsymbol{h} = 0$ (when $\boldsymbol{h} \neq 0$ this is a homogeneously magnetized phase); the SF phase, which is the so-called "spin-flop" phase with equal magnitudes of magnetization for different sublattices having equal angles with the applied field. The transitions PM – AF and PM – SF are second-order phase transitions but the transition AF – SF is a first-order phase transition. All three phase transition lines start from one point $(T_b,\ h_b^2)$, which according to the number of crossing lines is called the bicritical point. It is necessary to mention that the mean field approximation gives linear behavior, i.e.

$T \sim h^2$, for the second-order lines in the vicinity of the bicritical point. The influence of fluctuations leads to a violation of this condition along with specific dependence of the critical temperature on the value of applied field.[3] A detailed RG analysis in the vicinity of the bicritical point in the easy-axis AFM's was made by Fisher and Nelson (1974), Fisher (1975), and Kosterlitz *et al* (1976).

7.2.1 Ginzburg-Landau Functional

The Ginzburg-Landau functional for the system can be obtained following the general scheme developed in Chapter 2. We should start from the Heisenberg Hamiltonian with anisotropic interaction

$$H = -\frac{1}{2}\sum_{ll'}\sum_{\alpha=1}^{n} J_{ll'}^{\alpha} S_l^{\alpha} S_{l'}^{\alpha} - \boldsymbol{h}\sum_{l}\boldsymbol{S}_l. \tag{7.10}$$

Repeating the steps of the derivation for the isotropic case (see Sect. 2.2.2) one arrives at

$$Z \sim \int D\phi^{\alpha} \exp\left\{\frac{T}{2}\sum_{ll'}\sum_{\alpha=1}^{n}[J_{ll'}^{\alpha}]^{-1}\phi_l^{\alpha}\phi_{l'}^{\alpha} + \sum_{l}\ln[\chi_l^{1-n/2} I_{n/2-1}(\chi_l)]\right\}, \tag{7.11}$$

where components of the vector $\boldsymbol{\chi}_l$ are defined by

$$\boldsymbol{\chi}_l = \boldsymbol{\phi}_l - \boldsymbol{h}.$$

It is very important that for the AFM systems, the maxima of the Fourier transform J_q^{α} are positioned at the boundaries of the Brillouin zone ($\boldsymbol{q} = \boldsymbol{q}_{oi}$), while the Fourier transformant of the homogeneous field $\boldsymbol{h}$ has the only one component different from zero with $\boldsymbol{q} = 0$ ($\boldsymbol{h}_{q\neq o} = 0$). This leads to the Ginzburg-Landau functional for the fluctuating field $\boldsymbol{\phi}_q$ in the form which incorporates $\boldsymbol{h}$ as a parameter, influencing values of the trial vertices and not as an external field conjugated with the order parameter. As we have restricted consideration to easy-axis AFM's with the field $\boldsymbol{h}$ aligned along the easy-axis, it is convenient to separate the vector $\boldsymbol{\phi}$ into longitudinal and transverse components denoted below as $\boldsymbol{\sigma}$ and $\boldsymbol{S}$. Let us, as usual, divide small vicinities of the J_q^{α} maxima (essential modes) and expand all integrals to the $\boldsymbol{q}^2$ orders in these vicinities. We will mark the values $\boldsymbol{\sigma}_q$ and $\boldsymbol{S}_q$ which have the wave numbers belonging to these vicinities with the subscript 1 and the other modes with the subscript 2. Let us also expand

[3] This dependence was quite thoroughly experimentally studied (see Shapira (1984)).

the exponential term in Eq. (7.11) up to the fourth-order terms. Then the functional can be written as

$$\overline{\mathcal{F}} = \frac{1}{2}\int d\boldsymbol{r}\left\{(\nabla\boldsymbol{\sigma}_1)^2 + \overline{\tau}_1^{\parallel}\boldsymbol{\sigma}_1^2 + (\nabla\boldsymbol{S}_1)^2 + \overline{\tau}_1^{\perp}\boldsymbol{S}_1^2 + \overline{\tau}_2^{\parallel}\boldsymbol{\sigma}_2^2\right.$$

$$+ \overline{\tau}_2^{\perp}\boldsymbol{S}_2^2 - 2\boldsymbol{h}\boldsymbol{\sigma}_2 + \frac{1}{4}\sum_{i\leq j}^{2} u_{ij}\boldsymbol{\sigma}_i^2\boldsymbol{\sigma}_j^2 + \frac{1}{4}\sum_{i\leq j}^{2} v_{ij}\boldsymbol{S}_i^2\boldsymbol{S}_j^2$$

$$\left. + \frac{1}{4}\overline{v}_{12}(\boldsymbol{S}_1\boldsymbol{S}_2)^2 + \frac{1}{4}\sum_{i\leq j}^{2} w_{ij}\boldsymbol{\sigma}_i^2\boldsymbol{S}_j^2 + \frac{1}{4}\overline{w}_{12}(\boldsymbol{\sigma}_1\boldsymbol{\sigma}_2)(\boldsymbol{S}_1\boldsymbol{S}_2)\right\}. \tag{7.12}$$

The coefficients before the gradient terms in Eq. (7.12) are fixed at one with the help of the standard normalization replacement. The next step should be integration in the partition function over the unessential modes $\boldsymbol{\sigma}_2$ and $\boldsymbol{S}_2$. However, it is first convenient to eliminate the term linear with respect to $\boldsymbol{h}$. The latter can be done with the help of the replacement $\boldsymbol{\sigma}_2 \to \boldsymbol{\sigma}_2 + \boldsymbol{M}$ with the choice of $\boldsymbol{M}$ defined from the equation

$$\overline{\tau}_2^{\parallel}\boldsymbol{M} + \frac{1}{2}u_{22}\boldsymbol{M}M^2 = \boldsymbol{h}.$$

This leads to the renormalization of the vertices $\overline{\tau}_{1,2}^{\parallel}$ and $\overline{\tau}_{1,2}^{\perp}$ resulting in

$$\begin{aligned} \tau_1^{\parallel} &= \overline{\tau}_1^{\parallel} + \tfrac{1}{4}u_{12}M^2, & \tau_2^{\parallel} &= \overline{\tau}_2^{\parallel} + \tfrac{1}{4}u_{22}M^2, \\ \tau_1^{\perp} &= \overline{\tau}_1^{\perp} + \tfrac{1}{4}w_{12}M^2, & \tau_2^{\perp} &= \overline{\tau}_2^{\perp} + \tfrac{1}{4}w_{22}M^2, \end{aligned}$$

and to the appearance of a correction to the functional Eq. (7.12) ($\mathcal{F} = \overline{\mathcal{F}} + \Delta\mathcal{F}$) linear with respect to $\boldsymbol{M}$

$$\Delta\mathcal{F} = \frac{\boldsymbol{M}}{8}\int d\boldsymbol{r}\left[2u_{12}\boldsymbol{\sigma}_2\boldsymbol{\sigma}_1^2 + 4u_{22}\boldsymbol{\sigma}_2\boldsymbol{\sigma}_2^2 + 2w_{12}\boldsymbol{\sigma}_2(\boldsymbol{S}_1\boldsymbol{S}_2)\right.$$

$$\left. + \overline{w}_{12}\boldsymbol{\sigma}_1(\boldsymbol{S}_1\boldsymbol{S}_2) + w_{12}\boldsymbol{\sigma}_2(\boldsymbol{S}_1\boldsymbol{S}_2)\right]. \tag{7.13}$$

Not all the terms in the functional Eq. (7.12) and Eq. (7.13) are equally important in the critical region. Taking into account that the values $\overline{\tau}_2^{\parallel} \sim |T/J_q^{\parallel} - 1|$ and $\overline{\tau}_2^{\perp} \sim |T/J_q^{\perp} - 1|$ are quite large ($\boldsymbol{q}$ values are far from $\boldsymbol{q}_{oi}$), one can omit the terms containing powers of $\boldsymbol{\sigma}_2$ and $\boldsymbol{S}_2$ higher then two and as a result use the functional

$$\mathcal{F} \approx \frac{1}{2}\int d\boldsymbol{r}\left\{(\nabla\boldsymbol{\sigma}_1)^2 + \tau_1^{\|}\boldsymbol{\sigma}_1^2 + (\nabla\boldsymbol{S}_1)^2 + \tau_1^{\perp}\boldsymbol{S}_1^2 + \tau_2^{\|}\boldsymbol{\sigma}_2^2\right.$$

$$+\tau_2^{\perp}\boldsymbol{S}_2^2 + \frac{1}{4}u_{11}(\boldsymbol{\sigma}_1^2)^2 + \frac{1}{4}v_{11}(\boldsymbol{S}_1^2)^2 + \frac{1}{4}w_{11}\boldsymbol{\sigma}_1^2\boldsymbol{S}_1^2$$

$$\left.+u_{12}\boldsymbol{M}\boldsymbol{\sigma}_2\boldsymbol{\sigma}_1^2 + \frac{1}{4}w_{12}\boldsymbol{M}\boldsymbol{\sigma}_2\boldsymbol{S}_1^2 + \frac{1}{4}\boldsymbol{M}\boldsymbol{\sigma}_1(\boldsymbol{S}_1\boldsymbol{S}_2)\right\}.$$

The integration over the fields $\boldsymbol{\sigma}_2$ and $\boldsymbol{S}_2$ in the partition function Z renormalizes the vertices u_{11}, v_{11}, and w_{11} leading in the lowest over $1/\tau_2^{\|}$ order to

$$u_1 = u_{11} - u_{12}^2 M^2/\tau_2^{\|}, \qquad u_2 = v_{11} - w_{12}^2 M^2/4\tau_2^{\|},$$

$$v = w_{11}/2 - w_{12}u_{12}M^2/2\tau_2^{\|} - \overline{w}_{12}^2 M^2/32\tau_2^{\|}.$$

As a result, the RG Hamiltonian for the system is reduced to

$$H = \frac{1}{2}\int d\boldsymbol{r}\left\{(\nabla\boldsymbol{\sigma}_1)^2 + \tau^{\|}\boldsymbol{\sigma}_1^2 + (\nabla\boldsymbol{S}_1)^2 + \tau^{\perp}\boldsymbol{S}_1^2\right.$$

$$\left.+\ \frac{1}{4}u_1(\boldsymbol{\sigma}_1^2)^2 + \frac{1}{2}v\boldsymbol{\sigma}_1^2\boldsymbol{S}_1^2 + \frac{1}{4}(\boldsymbol{S}_1^2)^2\right\}. \tag{7.14}$$

Comparing this functional with the Hamiltonian Eq. (6.53) one can see that it represents a particular case of the latter. Accounting for this fact, below we will consider it in the notations accepted in Eq. (6.53). However, in this case the vertices τ_i are dependent on the value $\boldsymbol{h}$ (through M^2). The latter governs the system along the $(\tau_1 - \tau_2)$-axis allowing to move it in the vicinity of the bicritical point.

7.2.2 RG Analysis

Let us, following Kosterlitz *et al* (1976), analyze behavior and the phase diagram transformation of the system in the critical region. In Chapter 6 we have shown that critical fluctuations lead to the transformation of continuous phase transitions into discontinuous ones. However, even in the case when the type of a transition is not affected, the phase diagram can attain distortions of a principal character, which allow one to test experimentally the fluctuation theory predictions. These distortions are related to the fact that transition temperatures are renormalized differently for various values $\tau_1 - \tau_2$. The closer the system is to the bicritical point, the stronger the renormalization is. As a result, the lines of continuous phase transitions are distinctly bent to the direction of the bicritical point, forming the so-called umbilical curve. The asymptotic behavior of this line ($\tau_1 - \tau_2 \to 0$) can be obtained in an analytical form. In order to find this behavior one has to introduce a "field" θ, breaking symmetry of the Gaussian part of the Hamiltonian Eq. (6.53)

$$H_0 = \frac{1}{2} \int dr \left\{ \tau(\phi_1^2 + \phi_2^2) - \frac{\theta}{n}(n_1\phi_1^2 - n_2\phi_2^2) \right\}.$$

Now, a translation in the system along the axis $\tau_1 - \tau_2$, under the influence of field $\boldsymbol{h}$ corresponds to a change of the parameter θ. When $\theta = 0$, the full functional has fixed points considered in Sect. 6.3.4. By linearizing the RG equations in the vicinity of the stable fixed point one can obtain the (positive) eigenvalue λ_1 defining the correlation length critical exponent $\nu = 1/\lambda_1$. When $\theta \neq 0$, the Gaussian form becomes anisotropic which leads to the appearance of additional $n-1$ positive eigenvalues λ_α. These eigenvalues describe the motion of the system from the stable fixed point.[4] In order to find eigenvalues λ_α for the system of arbitrary symmetry it is convenient to make substitution $g_1^{\alpha\alpha} = \tau^{\alpha\alpha} - 3g_2^{\gamma\gamma\alpha\alpha}/2$ in the set Eq. (1.31). The value $\tau^{\alpha\alpha} \sim \epsilon^2$ so that in the first ϵ-order $g_1^{\alpha\alpha} \approx -3g_2^{\gamma\gamma\alpha\alpha}/2$. Taking this into account we have in the lowest order

$$\begin{aligned} \frac{d\tau^{\alpha\alpha}}{dl} &= 2\tau^{\alpha\alpha} - 3\epsilon g_2^{\gamma\gamma\alpha\alpha} - \frac{9}{2} g_2^{\gamma\gamma\beta\alpha} g_2^{\bar{\gamma}\bar{\gamma}\beta\alpha} - 6g_2^{\gamma\gamma\alpha\alpha}\tau^{\gamma\gamma}, \\ \frac{dg_2^{\alpha_1\cdots\alpha_4}}{dl} &= \epsilon g_2^{\alpha_1\cdots\alpha_4} - 3\left[g_2^{\beta\gamma\alpha_1\alpha_2} g_2^{\beta\gamma\alpha_3\alpha_4} + \text{permutations}\right]. \end{aligned}$$

As the second equation does not contain $\tau^{\alpha\alpha}$, it is quite sufficient to use only the first one in order to find eigenvalues λ_α. This equation leads to

$$\lambda_\alpha c_\alpha = 2c_a - c_\gamma g_2^{\gamma\gamma\alpha\alpha},$$

where c_α are components of an eigenvector $\boldsymbol{c}$. When $n \leq 3$, the isotropic fixed point is a stable one. The tensor $g_2^{\alpha\beta\gamma\delta}$ reduces at this point to

$$g_2^{\alpha\beta\gamma\delta} = \frac{u^*}{3}[\delta_{\alpha\beta}\delta_{\gamma\delta} + \delta_{\alpha\gamma}\delta_{\beta\delta} + \delta_{\alpha\delta}\delta_{\beta\gamma}],$$

where $u^* = 1/(n+8)$. In this case for λ_a we have the following equation

$$\lambda_a c_\alpha = (2 - 4u^*)c_\alpha - u^* \sum_\gamma c_\gamma.$$

[4] In principle, each of n directions of ϕ^α may correspond to its own eigenvalue λ_α.

The largest eigenvalue λ_1, in the first ϵ-approximation, is equal to

$$\lambda_1 = 2 - \frac{n+2}{n+8}\epsilon.$$

The eigenvector $\boldsymbol{c}$ corresponding to this eigenvalue has equal components c_a. Excepting this eigenvalue, we have also $n-1$ degenerate eigenvalues $\lambda_2 = 1 - 2u^*$, corresponding to orthogonal vectors, components of which satisfy the equality $\sum_\gamma c_\gamma = 0$. The critical exponent $\nu_2 = 1/\lambda_2$ describes the correlation length ξ behavior at $\tau = 0$ and $\theta \to 0$. When both values τ and θ are finite the correlation length satisfies the Fisher and Pfeuty (1972) scaling relation

$$\xi(\tau, \theta) = \tau^{-\nu} X(\theta \tau^{-\phi}).$$

Here X is an analytical function at $\theta \to 0$. The exponent $\phi = \nu/\nu_2$ is the so-called crossover exponent evidently defining the shape of the curve $\theta(\tau)$ when $\theta \to 0$ and $\tau \to 0$. For the isotropic fixed point one has

$$\phi = \frac{\nu}{\nu_2} \approx 1 + \frac{\epsilon n}{2(n+8)}.$$

More accurate calculations in higher orders of the perturbation theory (see Fisher and Nelson (1974) and Pfeuty *et al* (1974)) give $\phi(n=3) \approx 1.25$, and $\phi(n=2) \approx 1.175$.

Let us recall that for the functional under consideration the isotropic fixed point is stable only when $n < 4$. For $n \geq 4$ this point is replaced by the biconical stable point. In the first ϵ approximation the eigenvalues $\lambda_{b_{1,2}}$ were found by Kosterlitz *et al* (1976). They equal

$$\lambda_{b_1} = 2 + \left\{ -\frac{3}{8}u^* - \frac{n+1}{8}u_\alpha^* + \left[\left(\frac{3}{4}u_1^* - \frac{n+1}{4}u_2^* \right)^2 + (n-1)v^{*2} \right]^{1/2} \right\},$$

$$\lambda_{b_2} = 2 + \left\{ -\frac{3}{8}u^* - \frac{n+1}{4}u_\alpha^* - \left[\left(\frac{3}{4}u_1^* - \frac{n+1}{4}u_2^* \right)^2 + (n-1)v^{*2} \right]^{1/2} \right\}.$$

The value of the crossover exponent $\phi(n)$ for the biconical point can be obtained by means of numerical calculations. As a result, in the first approximation in ϵ we have

$$\phi(4) \approx 1.1667, \qquad \phi(5) \approx 1.1551.$$

Thus, with the motion away from the bicritical point ($\tau = 0$, $\theta = 0$) the lines of phase transitions on the plane (τ, θ) should have the shape characterized by the constant ratio $\theta/\tau^{\phi} \equiv w$. It is necessary to mention that the shape of the curve $\theta/\tau^{\phi} = w$ anticipated in the theory can be found only for the scaling fields θ and τ. These values do not coincide with the physically measured values $\tau_0 = (T-T_b)/T_b$ and h. They are equivalent only when we have zero coordinates for the bicritical point $\tau_0 = 0,\ h_b = 0$. The latter is possible only at infinitesimally small anisotropy,[5] when $h_b \to 0$. In the general case τ_0 and $\Delta h = h - h_b$ must be mixed to produce genuine scaling fields τ and θ.

Taking into account that the values τ and θ are dependent on h^2 (by means of M^2 dependencies in $\tau^{\parallel}$ and $\tau^{\perp}$) one can write in the vicinity of the bi-critical point (T_b, h_b)

$$\theta \approx h^2 - h_b^2 = 2h_b(h - h_b),$$

which means that the scaling variables can be found in terms of new fields τ' and θ'. In a quite general case, the variables τ' and θ' can be expressed with the help of equations $\tau' = \tau + q\theta,\ \theta' = \theta - p\tau$. The factors q and p are to be defined from the experimental phase diagram. In order to do this one can use the fact that the tangent of the spin-flop line $h_{\phi}^2(T)$ corresponds to the tangent of the axis of the zero "anisotropy field" θ, i.e. $\theta = 0$. From this fact it follows immediately that $p = T_b(dh_{\phi}^2/dT)_b$. The factor q requires more careful consideration. It was calculated by Fisher in 1975. He found that in the lowest ϵ-order for the case $n_1 = n - 1,\ n_2 = 1$ and $n < 4$, q is equal

$$q(n) = \frac{n+2}{3n} q_1 + \epsilon\theta, \tag{7.15}$$

where the parameter $q_1 = -(dT_c^{\parallel}/dh^2)|_{h=0} T_b^{-1}$ can be defined from the slope of the experimental curve along the phase transition line when $h \to 0$.

In conclusion we will discuss two more facts, important from the experimental point of view. The first: out of the immediate vicinity of the bicritical point, one of the fields is not fluctuating. As a result, the critical behavior is to be described either by the Ising model $n_{\parallel} = 1$ or by the model $n_{\perp} = n - 1$ with the consequent critical exponents.[6] The second fact is that the motion away from the bicritical point in either of the directions (to the Ising model $\theta = \theta_{\parallel}$ or in the other case $\theta = \theta_{\perp}$) is described by similar scaling relations $\theta_{\parallel}/\tau^{\phi} = w_{\parallel}$ and $\theta_{\perp}/\tau^{\phi} = w_{\perp}$. The difference here is related to the different number of components of the fluctuating field along the various sides of the bicritical point. This is manifested in the different values of the constants $w_{\parallel}$ and $w_{\perp}$. One can also show that the ratio of these constants is a universal value $Q(n_1, n_2) \equiv w_{\perp}/w_{\parallel}$. This

[5]This corresponds to the so-called degenerate bicritical point.

[6]When $n = 3$ this is the so-called XY-model.

value can be calculated and compared with experiment (see below). In the lowest ϵ approximation Q can be easily evaluated. One needs to use in this approximation only analysis based on the Gaussian Hamiltonian H_0. This Hamiltonian changes sign from one of the sides of the bicritical point where $\tau^{\|} = n_1\theta/n$, while from the other side we have $\tau^{\perp} = -n_2\theta/n$. Taking into account that $\phi = 1 + o(\epsilon)$ one can obtain the quite explicit relation

$$Q(n_1, n_2) = \frac{\tau^{\|}}{\tau^{\perp}} = \frac{w_{\perp}}{w_{\|}} = \frac{n_1}{n_2} + o(\epsilon),$$

or, when $n_1 = n - 1$ and $n_2 = 1$, $Q(n) = (n-1) + o(\epsilon)$, $Q(2) = 1$. As was shown by Fisher (1975) the obtained expression is valid even at higher ϵ-orders

$$[Q(n_1, n_2)]^{1/\phi(\epsilon)} = \frac{n_1}{n_2} + o(\epsilon^3),$$

and for the most interesting case, $d = 3$, $Q(2,1) \approx 2^{5/4} \approx 2.378$.

7.2.3 Experimental Data

Experimental research of antiferromagnets with weak anisotropy has more than three decades of history. Before the appearance of Kosterlitz *et al* (1976), most experimental results were interpreted in the framework of the mean field approximation. The theory by Kosterlitz *et al* (1975) created a new wave of experimental studies directed to recovering the role of fluctuations in the vicinity of the bicritical point. Soon after that it was shown that the second-order lines in the vicinity of the bicritical points were not straight lines (in the coordinates T, h^2), as it followed from the Landau theory, but rather formed umbilici anticipated by the fluctuation theory (see Fig 7.5 (b)). In previous studies umbilici, for the transition from the paramagnetic to an ordered phase, were not found. This owes to the fact that the region for such curves is, as a rule, quite small and their observation requires precise experiments specially designed for these purposes.

In order to find an umbilicus one should know that it exists. Below, we list some of the materials in which considered fluctuation effects were quite persuasively found. Essentially more detailed data can be found in the review by Shapira (1984).

In the first set of studied materials shown in the Table 7.2 we should include two component easy-axis antiferromagnets: $GdAlO_3$, $NiCl_6 \cdot 6H_2O$ and $CsMnBr_3 \cdot 2D_2O$. Two of them $GdAlO_3$ and $CsMnBr_3 \cdot 2D_2O$ have orthorhombic lattices, while $NiCl_6 \cdot 6H_2O$ has monoclinic lattice. The values

Material	T_N, *K*	$\phi_{\rm exp}$	$Q_{\rm exp}$
$GdAlO_3$	3.878	1.17 ± 0.02	0.9 ± 0.2
$NiCl_6 \cdot 6H_2O$	5.34	1.29 ± 0.07	1.06 ± 0.22
$CsMnBr_3 \cdot 2D_2O$	6.30	1.22 ± 0.06	-

Table 7.2: The values T_N, ϕ and Q for two-component antiferromagnets.

of the Neel's temperatures and the universal parameters ϕ and $Q = w_\perp/w_\parallel$ for these materials are shown in the Table 7.3.[7] For $CsMnBr_3 \cdot 2D_2O$ the value $Q_{\rm exp}$ is not shown in the table, because the phase diagram was fit to obtain the theoretical value $Q = 1$. In this fitting, instead of Q, the value of q was defined (see Eq. (7.15)). The value q was essentially different from the Fisher's estimate, which could be related to the quasi one-dimensional nature of the material. The Fisher estimate is based on the mean field approximation, which should be violated for such materials. Nonetheless, the critical exponents for $CsMnBr \cdot 2D_2O$, in the close vicinity of the transition line should demonstrate a crossover to the three-dimensional behavior. The latter accounts for ϕ being close to the theoretically anticipated value. Typical experimental curves are very much like the curve shown in Fig. 7.5 (b).

The second set of materials with the bicritical umbilici include three-component antiferromagnets MnF_2 and Cr_2O_3. The first of these materials has a tetragonal symmetry with the easy-axis parallel to the *c*-axis, while the second is a tetragonal crystal. The comparison of the theoretical and experimental values ϕ and Q for this set is given in the Table 7.4. From this table one can see that there are essential discrepancies with Fisher's estimate. The latter can be ascribed to some indefiniteness (in the determination of q) in the procedure of the correction of RG results by the mean field estimates. Nonetheless, the fact that the umbilical curves (predicted by the fluctuation theory) were found experimentally, constituted by itself an important role of fluctuations in the vicinity of the bicritical point.

The last set of materials includes antiferromagnets which are practi-

[7] All these materials ($n = 2$) have $\phi_{\rm theor} = 1.175$ and $Q_{\rm theor} = 1.0$.

Material	T_N, *K*	ϕ_{theor}	$\phi_{\rm exp}$	Q_{theor}	$Q_{\rm exp}$
MnF_2	67.3	1.25	1.279 ± 0.031	2.34 ± 0.08	1.56 ± 0.35
Cr_2O_3	307.3	1.25	1.22 ± 0.15	2.51	-

Table 7.3: The values T_N, ϕ and Q for three-component antiferromagnets

Material	n	T_N, K	ϕ_{theor}	$\phi_{\exp}$
$RbMnF_3$	3	83	1.25	1.278 ± 0.026
$KNiF_3$	3	246.4	1.25	1.274 ± 0.045
$CsMnF_3$	2	51.4	1.175	1.185 ± 0.03

Table 7.4: The values T_N and ϕ for antiferromagnets with degenerate bi-critical point.

cally isotropic. Two of them, $RbMnF_3$ and KN_iF_3, are cubic crystals with very small ratios the field of cubic anisotropy h_a to the exchange field h_e. The first one has this ratio less than 2×10^{-5} and the other — 7×10^{-7}, which shows that they are quite isotropic magnetic materials. $CsMnF_3$ is an easy-plane hexagonal antiferromagnetic. It has $h_a/h_e < 10^{-5}$, so that when the external magnetic field is in the easy plane, the bi-critical point is practically degenerate (as well as in $RbMnF_3$ and KN_iF_3). In Table 7.4 the data T_N and $\phi(n)$ are given for these materials.

In conclusion it is necessary to mention experimental studies on bicritical points in uniaxially pressured cubic crystals $LaAlO_3$ and $SrTiO_3$ (see Bruse and Cowley (1981)). The experimental estimates for the exponent ϕ are 1.31 ± 0.07 in $LaAlO_3$ and 1.27 ± 0.06 in $SrTiO_3$. These data are fairly well in agreement with the theoretical value $\phi = 1.25$.

7.3 Dipole Interaction

As has already been shown, the cubic anisotropy of real crystals may essentially influence the phase diagram of these materials. This is especially true, when this changes the phase diagram owing to the fluctuational instability of the second-order phase transition (see Sect. 6.3.3). Unfortunately, the isotropic second-order form of the functional cannot always be considered as a quite reliable approximation. In particular, this form cannot be applied to systems with essentially long-ranged dipole forces. Such interactions reduce the degeneracy of systems with respect to longitudinal and transverse long-wave length fluctuations. Such types of interactions in the RG approach were first considered by Aharony and Fisher in (1973).

7.3.1 Dipole Hamiltonian

In order to take into account dipole interactions, one has to add to the isotropic Heisenberg Hamiltonian

$$H_0 = -\frac{1}{2}\sum_{ll'} J_{ll'} \boldsymbol{S}_l \cdot \boldsymbol{S}_{l'}, \tag{7.16}$$

the Hamiltonian with the structure

$$H_{dd} = -\frac{1}{2}(g_s\mu_B)^2 \sum_{l \neq l'} \frac{\partial^2}{\partial l^\alpha \partial l^\beta}(|\boldsymbol{l}-\boldsymbol{l}'|^{2-d} S_l^\alpha S_{l'}^\beta), \tag{7.17}$$

where μ_B is the Bohr's magneton, and g_s is Lande's factor. It is necessary to notice that the scalar product $(\boldsymbol{lS})$ in Eq. (7.17) is only defined when $n = d$ (i.e. in the three-dimensional case $n = 3$). The derivation of the Ginzburg-Landau functional from the Hamiltonian $H_0 + H_{dd}$ is practically the same as for the isotropic Hamiltonian H_0. The difference appears only in the structure of expansion of the Fourier transform $J^{\alpha\beta}(\boldsymbol{q})$ which now becomes more complicated

$$J^{\alpha\beta}(\boldsymbol{q}) = J^{\alpha\beta}(0) - c\boldsymbol{q}^2 - (\Delta^2 + h\boldsymbol{q}^2)\frac{q_\alpha q_\beta}{\boldsymbol{q}^2} - f q_\alpha q_\beta \tag{7.18}$$

This expansion, naturally, leads to a more complicated zero-order Green's function $G_0^{\alpha\beta}(\boldsymbol{q})$ and, as a consequence, to more complicated RG equations. Difficulties in the solution of RG equations compelled Aharony and Fisher (1973) to restrict their considerations to the limit of the small anisotropy parameter f. Linear expansions, with respect to f, were not considered as a satisfactory approximation, and in 1979 Sokolov and Tagantsev found an approximation which allowed reduction of the number of RG equations yet preserved in them the arbitrary value of the anisotropy parameter f. They applied this approximation to the case of cubic ferroelectrics. The essence of this approximation is the following: All the factors at $\boldsymbol{q}^2$ and $q_\alpha q_\beta$ in the bilinear form of the functional are renormalized to the order of the exponent η which is small both theoretically ($\eta \sim \epsilon^2$) and experimentally ($\eta \approx 0.07$). Therefore in the lowest ϵ-approximation they do not change. It looks like this approximation is well justified within the accuracy of contemporary experiments. It is also important that the value of the dipole gap Δ in ferroelectrics is large and has the same order of value as the Debye's momentum. A renormalization $\delta\Delta$ is of the order of the renormalization of the trial value τ and, in the critical region, this value is small in comparison with Δ. This means that Δ can also be treated as a constant. As a result in the free energy functional for a cubic crystal with the dipole interaction

$$H = \frac{1}{2} \sum_{\alpha,\beta=1}^{3} \left\{ \int_q \left[(\tau + \boldsymbol{q}^2 + f q_\alpha^2)\delta_{\alpha\beta} + (\Delta + h\boldsymbol{q}^2) n_\alpha n_\beta \right] \right.$$

$$\times \phi^\alpha(\boldsymbol{q})\phi^\beta(-\boldsymbol{q}) + (u + v\delta_{\alpha\beta})$$

$$\left. \times \int_{q_1,\cdots,q_4} \delta(\boldsymbol{q}_1 + \cdots + \boldsymbol{q}_4)\phi^\alpha(\boldsymbol{q}_1)\phi^\alpha(\boldsymbol{q}_2)\phi^\beta(\boldsymbol{q}_3)\phi^\beta(\boldsymbol{q}_4) \right\} \tag{7.19}$$

($n_\alpha = q_\alpha/q$) there are only two renormalizable values left. They are the fourth-order vertices u and v. The function $G_0^{\alpha\beta}$ for the functional Eq. (7.19) is defined as an inverse with respect to

$$(G_0^{\alpha\beta})^{-1} = (\boldsymbol{q}^2 + f q_\alpha^2)\delta_{\alpha\beta} + (\Delta^2 + h\boldsymbol{q}^2) n_\alpha n_\beta$$

function and it can be written as

$$G_0^{\alpha\beta} = \frac{\delta_{\alpha\beta}}{\boldsymbol{q}^2 + f q_\alpha^2} - \frac{(\Delta^2 + h\boldsymbol{q}^2) n_\alpha n_\beta}{(\boldsymbol{q}^2 + f q_\alpha^2)(\boldsymbol{q}^2 + f q_\beta^2)}$$

$$\times \left[1 + (\Delta^2 + h\boldsymbol{q}^2) \sum_{\gamma=1}^{3} \frac{n_\gamma^2}{\boldsymbol{q}^2 + f q_\gamma^2} \right]^{-1} . \tag{7.20}$$

As the value Δ is quite large, in the critical region we can formally put $\Delta \to \infty$. In this limit Eq. (7.20) transforms to

$$G_0^{\alpha\beta} = \frac{\delta_{\alpha\beta}}{\boldsymbol{q}^2 + f q_\alpha^2} - \frac{n_\alpha n_\beta}{(\boldsymbol{q}^2 + f q_\alpha^2)(\boldsymbol{q}^2 + f q_\beta^2)} \left[\sum_{\gamma=1}^{3} \frac{n_\gamma^2}{\boldsymbol{q}^2 + f q_\gamma^2} \right]^{-1} . \tag{7.21}$$

7.3.2 RG Analysis

Using Eq. (7.21) and the general RG equations obtained in the loop approximation (see Eq. (1.31)), one can obtain RG equations for the vertices u and v. The combinatorial factors in these equations can be easily calculated owing to the simplification arising from$n = d = 3$.[8] As a result one arrives at

$$\begin{aligned} \frac{du}{dl} &= u - \tfrac{1}{2}(7J_1 + 22J_2 + 4J_3)u^2 - 3(J_1 + 4J_2)uv - \tfrac{9}{2}J_2 v^2, \\ \frac{dv}{dl} &= v - 2(J_1 - J_3 - 2J_2)u^2 - 6(J_1 - J_2)uv - \tfrac{9}{2}(J_1 - J_2)v^2. \end{aligned} \tag{7.22}$$

[8]The dipole interaction invariant can only exist when the dimension of space is equal to the number of components of the order parameter ϕ.

Due to the special form of the function $G_0^{\alpha\beta}(\boldsymbol{q})$ from all of the integrals

$$K_{\alpha\beta\gamma\delta} = \Lambda \frac{\partial}{\partial \Lambda} \int_q G_0^{\alpha\beta}(\boldsymbol{q}) G_0^{\gamma\delta}(\boldsymbol{q})$$

only the following three integrals have nonzero values

$$J_1 = K_{\alpha\alpha\alpha\alpha}, \qquad J_2 = K_{\alpha\beta\alpha\beta}, \qquad J_3 = K_{\alpha\alpha\beta\beta}. \tag{7.23}$$

In the general case the integrals, Eq. (7.23) are very cumbersome, but a number of interesting results can be obtained without detailed knowledge of their numerical structures.

7.3.3 Cubic Symmetry

First of all, for the isotropic function $G_0^{\alpha\beta}$ the integral J_2 is equal to zero and the other two coincide. The set Eq. (7.22) in this limit can be reduced to

$$\frac{du}{dl} = u(1 - \frac{11}{2}u - 3v), \qquad \frac{dv}{dl} = v(1 - 6u - \frac{9}{2}v),$$

which at $n = 3$ coincides with the set Eq. (6.45). However, in this case we have $J_2 \neq 0$. If one restricts consideration with $f = 0$ (i.e. in the case when a crystal is isotropic), then the parameters J_1, J_2, and J_3 are related to each other with the help of the equations

$$J_1 = 8J_2, \qquad J_3 = J_1 - 2J_2.$$

Having in mind these relations, let us introduce parameters ξ and σ defined by

$$\xi = \frac{1}{9}\left(\frac{J_1}{J_2} - 8\right) \quad \text{and} \quad \sigma = \frac{4}{9}\left(\frac{J_1 - J_2}{J_2} - 2\right).$$

When $f = 0$, the values ξ and σ are evidently equal zero. Next, as has already been mentioned, the structure of the ordered phase is defined not by the vertices u and v but rather by their ratio $z = v/u$. From the set Eq. (7.22) one can obtain a single equation

$$\frac{4}{J_2 u}\frac{dz}{dl} = z^3 + (1 - 3\xi)z^2 + (2 - \xi - \sigma)z - \xi. \tag{7.24}$$

One can verify that the straight line $u = 0$ is not a phase trajectory of the RG set. Even when the trial value $u = 0$, this vertex is generated by the term proportional to v^2. As a result all fixed points are roots of the polynomial

$$z^3 + (1 - 3\xi)z^2 + (2 - \xi - \sigma)z - \xi = 0. \tag{7.25}$$

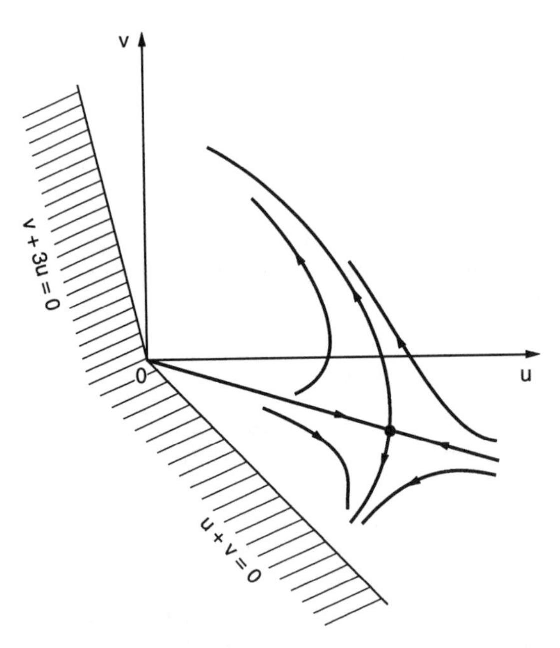

Figure 7.6: Phase portrait for the system with dipole interaction and cubic anisotropy. The shaded area denotes the region beyond the stability boundaries $u + v = 0$ and $v + 3u = 0$.

In the isotropic limit ($f = 0$) Eq. (7.25) reduces to

$$z(z^2 + z + 2) = 0$$

and it has a single real root $z = v/u = 0$. On the (u, v) plane this root corresponds to the Heisenberg straight line $v = 0$, which has a single (saddle) fixed point. In the general case, $v \neq 0$ and this line does not coincide with the Heisenberg line. The numerical solution by Sokolov and Tagantsev (1979) shows, that for physically reasonable f values ($-1 < f \leq 20$), there is only one saddle fixed point. The separatrix (see Fig 7.6) crossing this point has a tangent defined by the f sign: when $f < 0$, the angle is higher than 45°, while for $f > 0$ it is less than 45°. In any case, when trial vertices satisfy inequality $v_0 \neq (1 - z_0)u/z_0$, flow lines leave the stability region for the Ginzburg-Landau functional crossing the boundaries defined by the straight lines $u + v = 0$ and $v + nu = v + 3u = 0$. The straight lines $v = (1 - z_0)u/z_0$ and $v = 0$ do not coincide when $f \neq 0$. The latter means

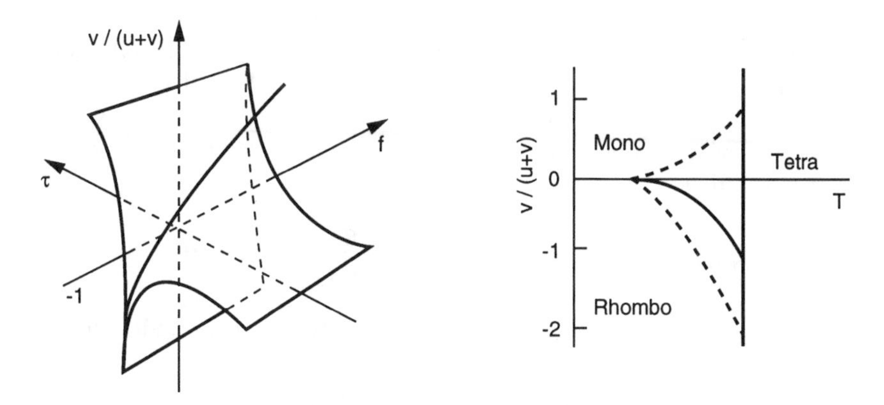

Figure 7.7: The phase diagram of a tetragonal (a) and cubic (b) ferroelectric.

that one can always find such initial values u_0, v_0 starting from which flow lines will cross the Heisenberg line $v = 0$. This line demarcates tetragonal ($v < 0$) and rhombohedral ($v > 0$) phases obtained in the Landau approximation. As a result a fluctuationally induced first-order transition occurs into the phase unrealizable in the mean field theory. The new phase is stabilized by critical fluctuations. Therefore, the second-order phase transition is split on two first-order transitions. The phase diagram in this case (for instance the P—T plane) has a typical "bird beak" constituted by the first-order transition lines. The value of prolongation of this beak and its direction are governed by the sign and value of the parameter of crystal anisotropy f. Such a phase diagram is shown in coordinates $(v/(u+v), \tau, f)$ in Fig 7.7 (b).

7.3.4 Tetragonal Symmetry

In tetragonal crystals of the easy-plane kind with dipole interaction, the situation to some extent reminds us of the easy-axes crystals considered above. Other than for the physical interest, this problem is also attractive for consideration owing to the existence of an analytical solution for the RG set (see Sokolov (1983)). The Ginzburg-Landau functional can also be represented in the critical region by Eq. (7.19). However, in this case the exponents α and β can accept only two values. When $f = 0$, the functional Eq. (7.19) does not change its structure under the rotation in (α, β) plane on 45° excepting change in the vertices u and v. This symmetry property essentially simplifies analytical consideration of the model. It helps to reduce consideration to only one of two possible ordered phases (rhombic or monoclinic). At the presence of the dipole forces ($f \neq 0$) the combinations like $\boldsymbol{q}\boldsymbol{\phi}$ do not conserve under the influence of the independent rotations of the variable $\boldsymbol{\phi}$. The full transformation, influencing along with $\boldsymbol{\phi}$ two components of the vector $\boldsymbol{q}$, but conserving the Hamiltonian Eq. (7.19) can be represented as

$$\boldsymbol{\phi}_{1,2} = \frac{\boldsymbol{\phi}_1 \pm \boldsymbol{\phi}_2}{\sqrt{2}}, \qquad \boldsymbol{q}_{1,2} = \frac{\boldsymbol{q}_1 \pm \boldsymbol{q}_2}{\sqrt{2(1+f)}}, \qquad \boldsymbol{q}_3 \to \boldsymbol{q}_3.$$

The vertices of the Hamiltonian Eq. (7.19) are transformed under the influence of this transformation as follows

$$f \to -\frac{f}{1+f}, \qquad h \to \frac{h+2f}{1+f}, \qquad \tau \to \tau,$$
$$\Delta \to \Delta, \qquad v \to -v, \qquad u \to \frac{3v+2u}{2}. \tag{7.26}$$

In the one-loop approximation the RG set for the two renormalizable vertices u and v reads as follows

$$\begin{aligned} \tfrac{du}{dl} &= u - 3(3-2K)(u^2+uv) - \tfrac{9}{2}(1-K)v^2, \\ \tfrac{dv}{dl} &= v - \tfrac{3}{2}(2K-1)(3v^2+4uv) - 2(4K-3)u^2, \end{aligned} \tag{7.27}$$

where $K = 1+(1-\sqrt{1+f})/2f$. From the set Eq. (7.27) one can easily obtain all fixed points of the system. Except for the trivial fixed point $u^* = 0$ and $v^* = 0$ we have only one additional fixed point with the coordinates

$$u^* = \frac{1-K}{12(4K^2-6K+3)}, \qquad v^* = \frac{4K-3}{36(4K^2-6K+3)}. \tag{7.28}$$

The symmetry discussed above should also be reflected in this point. The substitution $f \to -\bar{f}/(1+\bar{f})$ transforms the value K according to $K = 3/2 - \bar{K}$. It can be easily verified that

$$\begin{aligned} v^* &= (3-4\bar{K})/36(4\bar{K}^2-6\bar{K}+3) &= -\bar{v}^*, \\ (3v^*+2u^*)/2 &= 1-\bar{K}/36(4\bar{K}^2-6\bar{K}+3) &= \bar{u}^*. \end{aligned}$$

These relations evidently reflect the symmetry of the transformation Eq. (7.26).

The set Eq. (7.27) can be exactly integrated. Let us consider flow lines on the (u, v) plane. Thus we have to solve the equation

$$\frac{du}{dv} = \frac{2u - 6(3-2K)(u^2+uv) - 9(1-K)v^2}{2v - 3(2K-1)(3v^2+4uv) - 4(4K-3)u^2}.$$

The substitution $u = v\nu(v)$ transforms this equation into a linear one

$$[\rho(\nu) - \nu f(\nu)]\frac{dv}{d\nu} = vf(\nu) + 2,$$

where the functions $\rho(\nu)$ and $f(\nu)$ are defined by the relations

$$\begin{aligned} \rho(\nu) &= -6(3-2K)(\nu^2+\nu) - 9(1-K), \\ f(\nu) &= -4(4K-3)\nu^2 - 3(2K-1)(4\nu+3). \end{aligned}$$

The equations obtained above can be quite easily integrated with the help of the following factorization

$$\rho(\nu) - \nu f(\nu) = [3(\nu+1)^2 + 1][\nu(4K-1) + 3(K-1)] \equiv a(\nu)b(\nu).$$

As a result flow lines can be represented in the following parametric form

$$\begin{aligned} v(\nu) &= \tfrac{1}{b^3(\nu)}\left\{c_0 + \nu\left(4K^2 - 3K + \tfrac{3}{2}\right) + 5K^2 - 3K + \tfrac{3}{4}\right. \\ &\quad \left. + \tfrac{a(\nu)(3-6K+4K^2}{\sqrt{3}}\tan^{-1}\left[\tfrac{3+4\nu}{\sqrt{3}}\right]\right\}, \\ u(\nu) &= \nu v(\nu). \end{aligned}$$

Though the solution looks quite cumbersome, the flow line picture is simple. Topologically it is equivalent to the one shown in Fig. 7.6. As one can see from this figure, at any values of the initial vertices u and v the effective Hamiltonian is unstable. The separatrix, demarcating two different kinds of flows through the stability boundaries, crosses the saddle point Eq. (7.28) and is analytically defined by the line $v = -u(3-4K)/3(1-K)$. The stability regions of rhombic and monoclinic phases ($v > 0$ and $v < 0$) are separated by the Heisenberg line $v = 0$. It is important that for any value f (and hence practically for any K) the straight line $v = 0$ and the separatrix line never coincide. Only when $f \to 0$ the value $K \to 3/4$ these two lines approach each other. When $f > 0$, the separatrix is below the line $v = 0$ (this case is shown in Fig. 7.6). In the case $f < 0$, the separatrix is above the line $v = 0$. Thus one can always find such trial values u and v for which flow line will cross the phase equilibrium line $v = 0$. The latter means that the unfavorable phase, from the point of view of the Landau theory, is to be established through the first-order phase transition. For instance, a crystal with the initial anisotropy of constants corresponding to rhombic phase will be ordered in the monoclinic phase and *vice versa*. As usually, the stability of the Landau phase out of the critical region leads to a consequent transition in this domain. On the phase diagram this reflects as a beak discussed above and as shown in Fig. 7.7 (a). The dashed lines in this figure show limiting positions of the beak. One limit corresponds to $f \to \infty$, $K \to 1$, and $u/v \to 0$. The other limit can be seen from the following deliberation. Homogeneous states,[9] as is seen from the structure of $G_0^{\alpha\beta}(\boldsymbol{q})$, are thermodynamically favorable only when $f > -1$. This condition defines the second boundary for the beak $K \to 1/2$, $u/v \to -3/2$, and therefore, $v/(u+v) \to -2$. An arbitrary positioning of the beak is shown in Fig. 7.7 (a) by the solid line. The beak's turn to the side of one or the other phase is, naturally, defined by the equalities $f > 1$ or $f < 1$. The free energies for tetragonal and cubic ferroelectrics confirming RG considerations were calculated by Sokolov (1983) and Mayer (1984).

7.4 Impure Systems

In this section we will consider the influence of randomly distributed impurities on the critical behavior of fluctuating systems. It is commonly accepted to classify impurities in accordance with their distribution in the parent material. If impurities are in equilibrium with the material they are usually called "annealed". Thermodynamic behavior of such impurities is defined by the values of variables of the main system. Let us define these variables as before $\boldsymbol{\phi}(\boldsymbol{r})$, and variables describing the impurities' configuration as $\tau(\boldsymbol{r})$. The distribution function, describing realization of these configurations is defined by

$$\rho\{\boldsymbol{\phi}, \tau\} = \frac{1}{Z} \exp\left(-H\{\boldsymbol{\phi}, \tau\}\right) \tag{7.29}$$

[9] We do not consider nonhomogeneous ordering here.

where $H\{\phi,\tau\}$ is the Hamiltonian of the system, and Z is the partition function defined as

$$Z = \int D\phi D\tau \exp\left(-H\{\phi,\tau\}\right).$$

In order to obtain the distribution function relating to each particular configuration one has to carry out the functional integration over one of the functional variables $\phi(\boldsymbol{r})$ or $\tau(\boldsymbol{r})$, which leads to

$$\begin{aligned}
\rho\{\phi\} &= \tfrac{1}{Z}\exp\left(-H\{\phi\}\right) &&= \tfrac{1}{Z}\int D\tau \exp\left(-H\{\phi,\tau\}\right), \\
\rho\{\tau\} &= \tfrac{1}{Z}\exp\left(-H\{\phi,\tau\}\right) &&= \tfrac{1}{Z}\int D\phi \exp\left(-H\{\phi,\tau\}\right).
\end{aligned}$$

As a rule, physical systems are prepared in conditions when impurities cannot relax to equilibrium but are "frozen" as a particular configuration. This configuration results from a particular experimental procedure of sample preparation. Such impurities are traditionally called "frozen". A distribution density of the field configuration at a fixed $\rho\{\tau\}$ can be represented as

$$\rho\{\phi|\tau\} = \frac{1}{Z\{\tau\}}\exp\left(-H\{\phi|\tau\}\right),$$

where $H\{\phi|\tau\}$ and $Z\{\tau\}$ are the Hamiltonian and the partition function of the system at a fixed impurity distribution

$$Z\{\tau\} = \int D\phi \exp\left(-H\{\phi|\tau\}\right).$$

The resulting distribution function is defined as a product $\rho\{\tau\}\rho\{\phi|\tau\}$. In order to define the field probability density one has to carry out the integration over the impurity distribution

$$\rho\{\phi\} = \int D\tau \frac{\rho\{\tau\}}{Z\{\tau\}}\exp\left(-H\{\phi|\tau\}\right).$$

It looks quite natural to define free energy as a mean value of $\ln Z\{\tau\}$ with the distribution $\rho\{\tau\}$, i.e. $\mathcal{F} = -\int D\tau\rho\{\tau\}\ln Z\{\tau\}$. One can show that this definition leads to the genuine value for the mean magnetization $\phi_o = \langle\phi\rangle$, namely $\phi_o = \partial\mathcal{F}/\partial \boldsymbol{h}$. In order to prove this statement we need to use the definition

$$\phi_o = \int D\tau\rho\{\tau\}\langle\phi\rangle_\tau, \tag{7.30}$$

where $\langle\phi\rangle_\tau$ denotes mean value ϕ defined with a fixed impurity configuration, i.e.

$$\langle\phi\rangle_\tau = \int D\phi \frac{\phi}{Z\{\tau\}} \exp\left(-H\{\phi|\tau\}\right). \tag{7.31}$$

On the other hand we have

$$\langle\phi\rangle_\tau = \frac{\partial\mathcal{F}\{\tau\}}{\partial \boldsymbol{h}},$$

where $\mathcal{F}\{\tau\}$ is a free energy at a fixed configuration $\tau(\boldsymbol{r})$, i.e.

$$\mathcal{F}\{\tau\} = -\ln Z\{\tau\}. \tag{7.32}$$

Considering Eqs. (7.30) – (7.32) and taking into account the definition for the functional $Z\{\tau\}$ one can conclude that the free energy of the system should be defined by the expression

$$\mathcal{F} = -\int D\tau\rho\{\tau\}\ln Z\{\tau\}. \tag{7.33}$$

The fact that the free energy is dependent on the impurity distribution does not evidently lead to the change of critical behavior. The presence of impurities in the system introduces an additional parameter with the dimension of length. This parameter competes with the correlation length. In the critical region the correlation length will sooner or later exceed this parameter essentially. Therefore, one can expect that small scale impurity perturbations will only *unessentially* influence trial vertices in the system not affecting the asymptotical scaling behavior. This gives the impression that the presence of impurities does not affect the phase transition at all. This statement, however, is not always justified. The considered deliberations based on the disappearance of any characteristic scale excepting correlation length, completely ignore possible renormalization of effective interaction between impurities when $T \to T_c$. Nonetheless, such a renormalization may be very important. The physical reason for this effect can be understood if we interpret the random field as random variations of vertices in the Ginzburg-Landau functional

$$H = \int d\boldsymbol{r}\left\{\frac{\tau_0+\tau_T(\boldsymbol{r})}{2}\boldsymbol{\phi}^2(\boldsymbol{r}) + \frac{c_0+\tau_c(\boldsymbol{r})}{2}\left(\boldsymbol{\nabla}\boldsymbol{\phi}(\boldsymbol{r})\right)^2 + \frac{u_0+\tau_u(\boldsymbol{r})}{8}\left(\boldsymbol{\phi}^2(\boldsymbol{r})\right)^2\right\}, \tag{7.34}$$

which are being renormalized when the system tends to T_c. As a result of this renormalization an effective impurity interaction changes. If this interaction is increasing with the renormalization, the influence of the impurities becomes essential. As this interaction is induced by the critical process, its increase may change fixed points, regions of attraction to these points, critical exponents, and so on. Hence, to understand the role of impurities, one has to return to an RG analysis.

7.4.1 Harris's Criterion

It is possible to find quite a general criterion defining conditions when the influence of impurities becomes essential (see Harris (1974)). The dependence of the free energy functional on the impurity configuration can be characterized with the help of different moments of random functions

$$\Delta_{ij} = \int d\boldsymbol{r}\overline{\tau_i(\boldsymbol{r})\tau_j(\boldsymbol{r}')}, \quad \Delta_{ijl} = \int d\boldsymbol{r}d\boldsymbol{r}'\overline{\tau_i(\boldsymbol{r})\tau_j(\boldsymbol{r}')\tau_l(\boldsymbol{r}'')}, \quad \cdots \tag{7.35}$$

Here the bar over the products of $\tau_i(\boldsymbol{r})$ means configurational averaging. The impurities are unessential only if all of these moments tend to zero when $T \to T_c$. Let us first consider how they change with the RG transformation on the example of the average

$$\Delta = \int d\boldsymbol{r}\overline{\tau_\tau(\boldsymbol{r})\tau_\tau(\boldsymbol{r}')}$$

The scaling field τ is a linear function with respect to $T - T_c$ and under the change of the scale as $\bar{\boldsymbol{r}} = \boldsymbol{r}/l$ it acquires a factor $l^{1/\nu}$. The latter leads to the following Δ transformation

$$\bar{\Delta} = \int d\bar{\boldsymbol{r}}\overline{\bar{\tau}_\tau(\bar{\boldsymbol{r}})\bar{\tau}_\tau(\bar{\boldsymbol{r}}')} = l^{2/\nu - d}\int d\boldsymbol{r}\overline{\tau_\tau(\boldsymbol{r})\tau_\tau(\boldsymbol{r}')} = l^{2/\nu - d}\Delta. \tag{7.36}$$

The value $\bar{\Delta}$ is less than Δ when the following condition is fulfilled

$$\frac{2}{\nu} - d < 0. \tag{7.37}$$

Recalling the scaling relation $\alpha = 2 - \nu d$, one can see that impurities are essential to the critical behavior when the specific heat exponent α is a positive value. So far as the value τ is treated as a linear function of temperature, the random addition $\tau(\boldsymbol{r})$ can be considered as a local perturbation of temperature. This supports the following interpretation of the obtained criterion. The specific heat is a linear response on the temperature perturbation. If $\alpha > 0$, then at $T \to T_c$ the response of the system is infinitely increasing. It is natural that under this condition the influence of impurities is essential. If $\alpha < 0$, the response is finite and all deliberations about the suppression of impurity influence by the field fluctuations are valid. Therefore, the Harris criterion can be formulated as: *the presence of randomly distributed impurities essentially influences critical behavior when* $\alpha > 0$. This criterion can be easily generalized on arbitrary correlators, Eq. (7.35). Let λ_i be the eigenvalue corresponding to the RG transformation of the scaling field τ_i,[10] then, repeating the consideration that has led to Eq. (7.36) one can easily obtain for the moments Δ_{ij} and Δ_{ijl} the conditions

$$\lambda_i + \lambda_j - d \ < \ 0, \tag{7.38}$$

$$\lambda_i + \lambda_j + \lambda_l - d \ < \ 0, \tag{7.39}$$

which generalize the criterion Eq. (7.37). As in the isotropic case for nonmagnetic impurities (i.e. for the functional Eq. (7.34)) the exponent $\lambda = 1/\nu$ is the largest one, so the conditions Eq. (7.38) and Eq. (7.39) are less restrictive than for $\alpha < 0$, and they will not be considered further.

7.4.2 The Replica Method

In the case where Harris's criterion shows the essential influence of impurities on the critical behavior, one has to solve a problem of a selfconsistent treatment of all renormalizable vertices (including impurities) in the Hamiltonian Eq. (7.34). The solution of this problem is made more difficult by the fact that in order to obtain free energy, one has to average the logarithm of the partition function $Z\{\tau\}$ over all possible impurity configurations (see Eq. (7.33)). The problem of the logarithm averaging can be avoided with the help of the following mathematical trick (see Emery (1975) and Grinstein and Luther (1976)). Let us make use of the following limiting representation for logarithm

$$\ln x = \lim_{n\to 0} \frac{\partial e^{n \ln x}}{\partial n} = \lim_{n\to 0} \frac{\partial x^n}{\partial n}.$$

[10]Strictly speaking, the parameters τ_i may not be scaling fields but rather their combinations. In such a case λ_i is the eigenvalue corresponding to the main term in the expansion of τ_i in terms of scaling fields (see Chapter 3).

Consequently for the free energy one has

$$\mathcal{F} = -\lim_{n\to 0}\frac{\partial}{\partial n}\left\{\left(\int D\boldsymbol{\phi}\right)^n \left(e^{-H_0\{\boldsymbol{\phi}\}}\right)^n \right.$$

$$\left. \times \int D\tau\rho\{\tau\}\left(e^{-\Delta H_0\{\boldsymbol{\phi}|\tau\}}\right)^n\right\}, \tag{7.40}$$

where the independent on $\tau_i(\boldsymbol{r})$ part has been evidently separated

$$H\{\boldsymbol{\phi}|\tau\} = H_0\{\boldsymbol{\phi}\} + \Delta H\{\boldsymbol{\phi}|\tau\}.$$

Now the problem of impurity averaging is reduced to the problem of n identical subsystems ("replicas") interacting with the scalar field $\tau(\boldsymbol{r})$. After the solution of this problem one has to find $\mathcal{F}$ using the consequent limiting procedure. As was proved in the previous section, the most important term is the impurity dependent part of τ in the Hamiltonian, Eq. (7.34). Let us also simplify consideration assuming that the variable $\tau(\boldsymbol{r})$ is a δ-correlated Gaussian variable in the vicinity T_c, i.e.

$$\overline{\tau(\boldsymbol{r})} = 0, \qquad \overline{\tau(\boldsymbol{r})\tau(\boldsymbol{r}')} = \frac{\Delta}{4}\delta(\boldsymbol{r}-\boldsymbol{r}').$$

In this case one can evidently carry out integration over the impurity variables in Eq. (7.40)

$$\int D\tau\rho\{\tau\}\exp\left\{-\int d\boldsymbol{r}\frac{\tau(\boldsymbol{r})}{2}(\boldsymbol{\phi}_1^2 + \cdots + \boldsymbol{\phi}_n^2)\right\}$$

$$= \exp\left\{\frac{\Delta}{8}\int d\boldsymbol{r}(\boldsymbol{\sigma}^2)^2\right\}, \tag{7.41}$$

where the symbol $\boldsymbol{\sigma}$ denotes the supervector $\boldsymbol{\sigma} = \{\boldsymbol{\phi}_1, \boldsymbol{\phi}_2, \ldots, \boldsymbol{\phi}_n\}$ composed from n vectors $\boldsymbol{\phi}_i$. This term describes interaction between n identical replicas of the parent pure material. Substituting Eq. (7.41) into Eq. (7.40) we arrive at the expression

$$\mathcal{F} = -\lim_{n\to 0}\frac{\partial}{\partial n}\int D\boldsymbol{\sigma}\exp\left\{-H_{eff}\{\boldsymbol{\sigma}\}\right\},$$

where the effective Hamiltonian is

$$H_{eff} = \frac{1}{2}\int d\boldsymbol{r}\left\{\tau\boldsymbol{\sigma}^2 + (\boldsymbol{\nabla}\boldsymbol{\sigma})^2 + \frac{u}{4}\sum_{i=1}^{n}\sigma_i^4 + \frac{\Delta}{4}(\boldsymbol{\sigma}^2)^2\right\}. \tag{7.42}$$

If instead of the Gaussian δ-correlated impurity configurations one were to use the distribution

$$\rho\{\tau\} = (1-c)\delta(\tau) + c\delta(\tau - \Delta\tau)$$

then, the effective Hamiltonian would look like

$$H_{eff} = \frac{1}{2}\int d\boldsymbol{r}\left\{\tau\boldsymbol{\sigma}^2 + (\boldsymbol{\nabla}\boldsymbol{\sigma})^2 + \frac{u}{4}\sum_{i=1}^{n}\sigma_i^4 - \ln\left[(1-c) + c\exp(-\Delta\tau\boldsymbol{\sigma}^2)\right]\right\},$$

which in addition to the impurity vertex $\Delta \equiv \Delta_4$ has also higher order impurity vertices Δ_{2k} as a factors at $(\boldsymbol{\sigma}^2)^k$ powers. The latter may sometimes be very important (see Sect. 7.4.3).

The Hamiltonian Eq. (7.42) is equivalent to the system with a "hypercubic symmetry". This term means that Eq. (7.42) is a cubic functional with respect to the "vector " $\boldsymbol{\sigma}$, n components of which are in their turn m-component vectors. Such a functional can also be considered as a Hamiltonian of the system containing n interacting m-component fields.

7.4.3 RG Analysis

In order to obtain an RG set for this case one has to use Eq. (1.31) and a consequent representation for vertices $\hat{g}_2$

$$g_{2,s_1,\ldots,s_4}^{\alpha_1,\ldots,\alpha_4} = (-\Delta + uF^{s_1,\ldots,s_4})R_{s_1,\ldots,s_4}^{\alpha_1,\ldots,\alpha_4},$$

where

$$R_{s_1,\ldots,s_4}^{\alpha_1,\ldots,\alpha_4} = \frac{1}{3}(\Delta_{s_1s_2}^{\alpha_1\alpha_2}\Delta_{s_3s_4}^{\alpha_3\alpha_4} + \text{permutations}),$$

$$F^{s_1\ldots s_4} = \prod_{i=2}^{4}\delta_{s_1s_i}, \quad \Delta_{s_is_j}^{\alpha_i\alpha_j} = \Delta\delta_{\alpha_i\alpha_j}\delta_{s_is_j}.$$

Using this representation, one can obtain from Eq. (1.31) the resulting RG set, governing vertices u and Δ, in the form

$$\begin{aligned}\frac{du}{dl} &= \epsilon u - \frac{m+8}{2}u^2 + 6u\Delta,\\ \frac{d\Delta}{dl} &= \epsilon\Delta - (m+2)u\Delta + 4\Delta^2.\end{aligned} \tag{7.43}$$

Here the limit $n \to 0$ has already been taken. One has to be careful with this limit. The problem is that the calculations can only be performed for integer

n, but one needs to define all functions at arbitrary n. Such a procedure may not always be unique. However, in this case the set, Eq. (7.43) can also be obtained in the perturbation theory approach without use of the replica method (see Lubensky (1975)).

The renormalization of the critical temperature can be obtained from the first equation of the set Eq. (1.31) and is given by

$$\tau^* = \tau + \frac{m+2}{2} u - \Delta. \tag{7.44}$$

Excepting the trivial Gaussian point the set Eq. (7.43) has the following three fixed points

$$\begin{array}{lll} \mu_1^* : & u^* = 2\epsilon/(m+8), & \Delta^* = 0; \\ \mu_2^* : & u^* = \epsilon/(m-1), & \Delta^* = (4-m)\epsilon/8(m-1); \\ \mu_3^* : & u^* = 0, & \Delta^* = -\epsilon/4; \end{array}$$

The first of these points corresponds to a pure material. This point is stable only when $m > 4$. The specific heat critical exponent at this point is equal to

$$\alpha = \frac{4-m}{2(m+8)} \epsilon,$$

which is in agreement with Harris's criterion as we have stability only for $m > 4$. When $m < 4$, the stable fixed point is μ_2^*, which corresponds to an impure material as $\Delta^* \neq 0$. It is interesting to note that the point μ_3^* is also stable. The flow line portraits for both cases are shown in Fig. 7.8. However, one should not consider the state of the system at the point μ_3^* as a physically realizable situation. For a real temperature field $\tau(\boldsymbol{r})$ the value Δ is always positive. Hence, the state with $\Delta^* < 0$ is an unphysical state. When $m > 4$, the point μ_2^* also becomes unphysical. Therefore, when $m < 4$ we have only one physical stable fixed point μ_2^* with the critical exponents

$$\gamma = 1 + \frac{3m}{16(m-1)} \epsilon \quad \text{and} \quad \alpha = \frac{m-4}{8(m-1)} \epsilon,$$

which are different from the consequent exponents for pure systems.

A very interesting feature of the fixed point μ_2^* can be seen if one puts $m \to 1$. In this limit formally this point tends to infinity. This accounts for the following transformation of the set Eq. (7.43) at $m = 1$

$$\frac{du}{dl} = u \left[\epsilon - 3 \left(\frac{3}{2} u - 2\Delta \right) \right],$$

$$\frac{d\Delta}{dl} = \Delta \left[\epsilon - 2 \left(\frac{3}{2} u - 2\Delta \right) \right].$$

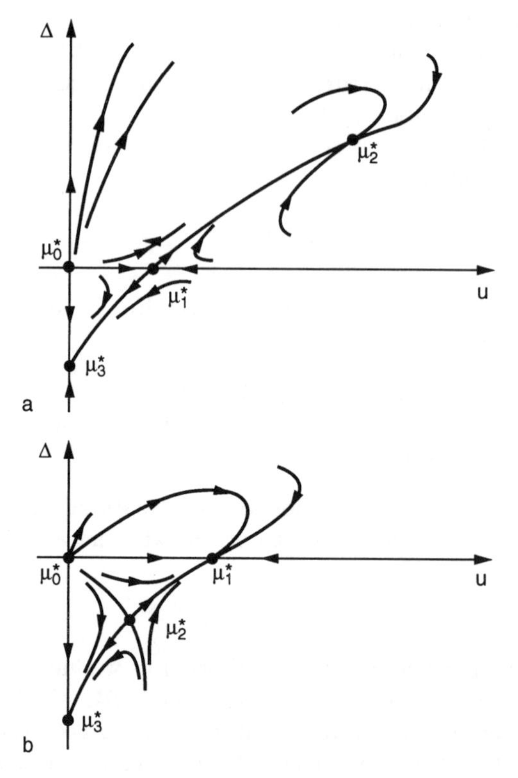

Figure 7.8: Flow lines for impure system at $\alpha > 0$ (a) and $\alpha < 0$ (b).

Notwithstanding the fact that the line $u = 4\Delta/3$ is an integral of the set corresponding to the integral line $u = 4\Delta/(m-1)$ when $m \neq 1$, the fixed point now cannot exist on this line. Khmelnitsky (1975) took into account a correction due to the next order terms in the RG set. In this order the fixed point satisfies the condition $u^* = 4\Delta^*/3$ as before. As a result the second-order terms, i.e. terms proportional to u^2, $u\Delta$, and Δ^2 compensate each other and the fixed point has order $\sqrt{\epsilon}$

$$u^* = \frac{4\Delta^*}{3}, \qquad \Delta^* = \left(\frac{6\epsilon}{53}\right)^{1/2}.$$

7.4.4 Flow Line Runaway

From Fig. 7.8 one can see also additional specific feature of flow lines for the impure system. The flow lines, starting in the vicinity of the axis Δ (small u), move along this axis towards quite large values of Δ. The latter resembles the situation when the continuous transition transforms into a discontinuous one. However, in this situation there is no physical reason for such a transformation. Intuitively, impurities would rather lead to the smearing of the transition than to making it more sharp. Considering that a more distinct runaway of flow lines can be seen at small u, let us analyze the limit $u \to 0$. Formally, a negative value of Δ in the Hamiltonian Eq. (7.42) should lead to the instability. As usual in such a case, one needs to keep in the functional terms like $\Delta_6(\boldsymbol{\sigma}^2)^3$, which should stabilize the expression for the partition function (at $\Delta_6 > 0$). It may seem that this leads to the functional corresponding to first-order phase transitions. However, this is not really the case. For instance, the above distribution $\rho\{\tau\} = (1-c)\delta(\tau)+c\delta(\tau-\Delta\tau)$ gives the order of signs ($\Delta_6 > 0$, $\Delta < 0$) appropriate, to the situation, but the whole function $\ln\{(1-c) + c\exp(-\Delta\tau\boldsymbol{\sigma}^2)\}$ is monotonic. It looks like such a monotonic dependence should take place for any physically realizable initial Hamiltonian H_{eff}. It is practically impossible to imagine a situation when impurities, of the kind considered here, will turn the continuous transition into a discontinuous one. That means that the whole sum $\sum_k \Delta_{2k}(\boldsymbol{\sigma}^2)^k$ may be monotonic notwithstanding any number of terms with negative coefficients Δ_k. The same example $\ln\{(1-c) + c\exp(-\Delta\tau\boldsymbol{\sigma}^2)\}$ shows that the series of Δ_{2k} with sign oscillating does not contradict the monotonicity and linear asymptotic increase of the function at $\boldsymbol{\sigma}^2 \to \infty$. If the monotonic nature of the function $\sum_k \Delta_{2k}(\boldsymbol{\sigma}^2)^k$ is not violated as a result of renormalizations when $T \to T_c$, then there is no reason for the discontinuous transition. Therefore, one has to find a new interpretation of the flow lines runaway to the side where $\Delta \to -\infty$. It looks like the most reasonable assumption for their runaway would be the smearing of

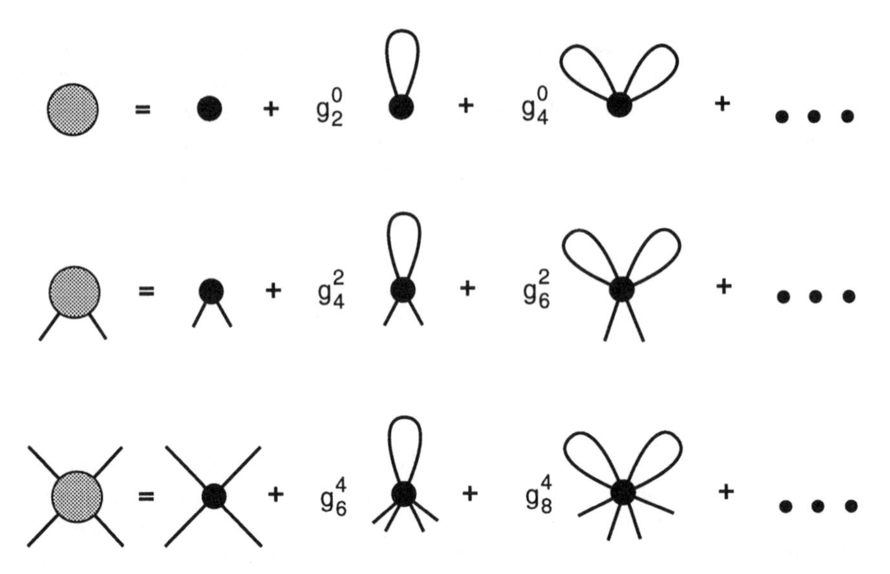

Figure 7.9: Loop contributions to renormalization of vertices. Combinatorial coefficients are equal to $g_{2k}^{2m} = (2k)!![(2k-2m)!!]^{-1} \prod_{i=m}^{k-1}(n+2i)$.

the transition.

One can easily verify that all vertices contain functions depending on c. If we restrict consideration to the small c values, then in the lowest c-order the renormalization of vertices will be linear with respect to Δ_{2k}. This means that the renormalization of this vertices is caused only by the loop contributions to Δ_{2k} at the expense of the higher-order vertices ($m > k$). Renormalizations of this type are shown in Fig. 7.9. From this figure one can see that in the first c-order the renormalization is coming from the maintenance of loops from the "legs" of the higher-order vertices. In the next section we will show that if the initial free energy functional is a monotonic function then it will always be such notwithstanding the loop renormalization of coefficients Δ_{2k}. Taking into account that the replica method is not a rigorous mathematical procedure, we will first prove the general theorem about the conservation of monotonicity for any number of components n. After that, this theorem will be applied to the interpretation of the results at $n \to 0$.

7.4.5 Loop Renormalizations

The importance of the summation of loop renormalizations in the Ginzburg-Landau functional is not restricted to the interpretation of the flow line runaway for impure systems. Namely, when Hamiltonians contain complex fourth-order invariants, the fluctuational renormalization of initial vertices may lead to the flow line runaway from the stability regions. Systems may manifest qualitatively new effects being sensitive to a particular realization of initial parameters. In such cases irrelevant parameters in the Ginzburg-Landau functional u_6, u_8, ... may essentially influence the system at early stages of the RG evolution by transforming initial fourth-order vertices with respect to separatrixes. As the separatrixes divide regions with different critical behavior, the renormalization caused by irrelevant vertices may affect the system catastrophically. If this is the case, it is necessary to obtain the "right" Hamiltonian which has incorporated all higher-order term renormalizations in the RG analysis. A similar problem was encountered by Fisher and Huse (1985) in relation to the wetting transitions. They found that all higher-order vertices were important for this physical situation. The summation of loop graphs was a quite good approximation for their case. Below we will follow the work by Lisyansky and Filippov (1986 b, 1987 a) in considering loop renormalizations of all local vertices.

Let us explicitly separate the term containing spatial derivatives in the free energy functional and represent the partition function in the form

$$Z = \int D\phi \exp\left\{\int d\boldsymbol{r}[(\boldsymbol{\nabla}\phi)^2 + F(\phi^2)]\right\},$$

where the function $F(\phi^2)$ is assumed to be representative in the form

$$F(\phi^2) = \sum_{k=0}^{\infty} u_{2k}(\phi^2)^k. \tag{7.45}$$

New vertices $\bar{u}_{2k}$ arise due to the summation of the loop graphs shown in Fig. 7.9. Let us denote the new function containing these vertices in the Taylor series as $\bar{F}(\phi^2)$. Though the series shown in the figure is reminiscent of a geometrical progression, this is not really the case. Each of the additives contains as a factor the value u_{2k} defined by Eq. (7.45). Naturally, in a general case all u_{2k} are different. It can be verified, however, that the differentiation of the series $\bar{u}_{2(k-1)}$, with respect to the value θ, equal to a one-loop graph, leads, except for a factor, to the series for the value $\bar{u}_{2k}$. A more accurate consideration recovers the factor leading to the following recursive relation

$$\bar{u}_{2k} = \frac{1}{k(n+2k-2)} \frac{\partial \bar{u}_{2(k-1)}}{\partial \theta}. \tag{7.46}$$

The constraint, Eq. (7.46) helps to find not only the sum of θ powers but also an analytical representation for the function $\bar{F}(\phi^2)$. With the help of new vertices $\bar{u}_{2k}$ one can define the renormalized functional which can be represented as

$$\bar{F}(\boldsymbol{\phi}^2) = \sum_{k=0}^{\infty} \bar{u}_{2k}(\boldsymbol{\phi}^2)^k. \tag{7.47}$$

Using Eq. (7.46) and Eq. (7.47) one can find the equation

$$\frac{\partial \bar{F}}{\partial \theta} = \frac{1}{2} \sum_{\alpha} \frac{\partial^2 F}{\partial \phi^\alpha \partial \phi^\alpha} \equiv \frac{1}{2} \nabla_\phi^2 \bar{F}. \tag{7.48}$$

This equation is a well known heat flow equation. Considering θ as effective "time" and taking into account that at $\theta \to 0$ the renormalized vertices are reduced to the initial ones, one can obtain the searched-for expression of $\bar{F}$ in terms of F and θ

$$\bar{F}(\boldsymbol{\phi}^2) = \frac{1}{(2\pi\theta)^{n/2}} \int d\boldsymbol{a} \exp\left\{-\frac{(\boldsymbol{\phi} - \boldsymbol{a})^2}{2\theta}\right\} F(\boldsymbol{a}^2) \equiv \hat{W}_\theta F(\boldsymbol{\phi}^2). \tag{7.49}$$

It is necessary to mention that the result obtained for the isotropic functional can be generalized to anisotropic systems. An equation in partial derivatives analogous to Eq. (7.48) can be generalized for the functional $\bar{F}[\boldsymbol{\phi}, \theta]$ to an arbitrary symmetry with any number of interaction fields. In a more general case Eq. (7.48) should be replaced by

$$\frac{\partial \bar{F}}{\partial \theta} = \frac{1}{2} \sum_{\alpha\beta} h^{\alpha\beta}(\boldsymbol{l} = 0) \frac{\partial^2 F}{\partial \phi^\alpha \partial \phi^\beta}. \tag{7.50}$$

Here, however, we restrict consideration with the case of the isotropic solution Eq. (7.49). This solution can be considered as an integral transformation satisfying the following relations

$$\frac{d^k \bar{F}(\boldsymbol{\phi}^2)}{d\phi^k} = \hat{W}_\theta \frac{d^k F(\boldsymbol{\phi}^2)}{d\phi^k}, \tag{7.51}$$

$$\hat{W}_{\theta_2} \left\{ \hat{W}_{\theta_1} F(\boldsymbol{\phi}^2) \right\} = \hat{W}_{\theta_1 + \theta_2} F(\boldsymbol{\phi}^2). \tag{7.52}$$

From Eq. (7.51) it follows that if the initial function posses a parity property, and/or is defined as a fixed sign function, and/or some of its derivatives are defined as fixed sign functions, then all these properties are conserved in the renormalized function. In other words, if the initial functional corresponds to the second-order transition, the same is true for the renormalized functional $\bar{F}(\boldsymbol{\phi}^2)$. Even from these general deliberations one can see that these statements are not valid if $F(\phi^2)$ would correspond to the first-order phase transition. With the increase of "time" θ the "temperature" $\bar{F}(\phi^2,\theta)$ tends to a smoother function of "position vector" ϕ. This means that notwithstanding the initial distribution $F(\phi^2)$ at a quite sufficient value θ all extrema will completely vanish. As a result a discontinuous phase transition turns into a continuous one. The conditions for such transformations will be considered later. Here, we discuss to some extent the reverse possibility.

Let us find such an initial form $F(\phi^2)$, which does not change under the influence of the transformation Eq. (7.49) (excepting multiplication on a constant factor). Mathematically this is an eigenfunction problem for the transformation Eq. (7.49), i.e. we have

$$\hat{W}_\theta F(\phi^2) = A(\theta)F(\phi^2),$$

or, taking into account the properties Eqs. (7.51) and (7.53), we can also write

$$\frac{dA(\theta)}{d\theta}F(\phi^2) = \frac{1}{2}A(\theta)\boldsymbol{\nabla}^2 F(\phi^2).$$

After separation of variables and integration of the consequent equation for the eigenvalue $A(\theta)$, one arrives at

$$A_\lambda(\theta) = \exp\left(-\frac{\lambda^2\theta}{2}\right),$$

where λ is an arbitrary separation constant. The function F satisfies the Helmholtz equation. Making use of the isotropy of the problem, one can write this equation in the form

$$\frac{d^2F}{d\phi^2} + \frac{n-1}{\phi}\frac{dF}{d\phi} + \lambda^2 F = 0. \tag{7.53}$$

The solution of this equation can be obtained in terms of Bessel's functions

$$F_\lambda \sim \phi^{-(n-2)/2} I_{n/2-1}(\lambda\phi).$$

Note, that the Neumann's function, also satisfying Eq. (7.53), has a singularity when $\phi \to 0$ and, therefore, it cannot give a physical representation of the Ginzburg-Landau functional. It is interesting to notice that, for n-component Heisenberg magnets, the initial expression $F(\phi^2)$ has the form (see Sect. 2.2.2, Eq. (2.22))

$$F(\phi^2) = \ln[\phi^{-(n-2)/2} I_{n/2-1}(\phi)],$$

i.e. the function $F(\phi^2)$ is a logarithm of the eigenfunction of the transformation Eq. (7.49). The latter means that the renormalization of this function cannot be reduced to multiplication by a constant as in the previous case.

In experiment only the effective form $\bar{F}(\phi^2)$ can be measured. Hence, the restoration of the initial functional $F(\phi^2)$ from the renormalized value is of major importance. In other words, we need to restore the vertices u_{2k} from the given parameters $\bar{u}_{2k}$. Having this in mind, let us, first of all, carry out the integration over the angular variables in Eq. (7.49). This procedure leads to

$$\bar{F}(\phi^2) = \frac{1}{\theta \phi^{n/2-1}} \exp\left(\frac{\phi^2}{2\theta}\right)$$

$$\times \int_0^\infty da a^{n/2} F(a^2) I_{n/2-1}\left(\frac{a\phi}{\theta}\right) \exp\left(-\frac{u^2}{2\theta}\right). \tag{7.54}$$

The function $I_\nu(a\phi/\theta)(a\phi/\theta)\exp(-\phi^2/2\theta)$ is a generating function for the generalized Laguerre's polynomials $L_k^{(\nu)}(a^2/2\theta)$. Taking this into account and introducing, for the sake of simplicity, the new variable $\phi^2/2\theta \to \phi^2$ one arrives at

$$\bar{F}(\phi^2) = \sum_{k=0}^{\infty} c_k \frac{(-1)^k}{k!} (\phi^2)^k = \sum_{k=0}^{\infty} \bar{u}_{2k} (\phi^2)^k,$$

where c_k are coefficients in the expansion of the function $F(\phi^2)$ over the Laguerre's polynomials

$$c_k = k! \int_0^\infty dt \frac{t^\nu e^{-t}}{\Gamma(k+\nu+1)} L_k^{(\nu)} F(t).$$

This formula leads to the inverse transformation of Eq. (7.49)

$$F(\phi^2) = \sum_{k=0}^{\infty} c_k L_k^{(\nu)}(\phi^2) = \sum_{k=0}^{\infty} (-1)^2 k! \bar{u}_{2k} L_{2k}^{(\nu)}(\phi^2). \tag{7.55}$$

For the scalar field $\nu = n/2 - 1 = -1/2$ and the polynomials $L_k^{(\nu)}$ can be reduced to the Hermite's polynomials

$$L_{2k}^{(-1/2)}(\phi^2) = \frac{(-1)^k}{k! 2^{2k}} H_{2k}(\phi),$$

so that the inverse transformation takes the form

$$F(\phi^2) = \sum_{k=0}^{\infty} 2^{-2k} \bar{u}_{2k} H_{2k}(\phi). \tag{7.56}$$

Now, the conditions for the smearing of extrema of the initial functional $F(\phi^2)$ can be found with the help of Eqs. (7.55) and (7.56). These conditions answer the question, when does the functional corresponding to a first-order phase transition turn into the functional corresponding to a continuous transition? The simplest case, for this problem, is the scalar field situation. Let us fix a value $\phi = \phi_0$ and transform $\bar{F}(\phi^2)$ as follows

$$\bar{F}(\phi^2) = \frac{1}{\sqrt{\pi}} \int_{-\infty}^{\infty} da F[(a+\phi_0)^2] \exp\{-[a - (\phi - \phi_0)]^2\}.$$

As the function $\exp[-(\phi-\phi_0)^2 + 2(\phi-\phi_0)a]$ is the generation function for the Hermite's polynomials $H_k(a)$, one can rewrite the previous formula as

$$\bar{F}(\phi^2) = \sum_{k=0}^{\infty} \frac{b_k(T)}{k!} (\phi - \phi_0)^k,$$

where $b_k = d^k \bar{F}/d\phi^k|_{\phi=\phi_0}$ are coefficients in the expansion of the function $F[(a+\phi_0)^2]$ over the Hermite's polynomials. These coefficients, but for a numerical factor, are equal to

$$b_k(\phi_0, T) \sim \frac{1}{\sqrt{\pi}} \int_{-\infty}^{\infty} da F[(a+\phi_0)^2] e^{-a^2} H_k(a).$$

When the function $\bar{F}$ has extremum at the point ϕ_0, the coefficient $b_1(\phi_0, T) = 0$. Monotonic and nonmonotonic functions $\bar{F}(\phi^2, T)$ are demarcated by functions with an inflection point at $\phi = \phi_0$ (see Fig. 7.10). The class of these functions can be defined as follows: If for the given function $F(\phi^2)$ one can find ϕ_0 to satisfy the equations

$$b_1(\phi_0, T_0) = b_2(\phi_0, T_0) = 0, \tag{7.57}$$

then $\bar{F}$ has at $\phi = \phi_0$ and $T = T_0$ an inflection point. The value ϕ_0, defined by Eq. (7.57), is equal to the discontinuous jump of the order parameter at the overheating lability boundary. For practical purposes, it is also useful to know the phase equilibrium surface (the value ϕ_1) and the overcooling lability surface with the order parameter $\phi = \phi_2$. These surfaces are defined in the parametric space constituted by vertices of the functional $F(\phi^2)$. They can be found from the equations (see also Fig. 7.10)

$$b_0(\phi_1, T_1) = b_1(\phi_1, T_1) = 0, \quad \text{and} \quad b_1(\phi_2, T_2) = b_2(0, T_2) = 0.$$

Hence, all required information concerning the structure of the function $\bar{F}$

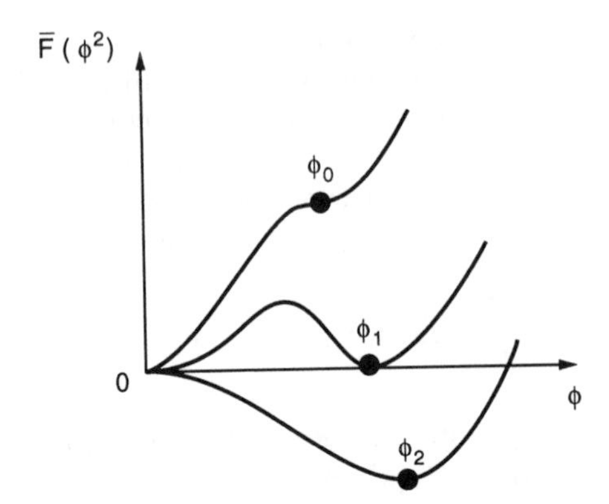

Figure 7.10: The structure of the function $\bar{F}(\phi^2)$ with the binodal and spinodal curves; and the discontinuous jumps of the order parameter on them (consequently ϕ_1 and ϕ_0, ϕ_2).

can be extracted from the structure of the expansion F over the Hermite's polynomials.

As has been shown above, the monotonic character of the free energy functional cannot be destroyed by the loop renormalization. Nonetheless, the full RG transformation also contains, in addition to this renormalization, the change of scale of all variables. This changes relations established between the vertices Δ_{2k}. In order to understand how important these deviations are, one has to use the exact RG equation for the full free energy functional. However, in the approximation linear over the concentration c only the terms linear with respect to F should be left in this equation. The generation of nonlocal vertices is also unimportant in this order of c and the function $\eta(\boldsymbol{q})$ should be neglected. Thus, taking into account only linear terms in the RG equation for the local functional (see Appendix B) we arrive at

$$\frac{\partial F}{\partial l} = dF - \frac{d-2}{2}\boldsymbol{\phi}\boldsymbol{\nabla}_\phi F + \boldsymbol{\nabla}_\phi^2 F.$$

This equation can be analytically solved. Using substitutions $\boldsymbol{x} = \xi(l)\boldsymbol{\phi}$ and $F = e^{dl}f$, with $\xi(l) = \exp[(2-d)l/2]$, one can find for the function f the equation

$$\frac{df}{dl} = \xi^2(l)\boldsymbol{\nabla}_x^2 f.$$

Let us also introduce the function $\theta(l)$ with the help of the equation $d\theta(l)/dl =$

$2\xi^2(l)$, then the equation for the function f can be written as

$$\frac{df}{d\theta} = \frac{1}{2}\nabla_x^2 f,$$

i.e. we have arrived again at the heat flow equation (see Eq. (7.48)). Finally, for the function $F(l, \boldsymbol{\phi}^2)$ one can write the following representation

$$F(l, \boldsymbol{\phi}^2) = \frac{1}{(2\pi\theta(l))^{n/2}} \int d\boldsymbol{a} \exp\left\{ -\frac{(\boldsymbol{\phi}\xi(l) - \boldsymbol{a})^2}{2\theta(l)} \right\} F(l = 0, \boldsymbol{a}^2) \quad (7.58)$$

Taking into account that when $l \to 0$ the function $F(l, \boldsymbol{\phi}^2)$ should turn into $F(l = 0, \boldsymbol{\phi}^2)$ the solution for θ should be taken in the form

$$\theta(l) = \frac{2}{2-d}(e^{(2-d)l} - 1)$$

satisfying the condition $\theta(0) = 0$.

Eq. (7.58), being a generalization of the transformation Eq. (7.49), allows us to derive the following important conclusions on the structure of the functional F at an arbitrary value of l. With the increase of l, the function not only conserves monotonicity but also becomes more smeared. Since at $l \to \infty$ the value $\xi(l) \to 0$, all higher-order vertices Δ_{2k} with $k > 2$ tend to zero too. The latter cannot be said about the factors at $\boldsymbol{\phi}^2$ and $(\boldsymbol{\phi}^2)^2$ owing to the coefficient e^{dl}. The vertex at $\boldsymbol{\phi}^2$ defines impurity renormalization of T_c and is not of much interest, as the value T_c is considered as a parameter of the experiment. On the contrary the vertex $\Delta_4 \equiv (-\Delta)$ is essential for critical behavior. In the linear RG approximation it increases as $\Delta_4 \sim e^{\epsilon l}$. In order to suppress this increase one has to keep terms of the order of Δ^2 and Δu, i.e. to "switch on" interactions with the traditional ϕ^4 vertex u. This vertex is also increasing (with the same rate as Δ) in the linear approximation. It is coupled to other vertices Δ_{2k} $(k > 1)$ in the higher orders of perturbation theory. In other words, we again return to the phase flow analysis on the (u, Δ) plane. Now, if the initial values of vertices u and Δ are of the same order, then the system, depending on the sign of the critical exponent α, will flow into a "pure" or "impure" stable fixed point. If the value $\Delta(l = 0)$ is sufficiently large, then we have flow line runaway, which should correspond to the smearing of the transition by impurities.

7.4.6 Anisotropic Systems

The study of critical behavior of a number of physical systems, as a rule, can be reduced to the consideration of the Ginzburg-Landau functional with a particular symmetry or with a number of interacting fluctuating fields. As has been shown above, the interaction of fluctuating fields (or components of vector field ϕ) can lead to the appearance of qualitatively new effects in the critical region (in comparison with the Landau theory approach). However, in reality we never have the ideal situation — different kinds of defects, impurities, and the like, are present in practically every experimental sample. Hence, the generalization of impurity influence on more complex systems yields substantial practical insight. As is shown in the previous sections, in isotropic systems impurities may change critical exponents or, at the most, smear the phase transition. In a more complex situation one may expect manifestations of qualitative deviations in comparison with an ideal system.

Here we will restrict consideration to an impure anisotropic system described by a single fluctuating field. This means that in the RG treatment only one additional vertex Δ is essential (see Bak (1976)). For the system with arbitrary symmetry, the Ginzburg-Landau functional contains N invariants composed from components ϕ_α. Let us denote p-th invariant as $B^p_{\alpha\beta\gamma\delta}\phi_\alpha\phi_\beta\phi_\gamma\phi_\delta$. For each of the invariants we have vertex u_p in the functional, i.e.

$$\sum_{\alpha\beta\gamma\delta} g_2^{\alpha\beta\gamma\delta}\phi_\alpha\phi_\beta\phi_\gamma\phi_\delta = \sum_{p=1}^{N} u_p \sum_{\alpha\beta\gamma\delta} B^p_{\alpha\beta\gamma\delta}\phi_\alpha\phi_\beta\phi_\gamma\phi_\delta. \tag{7.59}$$

The vertices u_p are convenient as variables for the RG analysis. As usual one can write the RG equations on the critical surface, in this case, defined by

$$\tau^* = -\sum_p c_p u_p^* + \Delta. \tag{7.60}$$

The RG set can be represented as

$$\begin{aligned} \frac{du_p}{dl} &= \epsilon u_p - \textstyle\sum_{nm} d^p_{nm} u_n u_m + 6\Delta u_p, \\ \frac{d\Delta}{dl} &= \epsilon\Delta - 2\textstyle\sum_p c_p u_p \Delta + 4\Delta^2, \end{aligned} \tag{7.61}$$

where the coefficients d^p_{nm} and c_p can be found with the help of Eq. (1.28) for a particular representation of the Ginzburg-Landau functional. The critical exponent ν for the "pure" system in the lowest order is defined (see Eq. (7.60)) as

$$2\nu = 1 - \sum_p c_p u_p^*.$$

Consequently, the specific heat exponent is equal to

$$\alpha = 2 - \nu d = 2\left(\epsilon - 2\sum_p c_p u_p^*\right).$$

Comparing this expression with the last equation of the set, Eq. (7.61), one can see that in order to obtain a stable fixed point with $\Delta^* = 0$ it is necessary to have α negative. The latter confirms Harris's criterion for anisotropic systems. From the set Eq. (7.61) we have the set defining all fixed points in the system as

$$\begin{aligned} \epsilon u_p^* &= \textstyle\sum_{nm} d_{nm}^p u_n^* u_m^* - 6\Delta^* u_p^*, \\ \epsilon\Delta^* &= 2\Delta^* \textstyle\sum_p c_p u_p^* - 4(\Delta^*)^2. \end{aligned} \tag{7.62}$$

Let us assume that when $\Delta^* = 0$ the "pure" system has a fixed point $u_p^* \equiv \epsilon x_p$. Let us introduce new variables $x_p = (1+6y)x_p'$ where $y \equiv \Delta^*/\epsilon$, then the set Eq. (7.62) can be rewritten as

$$x_p' = \sum_{nm} d_{nm}^p x_n' x_m', \tag{7.63}$$

$$y = 2\sum_p c_p(1+6y)x_p' y - 4y^2. \tag{7.64}$$

Eq. (7.63) is exactly the same as the one for the "pure" system. Hence it has the same solutions. Excepting the trivial solution $y = 0$, Eq. (7.64) also has a solution with $y \neq 0$. As a result, the number of fixed points is duplicated. For every fixed point of the pure system x_p one can find an additional impure fixed point

$$x_p^i = \frac{1}{1+6y} x_p, \quad y = -\frac{1}{4} + \sum_p \frac{c_p}{2} x_p^i.$$

Naturally, one has to study the stability of these points. As we discussed in Sect. 6.3.3, for pure systems at $n < 4$ the stable fixed point is always isotropic. Let u_1 be isotropic invariant in the functional Eq. (7.59). Then explicitly separating this vertex from the set Eq. (7.61) we arrive at[11]

$$
\begin{aligned}
\frac{du_1}{dl} &= u_1 - \left\{ \tfrac{n+8}{2} u_1^2 + u_1 \textstyle\sum_{p\neq 1} d_{p1}^1 u_p + \sum_{nm\neq 1} d_{nm}^1 u_n u_m \right\} \\
&\quad +6u_1\Delta, \\
\frac{d\Delta}{dl} &= \Delta - (n+2)u_1\Delta + 4\Delta^2 - 2\textstyle\sum_{p\neq 1} c_p u_p \Delta, \\
\frac{du_p}{dl} &= u_p - d_{p1}^p u_1 - \textstyle\sum_{nm} d_{nm}^p u_n u_m + 6u_p\Delta.
\end{aligned}
\tag{7.65}
$$

Linearizing the set Eq. (7.65) in the vicinity of the isotropic fixed point

$$
x_1^i = \frac{1}{2(n-1)}, \quad x_{p\neq 1}^i = 0, \quad y = \frac{4-n}{8(n-1)}, \tag{7.66}
$$

we arrive at the matrix

$$
S = \begin{pmatrix} 1-(n+8)x_1^i + 6y & -d_{p1}^1 x_1^i & 6x_1^i \\ 0 & 1 - d_{p1}^p x_1^i + 6y & 0 \\ -(n+2)y & -2c_p y & 1 + 8y - (n+2)x_1^i \end{pmatrix}
$$

which should have negative eigenvalues for the stability of this point. For the pure system at $n < 4$ we have inequality

$$
1 - d_{p1}^p x_1 < 0. \tag{7.67}
$$

In the impure case, this inequality is replaced by

$$
1 - d_{p1}^p x_1^i + 6y < 0,
$$

which is automatically satisfied provided the inequality Eq. (7.67) holds. Now, the stability is defined by the eigenvalues of the submatrix

$$
S_1 = \begin{pmatrix} 1-(n+8)x_1^i + 6y & 6x_1^i \\ -(n+2)y & 1 + 8y - (n+2)x_1^i \end{pmatrix},
$$

which is the same as for the $\mathcal{O}(n)$-symmetric functional. Thus, when $n < 4$ the isotropic fixed point is always a stable fixed point. Hence, in this case we have a conventional second-order phase transition with the critical exponents not depending on the concentration of impurities. Naturally,

[11] For the isotropic invariant the coefficients are equal: $d_{11}^1 = (n+8)/2$ and $c_1 = (n+2)/2$.

this statement should be considered as proven only in the limit of small concentrations.

Now, let us consider the case $n \geq 4$. As it was shown by Bak *et al* (1976 a,b), for the most physically realizable cases of pure system functionals, they do not have stable fixed points. Now, we are going to prove that this statement is true for the same impure systems. In other words, if a pure system has only unstable fixed points, points complementary to them are also unstable in the impure case. Bak *et al* (1976 a,b) found by direct verification that in any system with $n \geq 4$ (considered in their work,) pure fixed points possess the following property. For each of the fixed points $(u_1^*, \ldots, u_N^*)$ one can compose such a linear combination, satisfying the condition $\sum_p A_p u_p^* = 0$, that the RG equation for the variable $\sum_p A_p u_p^*$ in the vicinity of this point has the form

$$\frac{d}{dl}\left(\sum_p A_p u_p\right) = \lambda \left(\sum_p A_p u_p\right), \tag{7.68}$$

where the parameter λ is a strictly positive value defined by

$$\lambda = 1 - \sum_p a_p x_p > 0$$

(parameters a_p are different for different fixed points). This positiveness of λ constitutes an instability of a given fixed point with respect to the motion along the vector $\sum_p A_p u_p^*$. One can obtain an equation for the impure case, corresponding to Eq. (7.68) using the substitution $x_p \to x_p'$ $(x_p = (1+6y)x_p')$. As a result we have

$$\frac{d}{dl}\left(\sum_p A_p u_p'\right) = \lambda' \left(\sum_p A_p u_p'\right),$$

where $\lambda' = 1 - \sum_p a_p x_p^i + 6y$ and the values x_p^i and y are defined by the impure fixed point. One can transform the expression for λ' as follows

$$\lambda' = 1 - \sum_p a_p x_p (1+6y) + 6y = \left(1 - \sum_p a_p x_p\right)(1+6y) = \lambda(1+6y).$$

Recalling that $y = \Delta^*/\epsilon > 0$ and $\lambda > 0$ we have $\lambda' > 0$. Thus, if at $n \geq 4$ the pure fixed point is unstable, then the complementary impure point is also unstable, i.e. the appearance of stable fixed points under the influence of impurities is impossible.

7.4.7 Systems with Competing Interactions

Here we discuss the influence of impurities on the critical behavior in the systems with two interacting fields (see Lisyansky and Fillipov (1987 b)). Let us assume that the number of components is different for each of the fluctuating fields ϕ_1 and ϕ_2, so that the free energy functional for a pure system should be written as

$$H_1 = \int d\boldsymbol{r} \left\{ \frac{\tau_1}{2} \phi_1^2(\boldsymbol{r}) + \frac{\tau_2}{2} \phi_2^2(\boldsymbol{r}) + \frac{1}{2} \left(\nabla \phi_1(\boldsymbol{r})\right)^2 + \frac{1}{2} \left(\nabla \phi_2(\boldsymbol{r})\right)^2 \right.$$

$$\left. + \frac{u_1}{8} \left(\phi_1^2(\boldsymbol{r})\right)^2 + \frac{u_2}{8} \left(\phi_2^2(\boldsymbol{r})\right)^2 + \frac{v}{4} \phi_1^2(\boldsymbol{r}) \phi_2^2(\boldsymbol{r}) \right\}. \tag{7.69}$$

In principle, one can consider the case when the functional Eq. (7.69) also contains (at $n_1 = n_2$) the invariant $(\phi_1 \cdot \phi_2)^2$. However, this will additionally complicate the RG analysis, but does not qualitatively change the set of obtained data. In view of the results discussed in the previous sections, one can see that introduction of impurities can be made through the addition to the Hamiltonian Eq. (7.69) of the term

$$H_{imp} = \int d\boldsymbol{r} \left\{ \frac{\tau_1(\boldsymbol{r})}{2} \phi_1^2(\boldsymbol{r}) + \frac{\tau_2(\boldsymbol{r})}{2} \phi_2^2(\boldsymbol{r}) \right\},$$

where $\tau_1(\boldsymbol{r})$ and $\tau_2(\boldsymbol{r})$ are random Gaussian functions with the correlators

$$\overline{\tau_1(\boldsymbol{r})} = \overline{\tau_2(\boldsymbol{r})} = 0 \qquad \overline{\tau_i(\boldsymbol{r}) \tau_j(\boldsymbol{r})} = \Delta_{ij} \delta(\boldsymbol{r} - \boldsymbol{r}').$$

Here, we have left the correlator Δ_{12} different from zero, though the local change of the trial critical temperature for one of the fields may not influence the trial temperature for the other field. Such a situation can be realized, for instance, when fluctuations of each of the fields are interacting with different kinds of defects. Thus, the trial value Δ_{12} should be small. However, as is shown below, the vertex Δ_{12} in the general case is generated in RG equations and it cannot be put to zero. Fortunately, in many cases the generated value of Δ_{12} is small and in the stable fixed point is equal to zero. This means that in a number of cases one can treat the vertex Δ_{12} as irrelevant.

The critical thermodynamics for the system with the functional

$$H = H_1 + H_{imp} \tag{7.70}$$

is governed by RG equations for the vertices u_1, u_2, v, Δ_{11}, Δ_{12}, and Δ_{22}. These equations can be obtained with the help of the general procedure of the replica method (see Sect. 7.4.3), which results in

$$
\begin{aligned}
\frac{du_i}{dl} &= \epsilon u_i - \frac{n_i+8}{2}u_i^2 - \frac{n_{j\neq i}}{2}v^2 + 6u_i\Delta_i, \\
\frac{dv}{dl} &= v\left[\epsilon - \sum_{i=1}^{2}\left(\frac{n_i+2}{2}u_i - \Delta_{ii}\right) - 2v + 4\Delta_{12}\right], \\
\frac{d\Delta_{ii}}{dl} &= \Delta_{ii}\left[\epsilon - (n+2)u_i + 4\Delta_{ii}\right] - n_{j\neq i}\Delta_{12}v, \\
\frac{d\Delta_{12}}{dl} &= \Delta_{12}\left[\epsilon - \sum_{i=1}^{2}\left(\frac{n_i+2}{2}u_i - \Delta_{ii}\right) + 2\Delta_{12}\right] \\
&\quad -v\sum_{i=1}^{2}\frac{n_i}{2}\Delta_{ii}.
\end{aligned}
\tag{7.71}
$$

A comprehensive analysis of the set, Eq. (7.71), has not yet been done. Here we apply the approach already tried in previous sections and study this set with the help of confining (flow motion) hyperplanes. Then, using continuity of the integrals for a set of differential equations, we will extrapolate, so far as possible, the solutions to the rest of parametric space.

Symmetric Case $\Delta_{11} = \Delta_{22}$, $u_1 = u_2$, $n_1 = n_2$

These conditions reduce the set Eq. (7.71) to

$$
\begin{aligned}
\frac{du}{dl} &= u\left[\epsilon - \frac{n+8}{2}u + 6\Delta\right] - \frac{n}{2}v^2, \\
\frac{dv}{dl} &= v[\epsilon - (n+2)u - 2v + 2\Delta + 4z], \\
\frac{dz}{dl} &= z[\epsilon - (n+2)u + 2\Delta + 2z] - nv\Delta, \\
\frac{d\Delta}{dl} &= \Delta[\epsilon - (n+2)u + 4\Delta] - nvz,
\end{aligned}
\tag{7.72}
$$

where we introduce the notation

$$\Delta_{11} = \Delta_{22} \equiv \Delta, \qquad u_1 = u_2 = u, \qquad \Delta_{12} = z.$$

It is evident that the symmetry restrictions confine flow lines to the chosen hyperplane, so that the set Eq. (7.72) gives quite exhaustive characteristics of the phase portrait. As usual, let us switch to ratios of vertices Δ, u, z, and v with the help of the following normalization

$$f_1 = u/\Delta, \qquad f_2 = v/\Delta, \qquad f_3 = z/\Delta.$$

As a result, the set Eq. (7.72) is reduced to the following three equations

$$
\begin{aligned}
\frac{1}{\Delta}\frac{df_1}{dl} &= 2f_1 + nf_1f_2f_3 - \frac{n}{2}f_2^2 - \frac{4-n}{2}f_1^2, \\
\frac{1}{\Delta}\frac{df_2}{dl} &= -2f_2(1+f_2) + nf_2^2f_3 + 4f_2f_3, \\
\frac{1}{\Delta}\frac{df_3}{dl} &= (f_3-1)(nf_2(1+f_3)+2f_3).
\end{aligned}
\tag{7.73}
$$

Owing to the cancellation of terms containing f_1, the two last equations of the set Eq. (7.73) can be considered independently of the first one. On the hyperplane (f_2, f_3) we have the following fixed points

	f_2^*	f_3^*
μ_1^*	0	0
μ_2^*	0	1
μ_3^*	$-2/(n-2)$	1
μ_4^*	$-2(1-f_3^*)/(n+2)$	$[-(n+2)+\sqrt{(n+2)^2+4n^2}]/2n$
μ_5^*	$-2(1-f_3^*)/(n+2)$	$[-(n+2)-\sqrt{(n+2)^2+4n^2}]/2n$

one of which (μ_1^*) is stable. The flow line portrait on the plane (f_2, f_3) is shown in Fig. 7.11 (a). In this figure one can see that flow lines starting on the left hand side of the straight line $f_3 = 1$ end up at the stable fixed point μ_1^*. At this point we have $f_2^* = f_3^* = 0$ which means that interaction of the fields (direct with the vertex v and indirect by means of the vertex Δ_{12}) is absent. This important property will be used a little later. When $n < 2$ the phase portrait is different (see Fig. 7.11 (b)); however, the fixed point μ_1^* is a stable point as before. All the differences between these two cases occur in the flows along the separatrix line $f_3 = 1$. Let us study RG flows on this line. This line is really a special line only because of the conditions $\Delta_{11} = \Delta_{22} = \Delta_{12}$ occurring along it. This means that if, for some or another reason, both of the fluctuating fields have equal trial correlators Δ_{ij}, this equality will be conserved in the whole critical region. For instance, such a situation can be realized in magnets with competing interactions where both of the fluctuating fields have the same physical nature. However, even for the cases where this condition is not fulfilled it makes sense to analyze this case owing to its mathematical simplicity and

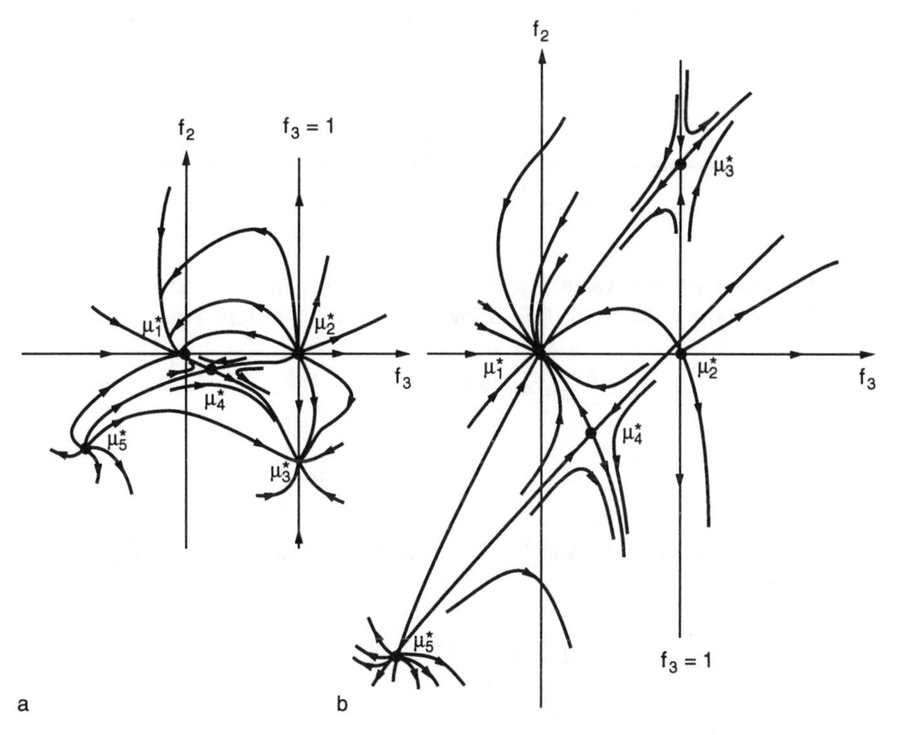

Figure 7.11: The flow line portrait on the (f_2, f_3) plane when $n > 2$ (a) and $n < 2$ (b).

instructiveness. Inserting $\Delta = z$ in the set Eq. (7.72) we arrive at

$$\begin{aligned} \frac{du}{dl} &= \epsilon u - \frac{n+8}{2}u^2 - \frac{n}{2}v^2 + 6u\Delta, \\ \frac{dv}{dl} &= v[\epsilon - (n+2)u - 2v + 6\delta], \\ \frac{d\Delta}{dl} &= \Delta[\epsilon - (n+2)u + 4\Delta - nv]. \end{aligned} \tag{7.74}$$

If one chooses as new variables the ratios $x = u/v$ and $y = \Delta/v$, then the set Eq. (7.74) is reduced to the following two independent equations

$$\begin{aligned} \frac{1}{v}\frac{dx}{dl} &= \frac{n-4}{2}x^2 + 2x - \frac{n}{2}, \\ \frac{1}{v}\frac{dy}{dl} &= -y[2y + (n-2)], \end{aligned} \tag{7.75}$$

which can be integrated exactly. The integral curves for the set Eq. (7.75) are given by

$$\begin{aligned} y &= (n-2)\left[c^{\frac{(n-4)(x-1)}{(n-4)x+n}} - 2\right]^{-1}, & n &\neq 2, \\ y &= \frac{x-1}{2+c(x-1)}, & n &= 2, \end{aligned}$$

where c is an arbitrary constant. The full list of fixed points for the set Eq. (7.75) contains, excepting the trivial Gaussian point $\mu_0^* = (0,0,0)$, the following seven points

	u^*	v^*	Δ^*
μ_1^*	$2\epsilon/(n+8)$	0	0
μ_2^*	$\epsilon/(n+4)$	$\epsilon/(n+4)$	0
μ_3^*	$n\epsilon/(n^2+8)$	$(4-n)\epsilon/(n^2+8)$	0
μ_4^*	0	0	$-\epsilon/4$
μ_5^*	$\epsilon/2(n-1)$	0	$(4-n)\epsilon/8(n-1)$
μ_6^*	$\epsilon/2(2n-1)$	$\epsilon/2(2n-1)$	$(2-n)\epsilon/4(2n-1)$
μ_7^*	$n\epsilon/2a_n$	$(4-n)\epsilon/4a_n$	$(n-4)(n-2)\epsilon/4a_n$

where $a_n = (n-1)(8-n)$. The points μ_2^*, μ_3^*, μ_6^*, and μ_7^* with $v^* \neq 0$ are seen in Fig. 7.12 (a), (c). When $n \to 1$ the point μ_7^* tends to infinity and is shown in Fig. 7.12 only conditionally. It is interesting to notice that stability of the points μ_3^* and μ_6^* are exchanged when the number

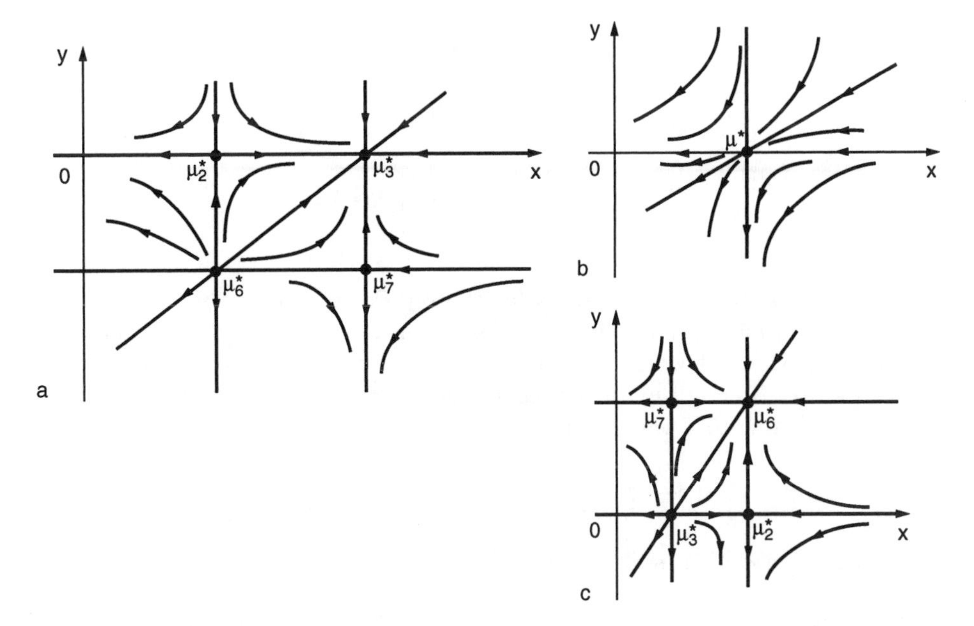

Figure 7.12: The flow line portrait on the hyperplane $f_3 = 1$ when $n = 1$ (a), $n = 2$ (b) and $n = 3$ (c).

of components crosses $n = 2$. If at $n = 1$ we have the pure fixed point with $\Delta^* = 0$ as a stable point, then at $n = 3$ we have in its place the point μ_6^* with $\Delta^* \neq 0$. The stable fixed point at $n = 3$ is present in the first ϵ-approximation and the vertices are of the order of ϵ. In the similar situation, the Ising magnet has vertices of the order $\sqrt{\epsilon}$ (see Sect. 7.4.3).

Small Δ_{12}

Let us now return to the set Eq. (7.73). As is seen in Fig. 7.11, the flow lines beginning in the vicinity of the vertical line $f_3 = 0$ are sharply bent down to the fixed point μ_1^*. That means, that though the vertices $z = \Delta_{12}$ are being generated at the trial value $z = 0$, their generation is a small value of the order of $v\Delta$, i.e. of the order of ϵ^2. Therefore, in this case one may neglect contributions of the order of zv in equations describing RG evolution of other vertices. As a result the set, Eq. (7.71) can be reduced to

$$
\begin{aligned}
\frac{du_i}{dl} &= \epsilon u_i - \frac{n_i+8}{2} u_i^2 - \frac{n_{j \neq i}}{2} v^2 + 6 u_i \Delta_i, \\
\frac{dv}{dl} &= v \left[\epsilon - \sum_{i=1}^{2} \frac{n_i+2}{2} u_i - \Delta_{ii} \right], \\
\frac{d\Delta_{ii}}{dl} &= \Delta_{ii} \left[\epsilon - (n+2) u_i + 4\Delta_{ii} \right],
\end{aligned}
\tag{7.76}
$$

where $i = 1, 2$.

When $n_1 = n_2$ fixed points for the set Eq. (7.76) can be found with the help of algebraic calculations. As a result, one can find 16 fixed points ($n > 1$), three of which are unphysical ($\Delta_{11}^* < 0$, or $\Delta_{22}^* < 0$, or $\Delta_{11}^* = \Delta_{22}^* < 0$). Among the other points, when $n > 4$ as before, only the point corresponding to a pure material is stable

$$
u_1^* = u_2^* = \frac{2\epsilon}{n+8}, \qquad v^* = \Delta_{11}^* = \Delta_{22}^* = 0,
$$

when $1 < n < 4$ the stable point is

$$
u_1^* = u_2^* = \frac{2\epsilon}{2(n-1)}, \qquad \Delta_{11}^* = \Delta_{22}^* = \frac{\epsilon(4-n)}{8(n-1)}, \qquad v^* = 0. \tag{7.77}
$$

This point has the symmetry $\mathcal{O}(n) \oplus \mathcal{O}(n)$ and corresponds to noninteracting fields $\boldsymbol{\phi}_1$ and $\boldsymbol{\phi}_2$. The critical state of these fields is defined by the critical exponents corresponding to the impure stable fixed point. The existence of this point may lead to a specific "impure-fluctuational" suppression of interaction between subsystems. The latter in its turn suppresses development of the fluctuational instability, which otherwise would lead to the first-order transitions. On the other hand, at a substantial strength of the initial coupling between fields $\boldsymbol{\phi}_1$ and $\boldsymbol{\phi}_2$ and small impurity concentrations (small Δ_{ij}) the presence of impurities may be insufficient to suppress the coupling between the subsystems. As a result the critical behavior of the system should remain similar to the pure case.

Flow portrait at $\Delta_{11} = \Delta_{22}$, $u_1 = u_2$, and $\Delta_{12} = 0$

In order to find the region of small Δ_{12} values, one has to consider another mathematical question. This problem is equivalent to the definition of the separatrix demarcating attraction regions to the fixed point Eq. (7.77). Usually, such a problem can be solved numerically. However, in this case it is possible to find a class of solutions for the set Eq. (7.76), with the help of which the separatrix can be found analytically. These solutions can be obtained on the hyperplane $\Delta_{11} = \Delta_{22} = \Delta$, $u_1 = u_2 = u$, and

$\Delta_{12} = 0$. The set Eq. (7.76) for this hyperplane is reduced to the system of three equations for the vertices u, v, and Δ or two equations for the ratios $f_1 = u/\Delta$ and $f_2 = v/\Delta$

$$\begin{aligned} \frac{1}{\Delta}\frac{df_1}{dl} &= \frac{n-4}{2} f_1^2 + 2f_1 - \frac{n}{2} f_2^2, \\ \frac{1}{\Delta}\frac{df_2}{dl} &= -2f_2(f_2+1). \end{aligned} \tag{7.78}$$

This set has the following four fixed points

	f_1^*	f_2^*
μ_a^*	0	0
μ_b^*	$4/(4-n)$	0
μ_c^*	1	-1
μ_d^*	$n/(4-n)$	-1

In any case ($n < 4$ or $n > 4$) the point μ_b^* is a stable fixed point. Five topologically different flow line portraits are shown in Fig. 7.13. On each of these portraits one can see the separatrix crossing points μ_a^*, μ_c^*, and μ_d^*, which demarcates the region of attraction from the instability region. The specific form of the set Eq. (7.78) allows us not only to find an analytical solution for the separatrix, but also to integrate the whole set. The solution for f_2 can be found in the form

$$f_2(l) = \frac{ce^{-2l}}{1 - ce^{-2l}}. \tag{7.79}$$

Substituting in the first equation of the set Eq. (7.78) the expression for $f_2(l)$ from Eq. (7.79) we arrive at Riccarti's differential equation. The solution for this equation can be found in elementary functions with one unknown partial solution. Here we are not going to write the explicit expression for the function $f_1(l)$ but rather represent it as a function of f_2. The partial solutions in this case can be easily found. In Fig. 7.13 one can see, that when $n \neq 4$ the following four straight lines are partial integrals of the set Eq. (7.78)

$$f_1 = f_2 + \frac{4}{4-n}, \quad f_1 = \frac{nf_2 + 4}{4-n}, \quad f_2 = -1, \quad f_2 = 0. \tag{7.80}$$

The case $n = 4$ is a special case in the sense that the first two integrals in Eq. (7.80) do not exist. The specific nature of this case can be also seen from the set Eq. (7.78) because at $n = 4$ the factor at f_1^2 is equal to zero and

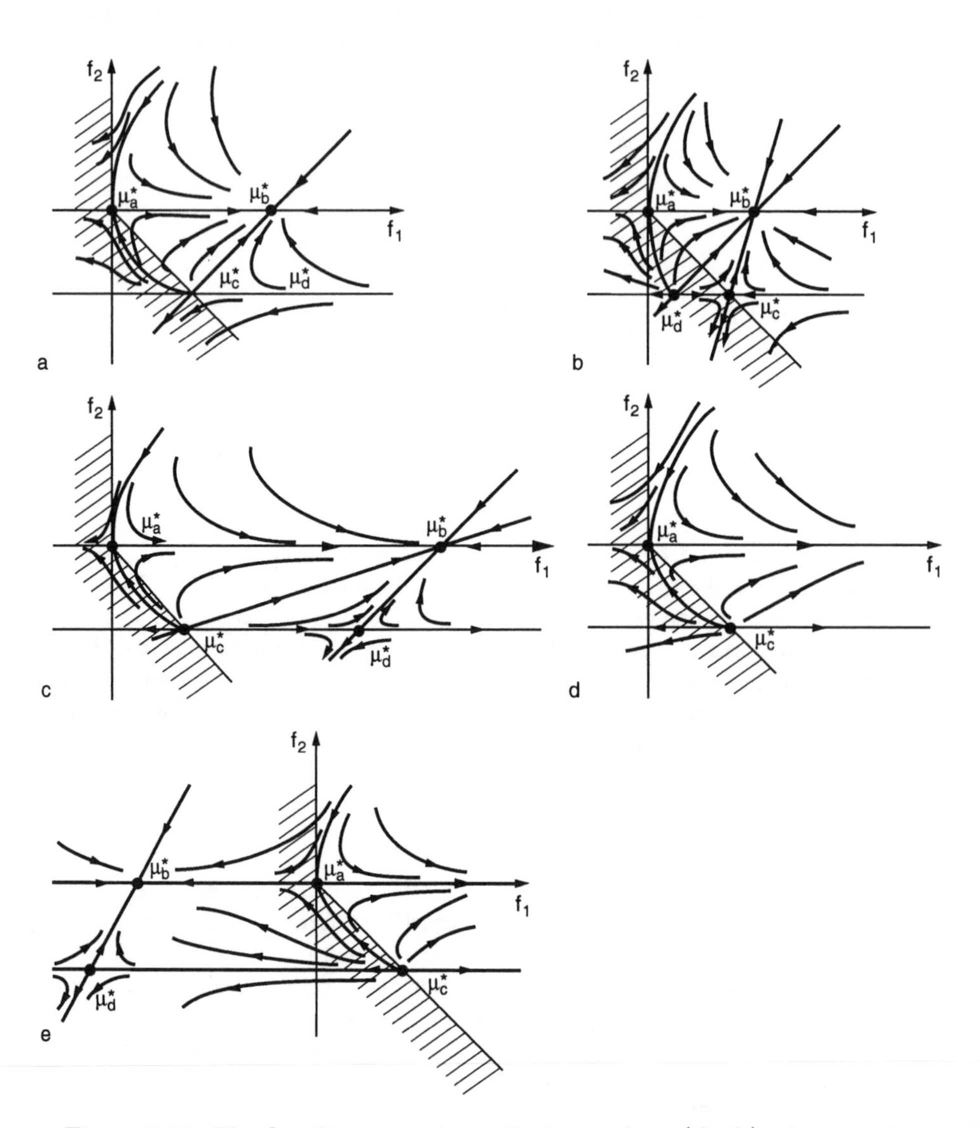

Figure 7.13: The flow line portrait on the hyperplane (f_1, f_2) when $n = 1$ (a), $n = 2$ (b), $n = 3$ (c), $n = 4$ (d), and $n = 5$ (f).

the first equation of this set reduces to the linear one. As a result of these deliberations the following equations for the flow lines can be obtained

$$
\begin{aligned}
f_1 &= f_2 + \tfrac{4}{4-n} + \tfrac{2(n-2)f_2}{(4-n)[1+(\alpha-1)|1+f_2|^{(n-2)/2}}, & n &\neq 2,4, \\
f_1 &= f_2 + 2 + \tfrac{f_2}{\alpha - \ln\sqrt{|1+f_2|}}, & n &= 2, \\
f_1 &= \tfrac{1+f_2}{f_2}\left[\alpha + 1 + f_2 - 2\ln|1+f_2| - \tfrac{1}{1+f_2}\right], & n &= 4,
\end{aligned}
$$

where α is an arbitrary constant. The separatrix consists from the flow line at $\alpha = 0$ and the ray of the straight line $f_2 = -1$ coming out of the repulsive center. The stability region of the functional is defined by the straight lines $f_1 = 0$ where $f_2 > 0$ and $f_1 = -f_2$ where $f_2 < 0$. As a result of the cross-section of the separatrix and stability boundary two regions of flow line runaway occur.

It is interesting to compare the picture of the integral curves with the analogous picture for the pure case. When $\Delta \equiv 0$ the set Eq. (7.76) at $u_1 = u_2$ reduces to the following two equations

$$
\begin{aligned}
\tfrac{du}{dl} &= \epsilon u - \tfrac{n+8}{2}u^2 - \tfrac{n}{2}v^2, \\
\tfrac{dv}{dl} &= v[\epsilon - (n+2)u - 2v].
\end{aligned}
\tag{7.81}
$$

The portrait of flow lines for this set is shown in Fig. 7.14. The separatrixes in this case are defined by the following evident equations

$$
\begin{aligned}
u &= \tfrac{nv}{4-n}, & v &= 0, & n &< 2, \\
u &= v, & v &= 0, & 2 < n &< 4, \\
u &= v, & u &= \tfrac{nv}{4-n}, & n &> 4.
\end{aligned}
$$

Comparison of pure and impure cases shows that when $n < 4$, impurities lead to the increase of the region of initial parameters which results in a second-order phase transition. This can be interpreted as suppression of the fluctuational instability by impurities. The case $n > 4$ should be discussed separately. In Fig. 7.14 (c) one can see the phase portrait for $n = 5$. Flow line portraits for different n for this case ($n > 4$) are topologically equivalent to Fig. 7.14 (c). The stable fixed point in this case corresponds

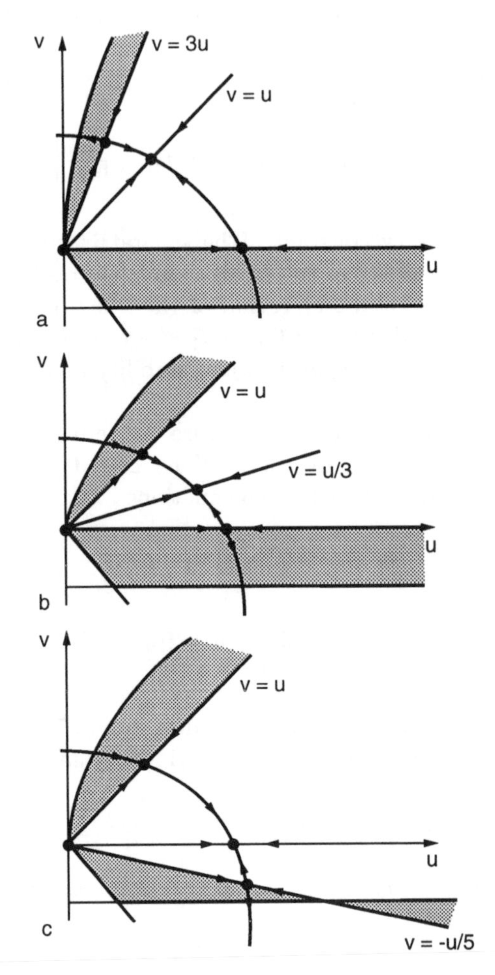

Figure 7.14: Comparison of the attraction regions in pure and impure cases when $n = 1$ (a), $n = 3$ (b), and $n = 5$ (c). The vertically-shaded regions are additions caused by impurities, while the slant-shaded region is a consequent contraction.

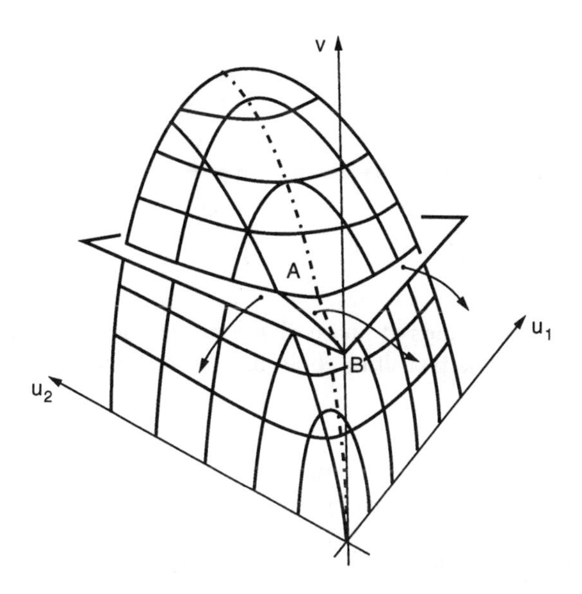

Figure 7.15: Schematic depiction of the cross-section of the separatrix surface given by Eq. (7.71) with the hyperplane constituted by $\Delta_1 = const$, and $\Delta_2 = const$.

to the splitting of the fields ($v^* = 0$). The flow line runaway in the region $f_1 \to \infty$ is caused by the normalization all of vertices in Eq. (7.78) by Δ and in reality it corresponds to $\Delta \to 0$.[12] As before, impurities lead to the change of the region of attraction. However, this case is different from the case $n < 4$. When $v > 0$ the attraction region is increased as before, while in the contrary case $v < 0$, depending on the ratio of the trial vertices, continuous as well as discontinuous phase transitions are possible in regions additional to the case $\Delta = 0$.

Phase Diagrams

Let us now return to the case $u_1 \neq u_2$, $\Delta_{11} \neq \Delta_{22}$ remaining with $n_1 = n_2$ for some time (as before $\Delta_{12} \ll \Delta_{ii}, u_i$). In this case an analytical solution cannot be obtained. Results of numerical calculations are shown in Fig. 7.15. In this figure one can see the cross-section of the separatrix surface with the hyperplane defined by the conditions $\Delta_1 = \overline{\Delta}_1$ and $\Delta_2 = \overline{\Delta}_2$ where $\overline{\Delta}_1$ and $\overline{\Delta}_2$ are some constants. In the mean field approximation

[12]As along with Δ the vertex v also tends to zero, the runaway occurs only along the axis f_1.

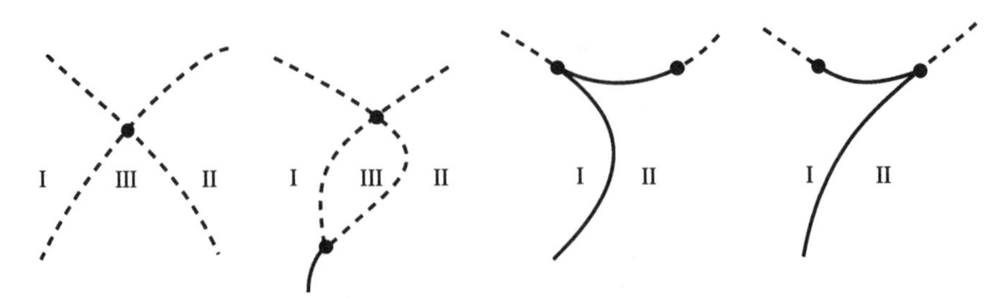

Figure 7.16: Possible kinds of phase diagrams. Dashed lines correspond to the second-order phase transition; bold lines – to the first-order transition.

when $v^2 < u_1 u_2$ the phase diagram with a tetracritical point is obtained. The separatrix surface is crossed by the surface $v^2 = u_1 u_2$ only when $u_1 = 0$ or $u_2 = 0$. The latter means that if the initial vertices satisfy the inequality $v^2 < u_1 u_2$, the phase flow lines will always reach the stable fixed point. Since in this point we have $v^* = 0$ (and, therefore, the inequality is satisfied), the phase diagram is of a tetracritical kind just as in the mean field approximation (see Fig. 7.16 (a)). Owing to the same reason ($v^* = 0$), the appearance of phase diagrams with the bicritical points is impossible as in this case we should have to fulfill the inequality $(v^*)^2 > u_1^* u_2^*$. Nonetheless, for the initial vertices one can imagine a situation where they are positioned between the surface $v^2 = u_1 u_2$ and the separatrix. In this case we also have a second-order phase transition but with parameters $v^* \neq 0$, u_1^*, u_2^*, satisfying, as before, $(v^*)^2 < u_1^* u_2^*$. In this case the phase diagram corresponds to the one shown in Fig. 7.16 (b).

So far we have discussed only the cases when flow lines converge to the stable fixed point. Let us now consider the possible types of phase diagrams when continuous transitions are transformed into discontinuous ones. We assume, at first, that $\overline{\Delta}_1 = \overline{\Delta}_2$. In this case the numerical analysis shows, that for $v > 0$ the fluctuational renormalization essentially changes only the parameter u_i with the smaller initial value, i.e. if $u_1 > u_2$ at $l = 0$, then at approaching the critical point $u_2 \to 0$ and the field ϕ_2 will be condensed. The phase diagram in this case has a "beak" restricted by lines of first-order transitions. When $\overline{\Delta}_1 \neq \overline{\Delta}_2$, the situation is a little different. Let us assume, for instance, that $\overline{\Delta}_1 > \overline{\Delta}_2$. Now, instead of the bisectorial plane $u_1 = u_2$ demarcating regions of the first-order transitions to the phases $\phi_1 \neq 0$, $\phi_1 = 0$ and $\phi_2 = 0$, $\phi_2 \neq 0$, we have a surface in the u_1, u_2, v space. The section of this surface by the separatrix surface and the plane $v = const$ is shown in Fig. 7.15 by the bold line. The dashed-dotted line in this figure is the section of the same surface by the plane $u_1 = u_2$. When at a given value v the point (u_1, u_2) is located at the left-hand side from the curve AB (naturally $u_1 > u_2$), then independently of the sign of the difference[13] $\tau_1 - \tau_2$, the first-order phase transition occurs to the phase $\phi_1 = 0$, $\phi_2 \neq 0$. When the point (u_1, u_2) is at the right hand side[14] of the curve AB, then the first-order transition occurs to the phase $\phi_1 \neq 0$, $\phi_2 = 0$. These two cases correspond to the phase diagrams shown in Fig. 7.16 (c,d).

Let us now return to the set Eq. (7.71) and consider the situation with $n_1 \neq n_2$. In this case we have the stable fixed point $(1 < n_1, n_2 < 4)$ with

$$u_i^* = \frac{\epsilon}{2(n_i - 1)}, \quad \Delta_{ii}^* = \frac{4 - n_i}{8(n_i - 1)}\epsilon, \quad \Delta_{12}^* = v^* = 0, \tag{7.82}$$

which corresponds to the functional of two independent impure subsystems. When $n_1, n_4 > 4$ this point is changed for the "pure" point without interaction

$$u_i^* = \frac{2\epsilon}{n_i + 8}, \qquad \Delta_{ii}^* = \Delta_{12}^* = v^* = 0. \tag{7.83}$$

The mixed point can also be found, when $n_1 > 4$ and $1 < n_2 < 4$. This point is defined by

$$u_1^* = \frac{2\epsilon}{n_1 + 8}, \quad u_2^* = \frac{\epsilon}{2(n_2 - 1)}, \quad \Delta_{11}^* = 0,$$

$$\Delta_{22}^* = \frac{4 - n_2}{8(n_2 - 1)}\epsilon, \qquad \Delta_{12}^* = v^* = 0. \tag{7.84}$$

The difference between the values u_i^* and Δ_{ij}^* may lead to different renormalizations of T_1 and T_2. As a result the initial inequality $T_1 - T_2 > 0$ may be changed to the opposite $T_1^* - T_2^* < 0$. The renormalization of τ_1^* is greater then the renormalization of τ_2^* provided: a) $n_1 > n_2$ at the point defined by Eq. (7.82), b) $n_1 < n_2$ at the point defined by Eq. (7.83), c) $n_1 < 8(n_2 + 8)/(5n_2 - 8)$ at the point defined by Eq. (7.84). Hence, an appropriate relation between the parameters T_1, T_2, n_1, and n_2 leads to the situation when at $T_1 > T_2$ the field ϕ_2 is ordered. The anomalous phase, from the context of the Landau theory, occurs as a result of the second-order phase transition. The latter means that the tetracritical point on the phase diagram is translated closer to the region of stability of the phase $\phi_2 \neq 0$, $\phi_1 = 0$.

In conclusion, let us briefly repeat all the consequences resulting from

---

[13]The values τ_1 and τ_2 are assumed to be quite close to each other.

[14]This case is possible if either of the inequalities holds $u_1 > u_2$ or $u_1 > u_2$.

the presence of impurities in the critical region. First of all, a small concentration of impurities does not smear second-order phase transitions. More than that, according to Harris's criterion, the negative specific heat exponents ($\alpha < 0$) lead to the conservation of the "pure" system critical behavior. If we have $\alpha > 0$, then the second-order phase transition remains, but the critical exponents are different from the "pure" system. The presence of impurities in the system with two interacting fields, described by the Hamiltonian Eq. (7.70), leads to the splitting of the fluctuating fields. As a result, the appearance of the asymptotic symmetry at the crossing point of the lines corresponding to second-order transitions, becomes impossible.[15] In a number of cases, impurities lead to the suppression of the fluctuational instability, which transforms the first-order transition into a second-order one. Finally, instead of the phase diagram with the bicritical point, a new kind of phase diagram appears (see Fig. 7.16 (b)). This diagram is characterized by the presence of both, bicritical and tetracritical points.

[15] Such a situation is typical for "pure" systems.

Chapter 8

Exactly Solvable Models and RG

In previous chapters we have shown that fluctuations lead to a peculiar picture of phase transitions in systems with competing interactions. This picture may be qualitatively different from predictions of the Landau theory. However when one, in a more complex calculation, obtains results that contradict simple physical consideration (Landau theory), the natural question arises: whether the data obtained reflect genuine physical reality or resulted from the approximations made. This doubt may be especially justified in relation to the RG approach, as all practical calculations are usually made by expanding over the parameter of the order of unity. Even though critical exponents calculated, by means of the fluctuation theory agree well with experimental data, there is no evidence that qualitative effects also should be seen in experiment. Keeping this in mind, in this chapter we study critical behavior of systems with competing interactions in the framework of simple, exactly solvable models. In these models the interaction of fluctuations is only partially taken into account. The latter leads to exact solvability — including the influence of fluctuations on the phase transition. The results obtained in the framework of such an approach are very instructive and easily interpreted.

8.1 Fluctuation Effects in Spherical Model

The spherical model was advanced by Berlin and Kac in 1952. It is one of not many exactly solvable models demonstrating a phase transition in ferromagnets. Excepting its mathematical simplicity, the model has a number of peculiarities which lead it to the ranks of quite successful approximations. First of all, as was shown by Stanley (1968), this model can be considered as the limit of infinite spin dimensionality for Heisenberg's model. The main approximation of the spherical model is the replacement of a large number of independent conditions $\boldsymbol{S}_l^2 = 1$, of Heisenberg's model, by the single requirement $\sum_l \boldsymbol{S}_l^2 = N$. Such a replacement can be justified in the limit of an infinitely large number of spin components. In this case each component may essentially fluctuate, while the "length" only slightly deviates from the average value. Hence, the sum $\sum_l \boldsymbol{S}_l^2$ can be treated like a constant. Thus, the Heisenberg model can be considered as a model restricted from below by Ising's model (single component spins) and from above by the spherical model (infinite component spins).

One of the attractive features of the spherical model is the absence of phase transitions for dimensions $d \leq 2$.[1] The latter agrees with the general theorem by Mermin and Wagner (1966), according to which, in the plane and one-dimensional systems with continuous degeneracy, the mean value of the order parameter should be equal to zero. Nonetheless, owing to the mathematical simplicity, this model was successfully applied by Barber and Fisher in 1973 to the study of quasi two-dimensional systems having finite thickness.

Up to the 1980s, the main applications of the spherical model were restricted to considerations of cubic lattices with the nearest-neighbor interactions (see, for instance, the review by Joyce (1972)). As a result this model was mainly applied to the study of ferro- and antiferromagnets (see Theuman (1969)). In the following sections this model is applied to the study of magnetic structures resulting from an arbitrary space dispersion of exchange integrals, and symmetry of crystal lattice. The results obtained in these cases differ essentially from the results obtained for cubic lattices. First of all, this difference is seen in the retaining of phase transitions even in the presence of external magnetic fields. The consideration of the spin condensation in particular states below the critical point in explicit form helps to reveal the physical sense of the "sticking" to the saddle point found by Berlin and Kac (1952).

As has been already mentioned, fluctuations of the spin interactions are only partially taken into account in the spherical model. As a result, the critical exponent δ coincides with the scaling value (the exponent $\eta = 0$) and the critical temperature is close to the genuine value. However, in order to understand whether these coincidences are properties of the model or consequences of the fluctuation effects, one should be able to "switch off"

[1] This statement is true for finite-range interactions. For infinite-range interactions the phase transition is present even at $d \leq 2$ (see Joyce (1966)).

the fluctuations and control the width of the fluctuation region. Having this in mind, we consider the situation (see Ivanchenko *et al* (1986 a)) when the interaction radius c essentially exceeds the lattice spacing a. Now, out of the fluctuation region (defined by the Ginzburg criterion (see Sect. 1.6)) results of the fluctuation theory and the mean-field theory should coincide, as in this region mean-field theory can be considered as an exact theory. Thus, one can say that the spherical model more or less adequately takes into account fluctuations only in the case when the Ginzburg parameter appears in a natural manner, and results obtained in this theory and the mean-field theory are different only in the critical region with the width $G_i \sim (a/c)^6$.

8.1.1 Inhomogeneous Ordering

Let us return to the Heisenberg's Hamiltonian

$$H = -\frac{1}{2}\sum_{ll'} J_{ll'}\boldsymbol{S}_l\boldsymbol{S}_{l'} - \boldsymbol{h}\sum_l \boldsymbol{S}_l,$$

where the vectors $\boldsymbol{S}$ and $\boldsymbol{l}$ are defined in the n- and d-dimensional spaces. As usual, we are going to calculate the partition function

$$Z = \int D\boldsymbol{S}_l \exp\left\{-\frac{H}{T}\right\}.$$

The main condition of the spherical model can be taken into account by the introduction in the integral δ-function

$$\delta\left(\sum_l \boldsymbol{S}_l^2 - N\right) = \int_{-\infty}^{\infty}\frac{dx}{2\pi}\exp\left[ix\left(\sum_l \boldsymbol{S}_l^2 - N\right)\right].$$

After the replacement $x \to ix/2T$, the expression for the partition function takes the form

$$Z = \int_{-i\infty}^{i\infty}\frac{dx}{4\pi iT}\int_{-\infty}^{\infty} D\boldsymbol{S}_l \exp\left\{\frac{N}{2T}\left[x - \frac{1}{N}\sum_{ll'}(x\delta_{ll'} - J_{ll'})\boldsymbol{S}_l\boldsymbol{S}_{l'} \right.\right.$$
$$\left.\left. + \frac{2\boldsymbol{h}}{N}\sum_l \boldsymbol{S}_l\right]\right\}.$$

The bilinear form with respect to the vectors $\boldsymbol{S}_l$ can be diagonalized with the help of Fourier's transformation, i.e.

$$-\sum_{ll'}(x\delta_{ll'} - J_{ll'})\boldsymbol{S}_l\boldsymbol{S}_{l'} + 2\boldsymbol{h}\sum_l \boldsymbol{S}_l = -\sum_q (x - J(\boldsymbol{q}))|\boldsymbol{S}_q|^2 + 2\boldsymbol{h}\sqrt{N}\boldsymbol{S}_o,$$

where the Fourier transformation is defined by

$$\boldsymbol{S}_l = \frac{1}{\sqrt{N}} \sum_q \boldsymbol{S}_q e^{i\boldsymbol{q}\boldsymbol{l}}, \qquad J_{ll'} = \frac{1}{N} \sum_q J(\boldsymbol{q}) e^{i\boldsymbol{q}(\boldsymbol{l}-\boldsymbol{l}')}.$$

Now we have to take into account that the condensing mode at the transition point becomes nonergodic (see Sect. 1.4). Therefore, one should not integrate over this mode. As usual, the wave vector $\boldsymbol{q}_{oi}$ specifying the structure of the ordered phase is defined by the symmetry equivalent maxima of the function $J(\boldsymbol{q})$ (see Sect. 2.2.2). In the following we omit the index i at the vector $\boldsymbol{q}_{oi}$, considering the case where the maxima are reached at the symmetrical points on the boundary of the Brillouin zone. We also denote the value $J(\boldsymbol{q}_o)$ by the symbol J. As a result the partition function can be written as

$$Z \propto \int_{-\infty}^{\infty} dx \exp\left\{\frac{N}{T}\mathcal{F}(x, \boldsymbol{S}, T)\right\}, \tag{8.1}$$

where

$$\mathcal{F} = \frac{1}{2}\left\{x + (J - x)|\boldsymbol{S}|^2 - \frac{nT}{N}\sum_{q \neq q_0} \ln\frac{x - J_q}{T} + \frac{\boldsymbol{h}^2}{(x - J_0)}\right\}. \tag{8.2}$$

In this equation we have introduced the notation $\boldsymbol{S} = \boldsymbol{S}_{q_o}/\sqrt{N}$. In Eqs. (8.1) and (8.2) we assumed that $\boldsymbol{q}_o \neq 0$, i.e. $J \neq J_0 \equiv J(\boldsymbol{q} = 0)$.[2] The fact that the integral in the exponent of Eq. (8.1) is proportional to N ($N \to \infty$) accounts for the application of the saddle point method. In the thermodynamic limit the value $\mathcal{F}(x_s, \boldsymbol{S}, T)$ (here x_s is x at the saddle point) is an exact value of the free energy per magnetic moment in the system. The saddle point equation can be easily defined from Eq. (8.2) as

$$\left.\frac{\partial \mathcal{F}}{\partial x}\right|_{x=x_s} = 0 = 1 - |\boldsymbol{S}|^2 - Tf(x_s) - \frac{\boldsymbol{h}^2}{(x_s - J_0)^2}, \tag{8.3}$$

where the function $f(x)$ is defined by

$$f(x) = \frac{n}{N}\sum_{q \neq q_0} \frac{1}{x - J(\boldsymbol{q})}. \tag{8.4}$$

As the value $\boldsymbol{S}$ describes the nonergodic mode, which in the most general case may not be in a state of thermal equilibrium with the environment, one should also add to Eq. (8.3) the requirement of equilibrium

[2]The case $q_0 = 0$ is considered below in Sect. 8.1.2.

$\partial\mathcal{F}(x_s, \boldsymbol{S}, T)/\partial\boldsymbol{S} = 0$. This condition can also be treated as an equation of state which has the form

$$\frac{\partial\mathcal{F}}{\partial\boldsymbol{S}} = 0 = (J - x_s)\boldsymbol{S}. \tag{8.5}$$

From Eqs. (8.3) and (8.5) one can see that with the appearance the order parameter $\boldsymbol{S}$, the value x_s is "sticking" to the quantity $x_s = J$. The magnitude of the order parameter should be defined from the equation

$$|\boldsymbol{S}|^2 = 1 - Tf(J) - \frac{\boldsymbol{h}^2}{(J - J_0)^2}. \tag{8.6}$$

Let us now give the physical sense of the saddle point "sticking". The equation for the saddle point can also be obtained if one considers the grand partition function for Heisenberg's Hamiltonian interpreting the condition $\sum_q \boldsymbol{S}_q^2 = N$ as a mean value for the sum $\sum_q \boldsymbol{S}_q^2$, defined with the distribution function

$$\rho\{\boldsymbol{S}_q\} \sim \exp\left\{\frac{1}{T}\sum_q \left(\mu + \frac{J(\boldsymbol{q}) - J}{2}\right)|\boldsymbol{S}_q|^2 + \sqrt{N}\boldsymbol{S}_o\boldsymbol{h}\right\}.$$

In this equation we add the constant J to the definition of the Hamiltonian to make its eigenvalues positive

$$H = \frac{1}{2}\sum_q (J - J(\boldsymbol{q}))|\boldsymbol{S}_q|^2 - \sqrt{N}\boldsymbol{h}\boldsymbol{S}_o.$$

Such an addition accounts for the resemblance of the value μ to the chemical potential. The equation $N = T(\partial \ln Z/\partial\mu)_T$, defining this "chemical potential," leads to the equation for the saddle point x_s. Now, it is quite clear that when $\boldsymbol{h} = 0$ we have an analogy with the degenerate Bose gas. The "chemical potential" $\mu = (J - x)/2$ for this gas should always be negative, "sticking" to zero ($x = J$) at the transition point (see Sect. 1.3.1). At this point the condensation of the mode with the wave vector $\boldsymbol{q} = \boldsymbol{q}_{oi}$ occurs. This state will accumulate more and more "particles" with the decrease of temperature, so that at $T = 0$ all other states are unoccupied and the value $\boldsymbol{S}_{q_o}^2 = N$ is macroscopically large. Thus, the "sticking" of the saddle point x_s corresponds to the Bose-Einstein condensation of spin modes.

Let us now return to Eq. (8.6). It can be rewritten in the form

$$|\boldsymbol{S}|^2 = -\left[\frac{T - T_c}{T_c} + \frac{\boldsymbol{h}^2}{(J - J_0)^2}\right], \tag{8.7}$$

where

$$T_c \equiv T_c(\boldsymbol{h}=0) = \frac{1}{f(J)}. \tag{8.8}$$

In the thermodynamic limit for the critical temperature we have

$$T_c = \left[\frac{n\Omega}{(2\pi)^d}\int \frac{d\boldsymbol{q}}{J - J(\boldsymbol{q})}\right]^{-1}, \tag{8.9}$$

where Ω is the volume of the lattice cell. Using Eq. (8.9), one can see that, for the finite-range interactions, when in the vicinity of the point $\boldsymbol{q} = \boldsymbol{q}_{\text{o}}$ the value $J(\boldsymbol{q})$ can be represented as $J(\boldsymbol{q}) \approx J[1 - c(\boldsymbol{q} - \boldsymbol{q}_{\text{o}})^2]$, the integral is divergent for $d \leq 2$. This gives $T_c = 0$ and means that the phase transition is only possible for $d > 2$. In the case of interactions with an infinite radius, when in the vicinity of $\boldsymbol{q} = \boldsymbol{q}_{\text{o}}$ the exchange interaction can be represented as $J(\boldsymbol{q}) \approx J[1 - c(\boldsymbol{q} - \boldsymbol{q}_{\text{o}})^\sigma]$, where $\sigma < 2$, the lower critical dimension is decreased to $d = \sigma$.

In accordance with Eq. (8.7) the critical temperature in the presence of the external field is defined

$$T_c(\boldsymbol{h}) = T_c\left[1 - \frac{\boldsymbol{h}^2}{(J - J_0)^2}\right].$$

Below this temperature the spontaneous magnetization is given by

$$|\boldsymbol{S}| = \left[1 - Tf(J) - \frac{\boldsymbol{h}^2}{(J - J_0)^2}\right]^{1/2}, \tag{8.10}$$

with the component along the direction of the magnetic field equal to

$$S_\parallel = T\frac{\partial \ln Z}{\partial h} = \frac{h}{(J - J_0)}. \tag{8.11}$$

The longitudinal magnetic susceptibility in the paramagnetic phase increases with the decrease of temperature until it reaches $\chi_\parallel = 1/(J - J_0)$. After that it does not change, which means that at the critical point $\chi_\parallel$ has a bend. Thus, in this case when $\boldsymbol{h} \neq 0$ the phase transition remains up to the field $h_c = (J - J_0)$. The physical sense of the result is quite clear. When the temperature is being decreased the stability of the paramagnetic phase is at first destroyed for the modes $\boldsymbol{S}_q$ with the wave vectors located in the vicinity of the symmetry equivalent maxima $\boldsymbol{q}_{\text{o}i}$. Now, the system is condensed into a noncollinear magnetic structure with $\boldsymbol{m}(\boldsymbol{r}) \propto \sum_i \boldsymbol{S}_{q_{\text{o}i}} \exp(i\boldsymbol{q}_{\text{o}i}\boldsymbol{r})$. In the homogeneous case, the phase transition to the ferromagnetic state can be destroyed by any finite, small external field which formally introduces the "condensation" at any temperature. The decrease of temperature only changes the part of spins condensed in the state with $\boldsymbol{S}_{q=0}$. On the contrary, to suppress the transition into the structure of a general kind, one should apply the field higher than some critical value. This value is defined by the condition that at $T = T_c$ all spins are condensed in the state with $\boldsymbol{q} = 0$ and the condensation of the modes with $\boldsymbol{q}_{0i}$ is impossible.

8.1.2 Homogeneous Ordering

Let us now explicitly evaluate the influence of fluctuations on the critical behavior of thermodynamic values. We assume that the interaction radius of the exchange integral $J(\boldsymbol{r})$ essentially exceeds the lattice constant a and the interactions between magnetic moments are of the ferromagnetic type. This means that $J(\boldsymbol{q})$ has a maximum (with the width of the order of c^{-1}) at the center of the Brillouin zone. As in this case the nonergodic mode is the one with $\boldsymbol{q} = 0$, instead of Eqs. (8.3) and (8.5), we have

$$\boldsymbol{S}^2 = 1 - f(x_s)T, \tag{8.12}$$

$$0 = (J_0 - x_s)S + h, \tag{8.13}$$

where $S = S_0/\sqrt{N}$.

Using Eqs. (8.12) and (8.13) one can find that

$$S^2 = T_c\left[f(J_0) - \frac{T}{T_c} f\left(J_0 + \frac{h}{S}\right)\right]. \tag{8.14}$$

The inequality $(a/c)^d \ll 1$ allows to consider the Brillouin zone, in the calculations of the function $f(x_s)$, as spherical one. Taking also into account that $J_{q>c^{-1}} \ll J_0$, one can find for different dimensions d

$$S^2 = \frac{nT_c}{J_0}\left[\frac{\pi d}{2\sin(d/2-1)\pi}\left(\frac{a}{c}\right)^d\left(\frac{h}{J_0 S}\right)^{d/2-1} + \frac{h}{J_0 S}\right], \tag{8.15}$$

where d is in the region $2 < d < 4$. When d is equal to four one has instead of Eq. (8.15)

$$S^2 = \frac{nT_c}{J_0}\left[2\left(\frac{a}{c}\right)^4 \ln\left(\frac{SJ_0}{h}\right) + 1\right]\frac{h}{J_0 S}, \tag{8.16}$$

and consequently for $d > 4$

$$S^2 = \frac{nT_c}{J_0}\left[1 + 4\left(\frac{a}{c}\right)^d \frac{1}{d-4}\right]\frac{h}{J_0 S}. \tag{8.17}$$

Thus, as is seen from Eqs. (8.15) – (8.17), when $d > 4$ the dependence $S(h)$ has the traditional form $S \sim h^{1/3}$, i.e. the exponent $\delta = 3$ which, as it could be anticipated, has the value of the mean-field approximation. When $2 < d < 4$ the situation is more complex. At the region of quite low fields δ coincides with the scaling value $\delta = (d+2)/(d-2)$ ($\eta = 0$), but with the increase of field a crossover occurs to the mean-field theory result. Namely, when $h/J_0 \gg (a/c)^{3d/(4-d)} = G_i^{3/2}$ the magnetization behaves like $S \sim h^{1/3}$. The latter is in accordance with the results obtained by Levaniuk (1959), which showed that at $T = T_c$ the mean field theory can be justified when $h/J_0 > (a/c)^9$ ($d = 3$). In the case when $d = 4$ one should add logarithmic corrections to the dependence $S \sim h^{1/3}$, which are essential in the exponentially narrow fluctuation region.[3]

In the framework of the above approximations one can find the critical temperature, which is renormalized with respect to the trial value $T_0 = J_0/n$ on the value of the order of $T_0(a/c)^d$

$$T_c = T_0 \left[1 - \frac{2}{d-2} \left(\frac{a}{c} \right)^d \right], \qquad d > 2.$$

From this formula, one can see that the shift of the critical temperature is confined to the fluctuation region and when $(a/c)^d \to 0$ the value T_c is the same as in the mean field approximation $T_c = J_0/n$. Of course, one should remember that all these results are obtained in the limit $(a/c)^d \ll 1$. In the case of short-range interactions the fluctuation region formally becomes of the order of T_c and the value of the critical temperature in this case can only be obtained numerically. It is also necessary to mention that, as far as the shift of T_c is substantially dependent on the radius of interactions (between magnetic moments), one can expect the manifestation of unusual ordered phases (from the point of view of the mean field theory) in systems containing a number of competing interactions with different interaction radii. One of such possibilities is considered in the following section.

8.2 Inversion of Phase Transitions

In this section, following the work by Ivanchenko *et al* (1986 a), we consider phase transitions in magnets with two sublattices. This means that the $\boldsymbol{l}$-th crystal cell contains either two nonequivalent magnetic atoms or two identical atoms occupying nonequivalent positions. Between the magnetic moments we have exchange interactions corresponding to atoms of the same kind (exchange integrals $I_{ll'}$ and $J_{ll'}$) and to atoms of different kinds ($K_{ll'}$).

[3] The width of this region is of the order of $\exp(-(c/a)^4)$.

The Heisenberg Hamiltonian for this case can be written as

$$H = -1/2 \sum_{ll'} \{ J_{ll'} \boldsymbol{n}_l \boldsymbol{n}_{l'} + I_{ll'} \boldsymbol{m}_l \boldsymbol{m}_{l'} + 2K_{ll'} \boldsymbol{n}_l \boldsymbol{m}_{l'} \},$$

where $\boldsymbol{n}_l$ and $\boldsymbol{m}_l$ are n-component unit vectors. For the case of spherical model one has to replace the constraints $\boldsymbol{n}_l^2 = 1$ and $\boldsymbol{m}_l^2 = 1$ of Heisenberg's model by the conditions $\sum_l \boldsymbol{n}_l^2 = N$ and $\sum_l \boldsymbol{m}_l^2 = N$. These conditions lead to the modification of the spherical model incorporating the case of two interacting order parameters. The partition function in this situation can be written as

$$Z = \int \frac{dx_1 dx_2}{(4\pi T)^2} d\boldsymbol{n}_q d\boldsymbol{m}_q \exp \left\{ \frac{N}{2T} \left[x_1 + x_2 + \frac{1}{N} \sum_q [(J_q - x_1)|\boldsymbol{n}_q|^2 \right.\right.$$

$$\left.\left. + (I_q - x_2)|\boldsymbol{m}_q|^2 + 2K_q \boldsymbol{n}_q \boldsymbol{m}_{-q}] \right] \right\} \sim \exp \left[\frac{N}{T} \mathcal{F}(x_{s_1}, x_{s_2}) \right],$$

where $J_q \equiv J(\boldsymbol{q})$, I_q, and K_q are Fourier's transformants of the exchange interactions, x_{s_1} and x_{s_2} are the values of x_1 and x_2 at the saddle points defined by the equations $\partial \mathcal{F} / \partial x_{s_i} = 0$. The integration over the variables $\boldsymbol{n}_q$ and $\boldsymbol{m}_q$ can be carried out immediately after the diagonalization of the bilinear form

$$\sum_q [(J_q - x_1)|\boldsymbol{n}_q|^2 + (I_q - x_2)|\boldsymbol{m}_q|^2 + 2K_q \boldsymbol{n}_q \boldsymbol{m}_{-q}]$$

$$= \sum_q [\lambda_q^+ |\boldsymbol{\phi}_q^+|^2 + \lambda_q^- |\boldsymbol{\phi}_q^-|^2]. \tag{8.18}$$

The unitary matrix diagonalizing the form Eq. (8.18) can be written

$$\hat{S}_q = \sqrt{\frac{\lambda_q^+ - \bar{I}_q}{2\lambda_q^+ - \bar{J}_q - \bar{I}_q}} \begin{pmatrix} 1 & -\frac{K_q}{\lambda_q^+ - \bar{I}_q} \\ \frac{K_q^*}{\lambda_q^+ - \bar{I}_q} & 1 \end{pmatrix}, \tag{8.19}$$

where we have introduced the notations $\bar{J}_q = J_q - x_1$ and $\bar{I}_q = I_q - x_2$. The eigenvalues $\lambda_q^\pm$ are defined by

$$\lambda_q^\pm = \frac{\bar{J}_q - \bar{I}_q}{2} \pm \sqrt{\left(\frac{\bar{J}_q - \bar{I}_q}{2} \right)^2 + |K_q|^2}.$$

For the same reason as in Sect. 8.1.1, we have to omit the integration over the condensing modes $\boldsymbol{\phi}_q^\pm$. Now, we also should complete the set of equations for the saddle points with $\partial\mathcal{F}/\partial\phi^{\pm}_{q_i} = \lambda^{\pm}_{q_i}\phi^{\pm}_{q_i} = 0$. The wave vectors $\boldsymbol{q}_i$ $(i = 1, 2)$ defining the condensing modes correspond to maxima of the functions $\lambda^{\pm}_{q}$. The nontrivial solution for $\boldsymbol{\phi}^{\pm}$ appears when any of the values $\lambda^{\pm}$ is "stuck" to zero. Let us show that when the temperature is being decreased, starting from the paramagnetic phase, the equality $\lambda^{-}_{q_2} = 0$ can never be reached, where $\boldsymbol{q}_2$ is the maximum point for the function $\lambda^{-}_{q_2}$. The latter is true owing to the inequality $\lambda^{+}_{q_2} > 0$ eventually leading to the condensation of the field $\boldsymbol{\phi}^{+}$. The proof of this statement is as follows. Let us assume that at the point $\boldsymbol{q}_2$ the equality $\lambda^{-}_{q_2} = 0$ is fulfilled along with the condition $\lambda^{+}_{q_2} < 0$. Then the explicit relation

$$\bar{J}_{q_2}\bar{I}_{q_2} - |K_{q_2}|^2 = 0$$

leads to the equality

$$\lambda^{-}_{q_2} = 0 = \frac{\bar{J}_{q_2} + \bar{I}_{q_2}}{2} - \frac{|\bar{J}_{q_2} + \bar{I}_{q_2}|}{2},$$

from which one can see $\lambda^{-}_{q_2}$ is equal to zero only when $(\bar{J}_{q_2} + \bar{I}_{q_2}) \geq 0$. Let us first consider the case $(\bar{J}_{q_2} + \bar{I}_{q_2}) = 0$. Taking into account the equality $\bar{J}_{q_2}\bar{I}_{q_2} = |K_{q_2}|^2$, one can obtain from the condition $\lambda^{-}_{q_2} = 0$ the relation $[(\bar{J}_{q_2})^2 + |K_{q_2}|^2] = 0$, which is impossible if $|K_{q_2}|^2 \neq 0$. Therefore, $\lambda^{-}_{q_2} = 0$ could only be allowed when $(\bar{J}_{q_2} + \bar{I}_{q_2}) > 0$. However, this inequality leads to

$$\lambda^{+}_{q_2} = \frac{\bar{J}_{q_2} + \bar{I}_{q_2}}{2} + \frac{|\bar{J}_{q_2} + \bar{I}_{q_2}|}{2} > 0,$$

which contradicts the initial assumption $\lambda^{+}_{q_2} < 0$. The continuity of the function λ^{+}_{q} means that near the point $\boldsymbol{q} = \boldsymbol{q}_2$ there is a vicinity of vectors $\boldsymbol{q}$ where $\lambda^{+}_{q} > 0$. Thus, with the decrease of temperature the value λ^{+}_{q} will change sign in the region where the value of λ^{-}_{q} is negative. As we always should have satisfied the inequality $(\bar{J}_{q_2} + \bar{I}_{q_2}) < 0$, the value λ^{-}_{q} cannot reach zero. Hence, the field $\boldsymbol{\phi}^{-}_{q}$ can never be condensed. The latter is not true when $K_q = 0$. This case requires special consideration and will be studied below.

Carrying out the integration over all $\boldsymbol{\phi}^{\pm}_{q}$ modes, excepting, of course, the nonergodic mode $\boldsymbol{\phi}^{+}_{q_1}$ one arrives at

$$Z \propto \int dx_1 dx_2 \exp\left\{\frac{N}{2T}\left[(x_1 + x_2) - \frac{nT}{N}\sum_{q}\ln(\lambda^{+}_{q}\lambda^{-}_{q}) + \frac{\lambda^{+}_{q_1}}{N}|\boldsymbol{\phi}^{+}_{q_1}|^2\right]\right\}.$$

The equilibrium value $\boldsymbol{\phi}^{+}_{q_1}$ and parameters x_1 and x_2 are being defined from the equation of state $\partial\mathcal{F}/\partial\boldsymbol{\phi}^{+}_{q_1} = 0$ and the saddle point equations

$\partial\mathcal{F}/\partial x_1 = \partial\mathcal{F}/\partial x_2 = 0$ which can be written as

$$\begin{aligned} 0 &= \lambda^+_{q_1}\phi^+_{q_1}, \\ \tfrac{1}{T} &= \tfrac{n}{N}\sum_q \tfrac{-\bar{J}_q}{\lambda^+_q\lambda^-_q} - \tfrac{\bar{J}_{q_1}}{N\lambda^-_{q_1}}|\phi^+_{q_1}|^2, \\ \tfrac{1}{T} &= \tfrac{n}{N}\sum_q \tfrac{-\bar{I}_q}{\lambda^+_q\lambda^-_q} - \tfrac{\bar{I}_{q_1}}{N\lambda^-_{q_1}}|\phi^+_{q_1}|^2. \end{aligned} \tag{8.20}$$

In a quite general case the calculation of sums in the set Eq. (8.20) is a nontrivial problem. Below, as in Sect. 8.1.2, we restrict consideration to the situation with the exchange radii (c_1, c_2) essentially exceed the lattice parameter. In addition to this we also assume the weak interaction between sublattices in comparison with the interlattice exchange, i.e. $K^2 \ll JI$. Making use of these assumptions one can reduce the set Eq. (8.20) to

$$\begin{aligned} \tfrac{1}{T} &= \tfrac{n}{x_1}\left[1-\left(\tfrac{a}{c_1}\right)^2\right] + \tfrac{3\overline{K}^2 n}{x_1^2 x_2} \\ &\quad + \tfrac{3a^3}{Jc_1^3}\left[1-\sqrt{\tfrac{x_1-J}{J}}\tan^{-1}\sqrt{\tfrac{J}{x_1-J}}\right] - \tfrac{\bar{J}}{N\lambda^-_{q_1}}|\phi^+_{q_1}|^2, \\ \tfrac{1}{T} &= \tfrac{n}{x_2}\left[1-\left(\tfrac{a}{c_2}\right)^2\right] + \tfrac{3\overline{K}^2 n}{x_2^2 x_1} \\ &\quad + \tfrac{3a^3}{Ic_2^3}\left[1-\sqrt{\tfrac{x_2-I}{I}}\tan^{-1}\sqrt{\tfrac{I}{x_2-I}}\right] - \tfrac{\bar{I}}{N\lambda^-_{q_1}}|\phi^+_{q_1}|^2. \end{aligned} \tag{8.21}$$

In this set the following notations are introduced $\bar{J} = \bar{J}_{q_1}$, $\bar{I} = \bar{I}_{q_1}$, $J = J_{q_1}$, $I = I_{q_1}$, and $\overline{K}^2 = \int_q \boldsymbol{q}^2|K_q|^2$. For the sake of definiteness below we choose $J > I$. Let us consider the values x_1 and x_2 in the vicinity of the transition point from the paramagnetic side ($\phi^+_{q_1} = 0$). When $K \ll J,\ I$, near the transition point we also have $(x_1 - J) \ll J$ and $(x_2 - I) \ll I$, so that approximately the values x_1 and x_2 are defined by

$$\begin{aligned} x_1 &= nT\left[1+2\left(\frac{a}{c_1}\right)^3 + 3\frac{\overline{K}^2}{JI}\right] \equiv AT, \\ x_2 &= nT\left[1+2\left(\frac{a}{c_2}\right)^3 + 3\frac{\overline{K}^2}{JI}\right] \equiv BT. \end{aligned} \tag{8.22}$$

When $\lambda^+_{q_1} \neq 0$, this set completely defines the functions $x_1(T)$ and $x_2(T)$. The value $\lambda^+_{q_1}$ decreases with the decrease of temperature until it reaches

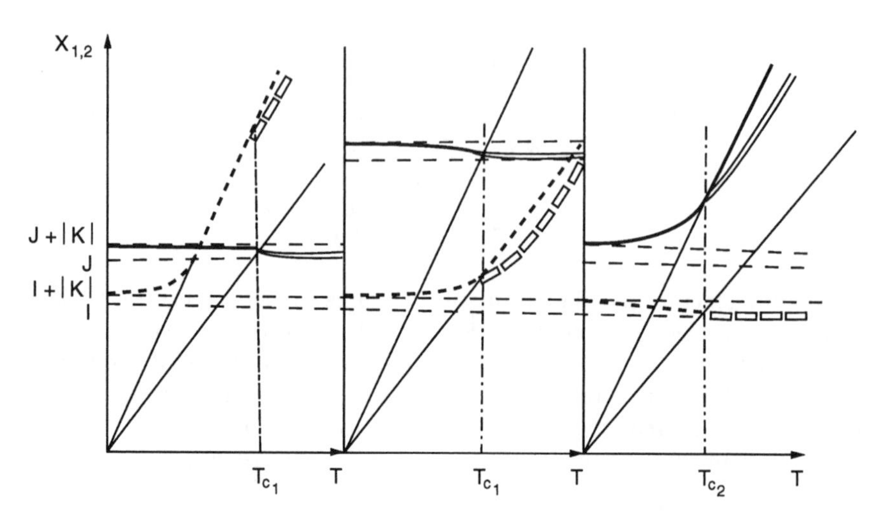

Figure 8.1: The temperature behavior of "chemical potentials" x_i at different relations between parameters $c_{1,2}$, J, and I.

zero, when inevitably a nontrivial solution $\phi^+_{q_1} \neq 0$ appears. This solution plays the role of an additional parameter in conditions where $\lambda^+_{q_1}$ is "stuck" to zero. The values x_1 and x_2 entering into the definition of $\lambda^+_{q_1}$ are now related by a single equation

$$\frac{\bar{J}-\bar{I}}{T} = \frac{n}{N}\sum_q \frac{\bar{I}\bar{J}_q - \bar{I}_q\bar{J}}{\lambda^+_q\lambda^-_q} = \frac{\bar{J}A}{x_1} + \frac{\bar{I}B}{x_2}.$$

The solutions of this equation can be obtained with the help of the relation $\bar{J}\bar{I} = |K|^2$ following from the requirement $\lambda^+_{q_1} = 0$. These solutions are shown in Fig. 8.1. The dependence of the solutions in the critical region can be described as follows. With the decrease of temperature from the paramagnetic phase the parameters x_1 and x_2 move along the lines $x_1 = AT$ and $x_2 = BT$ until they reach T_c, where $\lambda^+_{q_1} = 0$. So far as the partition function at $\lambda^+_{q_1} > 0$ is formally divergent, the motion along the lines $x_1 = AT$ and $x_2 = BT$ is prohibited. Therefore, the following decrease of temperature below T_c can only be allowed along the other branch of solutions $x_1(T)$ and $x_2(T)$ which appears owing to the condensation of the field $\phi^+_{q_1}$. Thus, the phase transition corresponds to the condensation of the single mode $\phi^+_{q_1}$. The latter, however, does not mean that there is an ordering in a single sublattice. The magnetizations of the sublattices are defined with the help of mean values of the fields $\phi_q^{\pm}$ as follows

$$\begin{pmatrix} \langle n_q \rangle \\ \langle m_q \rangle \end{pmatrix} = \hat{S} \begin{pmatrix} \langle \phi_q^+ \rangle \\ \langle \phi_q^- \rangle \end{pmatrix},$$

where the matrix $\hat{S}$ is given in Eq. (8.19). From this relation, taking into account that $\langle \phi_q^- \rangle = 0$, one arrives at

$$\langle n_{q_1} \rangle = S_{11}\phi_{q_1}^+, \quad \langle m_{q_1} \rangle = S_{21}\phi_{q_1}^+ = -\frac{K^*}{\bar{I}} S_{11}\phi_{q_1}^+. \tag{8.23}$$

This equation shows that at $T = T_c$ the magnetic order is established in both subsystems n and m. Even though Eq. (8.23) is applicable for any case, the situations shown in Fig. 8.1 are essentially different. Let us consider first the cases a and b. When $T = T_c$ the value x_1 is quite close to J, while the value x_2 is far from I, so that the inequality $K/(x_2 - I) \ll 1$ is well satisfied. Taking also into account the relation $\bar{I}\bar{J} = |K|^2$, one can see that in the vicinity of T_c the magnetizations satisfy the inequality $\langle m_{q_1} \rangle \ll \langle n_{q_1} \rangle$. When T tends to zero the value x_1 is slowly increasing to the limit value $x_1 = J + |K|$. So this value changes in a narrow region $J < x_1 < J + |K|$.[4] On the contrary, the value x_2 in a quite large range of temperatures moves along the line $x_2 \approx BT$, before it "sticks" to the limit $x_2 = I + |K|$. Since at $T = 0$ both parameters x_i differ from J and I by the same quantity ($|K|$), the magnetizations of the sublattices are of the same order of magnitude. When the inequality $JB < IA$ is fulfilled (see Fig. 8.1 (c)), the "chemical potentials" x_i are "stuck" in the reverse order. In the vicinity $T = T_c$ we have the inequality $(x_1 - J)/K \gg 1$, which means that $\langle m_{q_1} \rangle \gg \langle n_{q_1} \rangle$.[5] As in the previous case, with the decrease of temperature, the magnetizations are gradually becoming closer in magnitude. It is necessary to notice, the condition $JB < IA$ implies that the magnetization of the sublattice with a weak exchange becomes higher than the magnetization of the other sublattice. Of course, this statement is only justified in the vicinity close to T_c. The physical sense of this phenomenon can be clarified with the help of the consideration of the limiting case $|K_{q_0}|^2 = 0$, where q_0 is the point corresponding to the maxima of both functions $\lambda_q^{\pm}$. In such a situation the values $\lambda_{q_0}^+$ and $\lambda_{q_0}^-$ are reduced to $\bar{J}_{q_0}$ and $\bar{I}_{q_0}$. Now the statement that the field $\phi_{q_0}^-$ cannot be condensed is incorrect. In this case $\bar{J}_{q_0}$ and $\bar{I}_{q_0}$ contains the parameter x_1 and x_2 independently from each other and nullification of $\lambda_{q_0}^-$ does not lead to the positiveness of $\lambda_{q_0}^+$. In Fig. 8.2 one can see the behavior of the functions $x_1(T)$ and $x_2(T)$ for different ratios c_1/c_2 and J/I when $K_{q_0} = 0$. As is seen from this figure,

[4]One may say that x_2 "sticks" below T_c to the value of the order of J.

[5]The parameter x_2 in this region is close to J.

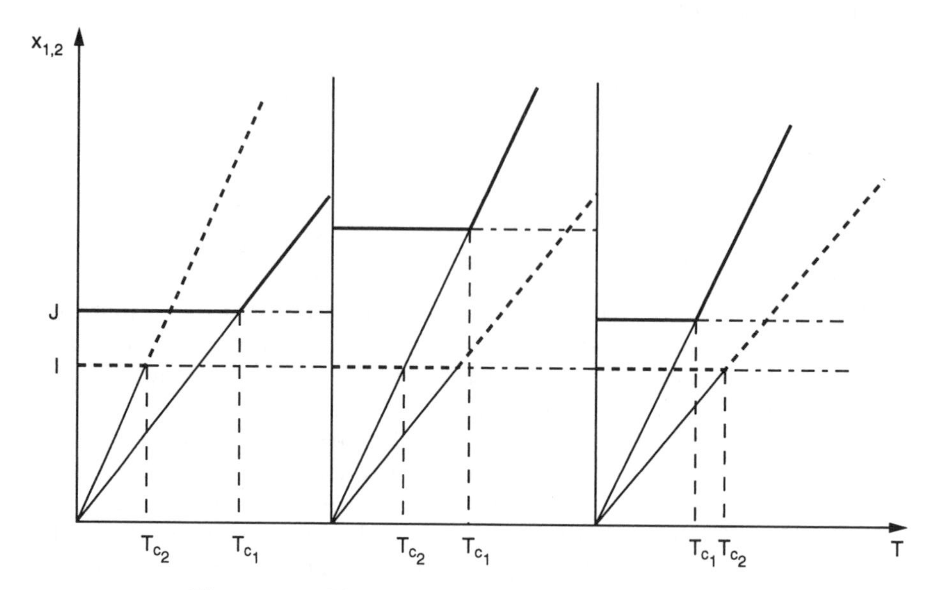

Figure 8.2: The same as in Fig 8.1, but $K_{q_0} = 0$.

with the decrease of temperature, the parameters x_1 and x_2 independently reach their minimum values $x_1 = J_{q_0}$ and $x_2 = I_{q_0}$ at temperatures T_{c_1} and T_{c_2} and "stick" to them down to $T = 0$. Now, in the region of temperatures between T_{c_1} and T_{c_2}, the intermediate phase with a single ordered sublattice exists.[6] Below the lowest of T_{c_1}, T_{c_2} both sublattices are ordered. According to the mean field theory, in the degenerate situation, serial ordering with two second-order transitions should also occur. However, in this case we will always have that the first ordered sublattice has the strongest interaction.[7] From this point of view the phase with $\langle \boldsymbol{m}_{q_o} \rangle \neq 0$, $\langle \boldsymbol{n}_{q_o} \rangle = 0$ realized at the condition $BJ < AI$, is an inverted phase in relation to the one anticipated by the mean-field theory approach. The manifestation of this phase is a result of the fluctuational shift of the temperature T_{c_1} below the renormalized value T_{c_2}. The condition $BJ < AI$ can be represented in the form

$$\frac{J}{I} \leq 1 + a^3 \left[\frac{1}{c_1^3} - \frac{1}{c_2^3} \right], \tag{8.24}$$

[6] The intermediate phase has $\langle n_{q0} \rangle \neq 0$, $\langle m_{q0} \rangle = 0$ for the cases a and b and $\langle m_{q0} \rangle \neq 0$, $\langle n_{q0} \rangle = 0$ for the case c.

[7] Since we have $J > I$, the first ordered is n-sublattice.

from which one can see an apparent relation of the inversion of the phase transitions to the difference in the exchange interaction radii ($c_1 < c_2$).

Let us now return to the case $K_{q_1} \neq 0$. In this case the structure corresponding to the condition Eq. (8.24) has $\langle \boldsymbol{n}_{q_o} \rangle \sim \langle \boldsymbol{m}_{q_o} \rangle K/J$, which is quite close to the inverted phase with $\langle \boldsymbol{n}_{q_o} \rangle = 0$. However, the finite value of K_{q1} leads to the smearing of the low-temperature transition. Nonetheless, the presence of the largest magnetization in the sublattice with the weaker exchange accounts for the interpretation of such a structure as a fluctuationally induced anomalous phase.

8.3 Model with Reduced Interaction

Let us consider one more exactly solvable model. The Ginzburg-Landau functional for this model was first advanced by Schneider *et al* in 1975 for the case of structural phase transitions

$$H = \frac{1}{2} \sum_{q} (\boldsymbol{q}^2 + \tau)|\phi_q|^2 + \frac{u}{V} \left(\sum_{q} |\phi_q|^2 \right)^2 \tag{8.25}$$

This functional differs from the ϕ^4-model owing to the reduction of the δ-function in a way providing for the momentum conservation law, i.e.

$$\delta \left(\sum_{i=1}^{4} \boldsymbol{q}_i \right) \to [\delta(\boldsymbol{q}_1 + \boldsymbol{q}_2)\delta(\boldsymbol{q}_3 + \boldsymbol{q}_4) + \ permutations].$$

Thus, in this model only interactions of fluctuations with oppositely directed momenta of equal magnitude are taken into account. Now, if the scalar ϕ^4-model belonged to the Ising-model universality class, the reduced model transforms to the spherical-model universality class. Sarbach and Schneider (1975) considered a modification of this model adding to the functional Eq. (8.25) the term with the power eight. At last, this model was modified to incorporate all even powers of the scalar order parameter (see Sarbach and Schneider (1977), and Ivanchenko *et al* (1986 a, 1987 b)).

8.3.1 General Consideration

The most general form of the Ginzburg-Landau functional with the scalar field appropriate for this model can be written as

$$H = \frac{\mathcal{B}}{2} \int d\boldsymbol{r}[(\boldsymbol{\nabla}\phi)^2 - 2h\phi + \overline{F}(\phi^2)], \tag{8.26}$$

where the space coordinates are made dimensionless with the help of the interaction radius c. The function $\overline{F}(\phi^2)$ is considered to be representative of Taylor's series defining dimensionless vertexes u_{2k}

$$\overline{F}(\phi^2) = \sum_{k=1}^{\infty} u_{2k}(\phi^2)^k. \tag{8.27}$$

The coefficient $\mathcal{B} = c^d T_c / a^d T \approx (c/a)^d$ occurred owing to the dimensionalization of the functional. Let us first show that this coefficient is directly related to the Ginzburg parameter G_i (see Sect. 1.6).

If one can neglect the influence of fluctuations, the value ϕ_0 can be found from the minimization of the functional Eq. (8.26)

$$\phi_0^2 = \left(-\frac{u_2}{k u_{2k}}\right)^{1/(k-1)} \Theta(-u_2) \equiv \left(-\frac{\tau}{k u_{2k}}\right)^{1/(k-1)} \Theta(-\tau), \tag{8.28}$$

where $\Theta(x)$ is the Heavyside step function, u_{2k} is the first finite vertex with $k \neq 1$. Eq. (8.28) is justified in the vicinity of second-order phase transitions, where $\phi_0 \ll 1$. Let us define correlation function in a traditional way as

$$G(\boldsymbol{r}, \boldsymbol{r}') = \frac{1}{\mathcal{B}} \frac{\delta \langle \phi(\boldsymbol{r}) \rangle}{\delta h(\boldsymbol{r}')} = \langle \phi(\boldsymbol{r}) \phi(\boldsymbol{r}') \rangle - \langle \phi(\boldsymbol{r}) \rangle \langle \phi(\boldsymbol{r}') \rangle. \tag{8.29}$$

In the limit of a small homogeneous field $h \to 0$, assuming that $\phi(\boldsymbol{r})$ only slightly deviates from the equilibrium value Eq. (8.28), one can obtain, using Eqs. (8.26) – (8.29), the following equation

$$\mu|\tau| G(\boldsymbol{r}) - \boldsymbol{\nabla}^2 G(\boldsymbol{r}) = \mathcal{B}\delta(\boldsymbol{r}), \qquad \mu = \begin{cases} 1, & \tau > 0, \\ 2(k-1), & \tau < 0, \end{cases} \tag{8.30}$$

which has the following solution

$$G(\boldsymbol{r}) \propto \frac{1}{\mathcal{B}} r^{(2-d)} e^{-r/r_c} \tag{8.31}$$

where $r_c = (\mu|\tau|)^{-1/2}$. This evaluation is justified only when fluctuations in the volume Ω_c of the order of r_c^d are small in comparison with the value ϕ_0, i.e.

$$\left\langle \int_{\Omega_c} d\boldsymbol{r} \left[\phi(\boldsymbol{r}) - \langle \phi(\boldsymbol{r}) \rangle\right] \right\rangle^2 = \int_{\Omega_c} d\boldsymbol{r} d\boldsymbol{r}' G(\boldsymbol{r}, \boldsymbol{r}') \ll (\Omega_c \phi_0)^2. \tag{8.32}$$

Using Eq. (8.31) one can obtain from Eq. (8.32) the following estimate

$$\tau^{(d_c - d)/2} \gg \frac{1}{\mathcal{B}}, \tag{8.33}$$

where $d_c = 2k/(k-1)$ is the highest dimension above which the theory for the Hamiltonian Eq. (8.26) with a finite number of vertexes u_{2k} becomes unrenormalizable (see Chapter 4). As the value $\mathcal{B} \geq 1$ and $\tau < 1$, from Eq. (8.33) it follows that when $d > d_c$, the mean-field theory is always applicable. On the contrary, when $d < d_c$ the mean-field theory is applicable only in the region where $\tau \gg G_i = \mathcal{B}^{2/(d-d_c)}$.

Now let us return to the above model. The main approximation of this model is replacement of all powers ϕ^{2k} in the functional Eq. (8.26), i.e. integrals of the kind $\int d\boldsymbol{r}\phi^{2k}$, by the following expressions

$$\int d\boldsymbol{r}\phi^{2k} \to V\left[\int \frac{d\boldsymbol{r}}{V}\phi^2(\boldsymbol{r})\right]^k \equiv Va^k[\phi^2]. \tag{8.34}$$

As there is $(2k-1)!!$ ways to combine $2k$ functions $\phi(\boldsymbol{r})$, the vertexes in Eq. (8.27) will change to

$$u_{2k} \to g_{2k} = (2k-1)!!u_{2k},$$

and finally transform the functional Eq. (8.26) as

$$H = \frac{\mathcal{B}}{2}\int d\boldsymbol{r}[(\boldsymbol{\nabla}\phi)^2 - 2h\phi + F(a[\phi^2])], \quad F(x) = \sum_{k=1}^{\infty} g_{2k}x^k.$$

For the following calculations it is convenient to represent the exponent $\exp[-\mathcal{B}VF/2]$ in the form

$$\exp\left[-\frac{\mathcal{B}V}{2}F(a)\right] = \frac{\mathcal{B}V}{4\pi i}\int_{-\infty}^{\infty} dx \int_{-i\infty}^{i\infty} dy$$

$$\times \exp\left\{-\frac{\mathcal{B}V}{2}[F(x) - y(x - a[\phi^2])]\right\}. \tag{8.35}$$

Now, one can see that the bilinear dependence of the functional $a = \int d\boldsymbol{r}\phi^2(\boldsymbol{r})V^{-1}$ on ϕ leads to the Gaussian integration over $\phi(\boldsymbol{r})$ in the partition function. In the ϕ^4-model in order to linearize the exponential expression with respect to $a[\phi^2]$ the Hubbard-Stratonovich identity is usually used. This identity is a particular case of the transformation Eq. (8.35). Now the partition function has the form

$$Z \propto \int D\phi_q dx dy \exp\left\{-\frac{\mathcal{B}V}{2}\left[\frac{1}{V}\sum_q(\boldsymbol{q}^2 + y)|\phi_q|^2 - xy + F(x) - 2h\frac{\phi_{q=0}}{V^{1/2}}\right]\right\}.$$

After carrying out the integration over ϕ_q, the partition function can be represented as

$$Z \sim \int dxdy \exp\left\{-\frac{\mathcal{B}V}{2}\left[F(x)+(\phi_0^2-x)-2h\phi_0+\frac{1}{\mathcal{B}V}\sum_q \ln(y+\boldsymbol{q}^2)\right]\right\}, \tag{8.36}$$

where, as usual, the integration over the nonergodic condensing mode $\phi_0 = \phi_{q=0}/\sqrt{V}$ is omitted. The integration in Eq. (8.36) can be carried out with the help of the saddle-point method, which gives an exact result in the limit $V \to \infty$. However, before the integration, we have to consider the following important feature of the model. The sum over the momenta $\sum_q \ln(y+\boldsymbol{q}^2)$ is divergent at large momenta. Strictly speaking this divergence has a formal nature. In solids we are always restricted in momenta by the value of the order of the inverse lattice cell $\Lambda \approx a^{-1}$. Therefore, one can always cut the integration at some cutoff momentum Λ. However, we are interested in the study of the universal critical behavior. Hence, it is more instructive to use the subtraction procedure. This procedure has been developed in the field theoretical RG (see Chapter 4). In other words, we are going to regularize the sum, absorbing, so far as possible, the dependence on the cutoff momentum into renormalized vertexes u_{2k}^*. For this purpose, when $2 < d < 4$, it is sufficient to perform the following two subtractions

$$\left(\sum \ln(y+\boldsymbol{q}^2)\right)_R = \sum\left[\ln(y+\boldsymbol{q}^2) - \ln \boldsymbol{q}^2 - \frac{y}{\boldsymbol{q}^2}\right]. \tag{8.37}$$

The regularized sum is quickly convergent and its dependence on Λ is unessential so that the cutoff value can be put to infinity. Both subtracted terms should be added to the exponential expression in Eq. (8.36). The first term, $\sum_q \ln \boldsymbol{q}^2$, is an additive correction to the free energy independent on thermodynamic variables. Hence, this correction is unessential and can be omitted. The second term, $y\sum_q \boldsymbol{q}^{-2} = yV\theta\mathcal{B}$, can be eliminated with the help of the substitution $x = x^* + \theta$. This substitution, naturally leads to the renormalization of all other vertexes in the expression

$$F = \sum_k g_{2k}x^k = \sum_k \sum_{m=0}^{k} C_k^m \theta^{k-m}(x^*)^m g_{2k}.$$

Changing the order of summations in this formula and taking into account that $g_{2k} = (2k-1)!!u_{2k}$, one arrives at

$$F(x^*) = \sum_k u_{2k}^*(x^*)^k, \qquad u_{2k}^* = \sum_{m=k}^{\infty} C_m^k \frac{(2m-1)!!}{(2k-1)!!} u_{2m}\theta^{(m-k)}. \tag{8.38}$$

One can show that the expression obtained is a renormalized function resulting from the loop renormalization shown in Fig. 7.9. Really, we have $(2k-1)!!$ ways leading to a creation of the loop from $2k$ "legs" for the vertex u_{2k}. In order to find a contribution from the vertex u_{2k} to the renormalization of the vertex u_{2m},[8] one has to keep from all possible loops only $(k-m)$ loop. The latter can be done by C_k^m ways. As a result we get the renormalization of Eq. (8.38). Let us recall that the contribution due to a loop is $\theta = \sum_q [\mathcal{B}V(\boldsymbol{q}^2 + \tau)]$ which coincides with the θ in Eq. (8.38) when $\tau = 0$. It was shown in Sect. 7.4.5 that the loop renormalization can be automatically taken into account with the help of the integral transformation $F(\phi^2) \to \bar{F}(\phi^2)$, which for the scalar order parameter reduces to Weierstrass's transformation

$$\bar{F}(x) = \frac{1}{\sqrt{2\pi\theta}} \int_{-\infty}^{\infty} dy \exp\left[-\frac{(u-\sqrt{x})^2}{2\theta}\right] F(u). \tag{8.39}$$

Substituting this function in Eq. (8.36), we obtain an expression which has incorporated all loop renormalizations along with the regularized value of the sum given by Eq. (8.37). As a result the free energy has the form

$$\mathcal{F} = \frac{\mathcal{B}V}{2}\left[\bar{F}(x) + y(\phi_0^2 - x) - 2h\phi_0 + \frac{1}{\mathcal{B}V}\left(\sum_q \ln(y + \boldsymbol{q}^2)\right)_R\right], \tag{8.40}$$

where the values x and y are defined from the saddle point equations $\partial\mathcal{F}/\partial x = \partial\mathcal{F}/\partial y = 0$, which defines them as functions of ϕ_0 and u_{2k}^*.[9] The substitution of the saddle point values x, y in Eq. (8.40) defines the free energy $\mathcal{F}(\phi_0)$ as a function of the order parameter. In order to find the equilibrium value ϕ_0 one needs to add the equation of state in the form $d\mathcal{F}(\phi_0)/d\phi_0 = 0$. Thus, the full set of equations can be written as

$$y = \bar{F}_x, \qquad \phi_0^2 - x - f(y) = 0, \qquad \phi_0 y = h. \tag{8.41}$$

In this set we have introduced the notations

$$\begin{aligned} \bar{F}_x &= \frac{d\bar{F}}{dx}, \quad f(y) = -\frac{\partial\Phi(y)}{\partial y}, \\ \Phi(y) &= \frac{1}{\mathcal{B}V}\left(\sum_q \ln(y + \boldsymbol{q}^2)\right)_R. \end{aligned} \tag{8.42}$$

[8]Naturally, the latter is only possible when $k \geq m$.

[9]The temperature dependence is introduced owing to $u_2 = \tau$ and, consequently, the renormalized value $u_2^* = \tau^*$.

Using Eq. (8.42) one can evaluate the function $f(y)$ in an explicit form

$$f(y) = \kappa y^{(d-2)/2}\Theta(y), \qquad \kappa = \frac{K_d \pi}{2\mathcal{B}\sin(\pi d/2)}. \tag{8.43}$$

8.3.2 Critical Exponents. Crossover

With the help of Eqs. (8.40) and (8.41), the free energy of the system is defined in the whole range of the parameters τ and h. In the following we are interested only in the critical behavior of the system. As one can see from Eq. (8.41), when $\phi_0 \to 0$ the variables x and y also tend to zero. For the system with a continuous phase transition, in the vicinity of the critical point one only has to keep two terms in the expansion of the function $\bar{F}(x)$, i.e.

$$\lim_{x\to 0} \bar{F}(x) = \tau^* x + u_{2m}^* x^m,$$

where u_{2m}^* is the first, non-zero vertex in the expansion Eq. (8.38) with a minimum value $m > 1$. For instance, in the ϕ^4-model $m = 2$, for the tricritical point $m = 3$, and so on. Now the set of Eq. (8.41) can be simplified to

$$y = \tau^* + Ax^{m-1}, \qquad \phi_0^2 - x - \kappa y^{(d-2)/2} = 0, \qquad y\phi_0 = h, \tag{8.44}$$

where $A = mu_{2m}^*$. In order to find critical exponents let us at first put $h = 0$ when $\tau^* < 0$. In this case the solution of the set of Eq. (8.44) can be written as

$$\phi_0^2 = x = \left(-\frac{\tau^*}{A}\right)^{\frac{1}{m-1}},$$

which means that the exponent $\beta = 1/2(m-1)$. The latter coincides with the mean-field value ($\beta = 1/2$ in the ϕ^4-model, $\beta = 1/4$ at the tricritical point, and so on). Substituting the value $x(\tau^*)$ in the expression of the free energy we arrive at

$$\mathcal{F}|_{h=0} \approx \tau^* x + \frac{A}{m}x^m \sim |\tau^*|^{\frac{m}{m-1}}. \tag{8.45}$$

A more interesting is the case when $\tau^* = 0$, $h \neq 0$. In this situation the set of Eq. (8.44) can be reduced to

$$\phi_0^2 - \left(\frac{y}{A}\right)^{\frac{1}{m-1}} - \kappa y^{\frac{d-2}{2}} = 0, \qquad y\phi_0 = h. \tag{8.46}$$

From this set one can see that when the dimension of space is less than the critical value $d_c = 2m/(m-1)$,[10] the exponent δ at $h \to 0$ happens

[10] This critical value coincides with the corresponding quantity obtained in the quantum field theory (see for instance Bogoliubov and Shirkov (1984)), i.e. $d_c = 4$ at the critical and $d_c = 3$ at the tricritical points.

to be equal to $(d+2)/(d-2)$. When $d > d_c$, then $\delta = 2m - 1$, which coincides with the mean-field theory result. It is necessary to notice that $\kappa \sim \mathcal{B}^{-1} = Gi^{(d_c-d)/2}$, which means that the term $\kappa y^{(d-2)/2}$ can be small when the Ginzburg parameter is small. In this case, until y reaches the value when the quantities $(y/A)^{1/(m-1)}$ and $\kappa y^{(d-2)/2}$ are of the same order, the index is given by $\delta = (2m-1)$, though later (at $y \to 0$) it turns out to be $\delta = (d+2)/(d-2)$. Thus, in this situation we have a natural crossover from the mean-field behavior to scaling. The fluctuation region boundary is now defined by the field $h^* \sim G_i^{(2m-1)/2(m-1)}$. In the ϕ^4-model this field is $h^* \sim G_i^{3/2}$.

In a similar manner, one can define the critical, susceptibility exponent γ: $\chi \sim \phi_0/h \sim |\tau^*|^{-\gamma}$. Noticing that $\chi^{-1} = y$, one can find from the set Eq. (8.41) that when $d > d_c$ $\gamma = 1$. On the other hand, when $d < d_c$ and $\tau^* \ll G_i$ we have $\gamma = 2/(d-2)(m-1)$. When $d < d_c$ and $\tau^* \sim G_i$ we again have a crossover to the mean-field behavior, so that when $|\tau^*| \gg G_i$ $\gamma = 1$.

It is not difficult to find that when $h = 0$ and $\tau^* > 0$ in the expression for the free energy, in addition to the term proportional to $(\tau^*)^{m/(m-1)}$, we have terms of the order of $(\tau^*)^{d/(d-2)(m-1)}$ if $d < d_c$, or terms of the order of $(\tau^*)^{d/2}$ for $d > d_c$. These terms define the critical exponent α:

$$\alpha = 2 - \frac{d}{(d-2)(m-1)} \quad (d < d_c) \quad \text{and} \quad \alpha = -\frac{4-d}{2} \quad (d > d_c).$$

The critical exponents obtained satisfy scaling laws (see Sect. 1.7.1). Naturally, the equation of state in this model also has the scaling form (see Sect. 4.4)

$$h = [(m-1)u_{2m}^* \kappa]2/(d-2) M^\delta H(\tau^* M^{-1/\beta}), \qquad H(x) = \left(1 + \frac{x}{u_{2m}^*}\right)^{2/(d-2)}.$$

An additional kind of crossover, that can be demonstrated in this model, is transition from tricritical to critical behavior when temperature approaches the critical point. Let us consider the following expansion for the function $\bar{F}_x$

$$\bar{F}_x = \tau^* + A_1 x + A_2 x^2 \qquad (A_1 = 2u_4^*, \;\; A_2 = 3u_6^*).$$

We have to keep the second term in the case when, owing to some reason, the coefficient of the first term may turn out to be zero (tricritical point). If the system is not exactly at the tricritical point, but quite close to it ($A_1 \ll A_2$), the inequality $A_2 x^2 \gg A_1 x$ can be satisfied in a quite broad temperature range. If the value A_1 is strictly equal to zero we have tricritical behavior. However, when A_1 is sufficiently small then at $x \to 0$ we inevitably have a transition from the tricritical to the critical branch of behavior (when $A_1 \approx A_2 x$). When the number G_i is small this transition occurs out of the fluctuation region and we have the following sequence for critical exponent δ: $\delta = 5 \rightarrow \delta = 3 \rightarrow \delta = (d+2)/(d-2)$, i.e. at d=3 $\delta = 5 \rightarrow \delta = 3 \rightarrow \delta = 5$. However, when the fluctuation region is large we have only one crossover: $\delta = 5 \rightarrow \delta = (d+2)/(d-2)$.

8.4 First-Order Transitions

As has been shown in Chapters 6 and 7, the transformation of the continuous phase transition into a discontinuous one is one of the most important qualitative effects of the fluctuational theory. In this theory, the fluctuational induction of a discontinuous transition is a result of either of the two following mathematical facts. The first is the absence of stable fixed points. However, even if such points are present, they frequently cannot be reached from the region of initial parameter of the Hamiltonian. In the second case the flow lines leave the stability region of the Hamiltonian having positive "masses," so that the correlation length cannot reach infinite values and the order parameter have to be established with a "jump." Strictly speaking, in all these cases one should introduce terms in the Hamiltonian higher order, then the fourth. However, in all those cases, when such calculations were done in the loop approximation (see Sect. 6.5) first-order transitions were confirmed. More than that, a number of materials, which according to the Landau theory should have continuous transitions, demonstrate first-order transitions. Nonetheless, there are materials, which in contradiction to the fluctuation theory, show second-order transitions. The latter may always be attributed to the absence of a small parameter in the theory. For instance, one cannot exterminate the transformation of stable points into unstable ones and *vice versa* when $\epsilon \rightarrow 1$. It is necessary to mention, however, that the interpretation of experimental results in such cases is a very subtle deal. One can always show reasons which may change the order of a phase transition in the Landau theory just as well as in the fluctuation theory. The latter is possible because, as a rule, reliable data concerning all possible interactions are absent. In any case, results obtained in expansions or in any other modification of RG analysis cannot be considered as a final answer concerning a fluctuationally induced first-order phase transition. In view of this controversy, it is very important to find fluctuationally induced transitions in the framework of exactly solvable models. Such a study is made in the following sections on the example of the Ginzburg-Landau functional for cubic symmetry (see Ivanchenko *et al* (1987 a,b)) and for interacting order parameters (see Lisyansky and Nicolaides (1993)).

8.4.1 Cubic Symmetry

As is shown in Sect. 6.3.3, the RG analysis leads to the fluctuational "plunge" to the first-order transition for the cubic Ginzburg-Landau functional in the ϕ^4-model

$$H = \int \frac{d^d\boldsymbol{r}}{2}\left\{\tau\boldsymbol{\phi}^2(\boldsymbol{r}) + (\boldsymbol{\nabla\phi})^2 + \frac{u}{4}(\boldsymbol{\phi}^2)^2 + \frac{v}{4}\sum_{\alpha=1}^{n}(\phi^\alpha(\boldsymbol{r}))^4\right\}, \tag{8.47}$$

Let us reduce the ϕ^4-terms according to

$$\int d\boldsymbol{r}[\phi^2(\boldsymbol{r})]^2 \to \frac{1}{V}\left[\int d\boldsymbol{r}\phi^2(\boldsymbol{r})\right]^2, \qquad \int d\boldsymbol{r}\phi_\alpha^4(\boldsymbol{r}) \to \frac{1}{V}\left[\int d\boldsymbol{r}\phi_\alpha^2(\boldsymbol{r})\right]^2.$$

This means that we have to use the following reduced functional

$$H = \frac{1}{2}\sum_{\alpha=1}^{n}\left[\sum_q \boldsymbol{q}^2|\phi_{\alpha q}|^2 + \frac{V}{2}\tau a_\alpha + \frac{vV}{4}a_\alpha^2\right] + \frac{uV}{4}\left(\sum_{\alpha=1}^{n} a_\alpha\right)^2,$$

where

$$a_\alpha = \frac{1}{V}\sum_q|\phi_{\alpha q}|^2 = \frac{1}{V}\int d\boldsymbol{r}\phi_\alpha^2(\boldsymbol{r}).$$

In order to linearize the exponential term in the partition function with respect to variables a_α, let us generalize the transformation Eq. (8.35)

$$\exp\left[-\frac{V}{2}F(a_\alpha)\right] = \left(\frac{V}{2\pi i}\right)^n\int_{-\infty}^{\infty}\prod_{\alpha=1}^{n}dx_\alpha\int_{-i\infty}^{i\infty}\prod_{\alpha=1}^{n}dy_\alpha$$

$$\times\exp\left\{-\frac{V}{2}[F(x_\alpha) - \sum_{\alpha=1}^{n}y_\alpha(x_\alpha - a_\alpha)]\right\}. \tag{8.48}$$

Using this representation, one can obtain the expression for the partition function in the form

$$Z \propto \int\left(\prod_{\alpha=1}^{n}D\phi_{\alpha q\neq 0}dx_\alpha dy_\alpha\right)\exp\left\{-\frac{V}{2}\sum_{\alpha=1}^{n}\left[\frac{1}{V}\sum_q(\boldsymbol{q}^2+\tau)|\phi_{\alpha q}|^2\right.\right.$$

$$\left.\left.+y_\alpha(\phi_\alpha^2 - x_\alpha) + \tau x_\alpha + \frac{u}{2n}\left(\sum_{\alpha=1}^{n}x_\alpha\right)^2 + \frac{v}{2}x_\alpha^2\right]\right\},$$

where as usual the condensed mode $\phi_{q=0}/\sqrt{V} = \boldsymbol{\phi}$ is eliminated from the integration. Here, as in the previous section, we have to apply the subtraction procedure (see Eq. (8.37)) which renormalizes the initial critical temperature

$$\tau \to \tau^* = \tau + \frac{nu+v}{V}\sum_q \boldsymbol{q}^{-2}$$

and vertexes after the substitution $x_\alpha = x_\alpha^* + \sum_q (V\boldsymbol{q}^2)^{-1}$. As a result, we have the final expression for the partition function in the form

$$Z \propto \int \left(\prod_{\alpha=1}^{n} dx_\alpha dy_\alpha\right) \exp\left\{-\frac{V}{2}\mathcal{F}(\phi_\alpha, x_\alpha, y_\alpha)\right\}, \tag{8.49}$$

where

$$\mathcal{F} = \frac{u}{2}\left(\sum_{\alpha=1}^{n} x_\alpha\right)^2 + \sum_{\alpha=1}^{n}\left[\frac{v}{2}x_\alpha^2 + \tau^* x_\alpha + y_\alpha(\phi_\alpha^2 - x_\alpha) + \Phi(y_\alpha)\right], \tag{8.50}$$

where the function $\Phi(y_\alpha)$ is defined by Eq. (8.42) when $\mathcal{B} = 1$. Using Eq. (8.49) one can find saddle point equations

$$\begin{aligned} \phi_\alpha^2 - x_\alpha - f(y_\alpha) &= 0, \\ u\textstyle\sum_{\beta=1}^{n} x_\beta + v x_\alpha + \tau^* - y_\alpha &= 0, \qquad \alpha = 1,\ldots,n, \end{aligned} \tag{8.51}$$

where $f(y_\alpha) = -d\Phi(y_\alpha)/dy_\alpha$. When $d = 3$ the functions $\Phi(y)$ and $f(y)$ are equal

$$\Phi(y) = -\frac{y^{3/2}\Theta(y)}{6\pi} \quad \text{and} \quad f(y) = \frac{\sqrt{y}\Theta(y)}{4\pi}.$$

One also has to add the equation of state to the set Eq. (8.51). This equation defines an equilibrium value of the order parameter $\boldsymbol{\phi}$ by

$$\frac{\partial \mathcal{F}}{\partial \phi_\alpha} = 2y_\alpha \phi_\alpha = 0. \tag{8.52}$$

In order to get an analytical solution of the sets Eqs. (8.51) and (8.52) we should make use of the symmetry of the initial Hamiltonian. The symmetry consideration allows the following two possibilities for the nontrivial solution $\boldsymbol{\phi}$: 1) the vector $\boldsymbol{\phi}$ is directed along a cube diagonal so that all components ϕ_α are equal; 2) the vector $\boldsymbol{\phi}$ is directed along a cube rib and all components ϕ_α excepting one (for instance ϕ_1) are equal to zero. All other solutions correspond to local minima of the function Eq. (8.50) and they give exponentially small contributions (of the order of $\exp(-aV)$, $a > 0$) to the free energy. Such solutions will not be considered in what follows.

Let us eliminate the variable x_α in the set Eqs. (8.51) and (8.52), then we arrive at

$$u\sum_{\beta=1}^{n}[\phi_\beta^2 - f(y_\beta)] + v[\phi_\alpha^2 - f(y_\alpha)] + \tau^* - y_\alpha = 0,$$

$$\phi_\alpha y_\alpha = 0, \qquad \alpha = 1, \ldots, n. \tag{8.53}$$

In the following we consider how solutions of the set Eq. (8.53) change with the change of v sign.

Positive v

In this case when $\tau^* > 0$ the free energy minimum corresponds to the solution $\phi = 0$ and

$$\sqrt{y_1} = \cdots = \sqrt{y_n} = -\frac{nu+v}{8\pi} + \sqrt{\left(\frac{nu+v}{8\pi}\right)^2 + \tau^*}, \tag{8.54}$$

and when $\tau^* < 0$ all y_i are equal to zero but the order parameter is defined from

$$\phi_1^2 = \cdots = \phi_n^2 = -\frac{\tau^*}{nu+v}. \tag{8.55}$$

Such a solution describes the second-order transition at $\tau* = 0$.

Negative v

Now, when $\tau^* \leq \tau_c^*$, where

$$\tau_c^* = -\frac{v(nu+v)^2}{(8\pi)^2(u+v)}, \tag{8.56}$$

in addition to the solution given by Eq. (8.54), a new, nontrivial solution appears with $y_1 = 0$, $\phi_2 = \cdots = \phi_n = 0$, and

$$\sqrt{y_{2\pm}} \;=\; \cdots \;=\sqrt{y_{n\pm}} = -\frac{v(nu+v)}{8\pi(u+v)}\left[1 \pm \sqrt{1-\frac{\tau^*}{\tau_c^*}}\right], \tag{8.57}$$

$$\phi_{1\pm}^2 \;=\; -\frac{\tau^*}{u+v} - \frac{(n-1)uv(nu+v)}{32\pi^2(u+v)^2}\left[1 \pm \sqrt{1-\frac{\tau^*}{\tau_c^*}}\right]. \tag{8.58}$$

One can verify that at $\tau^* = \tau_c^*$ the function $\mathcal{F}(\phi)$ has an inflection point at $\phi_1 = \phi_+ = \phi_-$. When $\tau^* < \tau_c^*$ this point is split into a maximum (at $\phi_1 = \phi_-$) and minimum (at $\phi_1 = \phi_+$). However, in the vicinity of the temperature τ_c^* the minimum $\mathcal{F}(\phi_+)$ is local so that $\mathcal{F}(\phi_1 = 0) < \mathcal{F}(\phi_+)$. With the decrease of τ^* the value $\mathcal{F}(\phi_+)$ decreases reaching the value $\mathcal{F}(\phi_1 = 0)$ at temperature $\tau^* = \tau_0^*$ defined from

$$n\sqrt{y_1}\left[\frac{y_1}{3} + \frac{nu+v}{8\pi}\sqrt{y_1} - \tau_0^*\right] = \sqrt{y_2}\left[\frac{\pi(u+v)}{v^2}y_2^{3/2}\right.$$
$$\left. + \frac{3nu+4v}{3v}y_2 + \left(\frac{nu+v}{8\pi} + \frac{4\pi\tau^*}{v}\right)\sqrt{y_2} - n\tau_0^*\right], \tag{8.59}$$

where the values $y_1(\tau_0^*)$ and $y_2(\tau_0^*)$ are defined by Eqs. (8.54) and (8.57). The following decrease of τ^* transforms $\mathcal{F}(\phi_+)$ to the absolute minimum. The maximum value $\mathcal{F}(\phi_-)$ moves with the decrease of temperature in the direction $\phi \to 0$ and at temperature $\tau^* = \tau_s^*$ this maximum annihilates with the local minimum $\mathcal{F}(\phi_1 = 0)$. The value τ_s^* can be defined from the condition $\phi_-^2 = 0$ which leads to

$$\tau_s^* = -\frac{(n-1)uv}{(4\pi)^2}. \tag{8.60}$$

The expression for the free energy can be constructed with the help of the definition Eq. (8.50) and the formulae, Eqs. (8.57) and (8.58). We do not write here the analytical expression which is very cumbersome, but the behavior of this function is plotted in Fig. 8.3. From this figure, one can see that the free energy transforms with temperature in a way typical for first-order transitions. It is important that this transition occurs well before the second-order transition expected from the mean-field theory. This statement follows from the condition that at $v = 0$ the value τ_s^* is positive (see Eq. (8.59)). The phase equilibrium curve is given by Eq. (8.59), while the surfaces $\tau_c^*(u, v)$ and $\tau_s^*(u, v)$, defined by Eqs. (8.56) and (8.60), give lability boundaries in the parametric space of the Hamiltonian.[11] The jumps of the order parameter at the lability boundaries are defined by

$$\phi_+(\tau_c^*) = \frac{-v(nu+v)[(n-2)u-v]}{[8\pi(u+v)]^2}, \quad \phi_+(\tau_s^*) = \frac{-(n-1)uv[(n-2)u-v]}{[4\pi(u+v)]^2}.$$

It is interesting to notice that the solution found from the set Eq. (8.53) does not have the limit $n = 1$. The latter is a consequence of the assumption $n > 1$ used in the derivation of Eqs. (8.57) and (8.58). Even though the solution, Eq. (8.58) at $n = 1$ formally coincides with Eq. (8.55), these solutions are different so far as $\phi_\pm^2$ appear when $\tau^* > 0$. In the limit $v \to 0$, the solution, corresponding to the first-order transition, as one may expect, vanishes.

[11] They are overheating and overcooling spinodales.

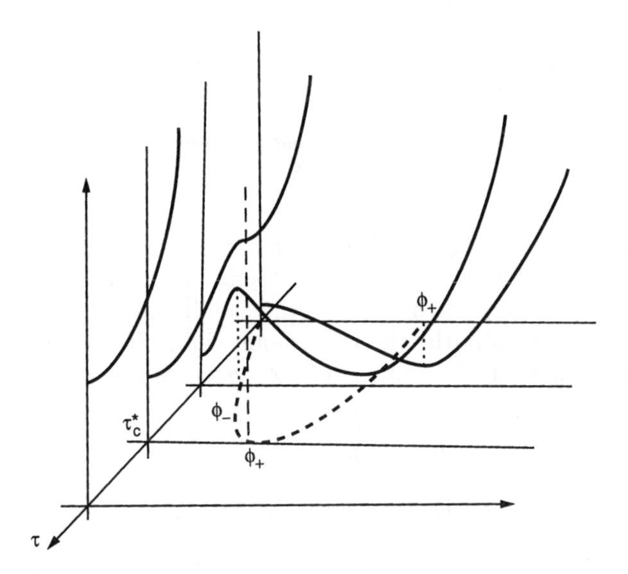

Figure 8.3: Transformation of the free energy function with temperature.

8.4.2 Interacting Fields

This case, in the framework of the model with reduced interactions, allows us to consider a quite general situation with m interacting fields. The model's Ginzburg-Landau functional can be written as

$$H = \frac{1}{2}\int d\boldsymbol{r}\left[\sum_{i=1}^{m} c_i(\nabla\boldsymbol{\phi}_i)^2 + \frac{V}{2}F(a_1,\ldots,a_m)\right], \tag{8.61}$$

where $F\{a_i\}$ is a quite arbitrary function of the variables $a_i = \int d\boldsymbol{r}\boldsymbol{\phi}_i^2(\boldsymbol{r})V^{-1}$ representable as Taylor's series with respect to the variables a_i. After the application of the transformation Eq. (8.48) and integration over the variables $\boldsymbol{\phi}_{q_i}$ we arrive at the expression for the partition function

$$Z \propto \int \left(\prod_{i=1}^{m} dx_i dy_i\right) \exp\left\{-\frac{V}{2}\left[F\{x_i\} + \sum_{i=1}^{m}[y_i(\phi_{0i}^2 - x_i + \theta_i) + \Phi_i(y_i)]\right]\right\},$$

where

$$\boldsymbol{\phi}_{oi} = \boldsymbol{\phi}_{q=o,i}/\sqrt{V}, \quad \Phi_i(y) = \frac{1}{V}\left(\sum_q \ln(y + c_i\boldsymbol{q}^2)\right)_R, \quad \theta_i = \frac{1}{Vc_i}\sum_q \boldsymbol{q}^{-2}.$$

Making use of the replacements $x_i \to x_i + \theta_i$ one finally arrives at

$$Z \propto \int \left(\prod_{i=1}^{m} dx_i dy_i \right) \exp\left\{ -\frac{V}{2} \mathcal{F}(\phi_{oi}, x_i, y_i) \right\},$$

$$\mathcal{F}(\phi_{oi}, x_i, y_i) = \bar{F}(x_i) + \sum_{i=1}^{m} [y_i(\phi_{oi}^2 - x_i) + \Phi_i(y_i)], \tag{8.62}$$

where the function $\bar{F}(x_i)$ is Weierstrass's transformation of the function $F(x_i)$ (see Eq. (8.39)). Again, the integration in the partition function can be carried out with the help of the saddle point method, leading to the exact expression in the thermodynamic limit. Hence, the whole set of equations defining the free energy of the system is given by

$$\frac{\partial \mathcal{F}}{\partial x_i} = \frac{\partial \mathcal{F}}{\partial y_i} = \frac{\partial \mathcal{F}}{\partial \phi_{0i}} = 0, \qquad i = 1, \ldots, m. \tag{8.63}$$

In explicit form the set Eq. (8.63) can be written as

$$y_i = \bar{F}_{x_i}, \qquad \phi_{0i}^2 - x_i - f_i(y_i) = 0, \qquad y_i \phi_{oi} = 0, \tag{8.64}$$

where

$$f_i(y) = -\frac{\partial \Phi_i(y)}{\partial y} = 2\kappa_i \sqrt{y}\Theta(y), \qquad \kappa_i = \frac{1}{8\pi c_i^{3/2}}.$$

The set Eq. (8.64) is quite complicated and its study is very cumbersome. However, the qualitative effects that we are interested in can be traced in the simplest case of two interacting fields. The presence of more than two fields can only diversify these effects. Hence, restricting consideration to the simplest case $m = 2$, we have to choose the function $F(x_i)$ in the ϕ^4-model as

$$F(x_i) = \sum_{i=1}^{2} \left(\tau_i x_i + \frac{1}{4} g_i x_i^2 \right) + \frac{1}{2} w x_1 x_2,$$

where, as usual, $\tau_i = (T - T_{c_i})/T$, T_{c_i} is the trial critical temperature of the i-th field. The function $\bar{F}(x_i)$ can be easily found with the help of the variables x_i shift. It has the same form as the function $F(x_i)$ but, instead of the trial critical temperatures we have renormalized values τ_i^*, i.e.

$$\tau_i \to \tau_i^* = \tau_i + \frac{1}{2} g_i \theta_i + \frac{1}{2} w \theta_{j \neq i}. \tag{8.65}$$

In Chapter 6 we have shown that according to the mean-field theory, disorder-order phase transitions are of the second order. When the parameter $\Delta = g_1 g_2 - w^2$ is negative, the mixed phase, $\phi_{o1} \neq 0$, $\phi_{o2} \neq 0$, is never realized, and the phase transition between the phases $\phi_{o1} \neq 0$, $\phi_{o2} = 0$ and $\phi_{o1} = 0$, $\phi_{o2} \neq 0$ is of the first-order. So, the point $\tau_1^* = \tau_2^*$ on the plane (τ_1^*, τ_2^*) is a bicritical point. The RG analysis shows that for certain values of the vertices g_1, g_2 and w disorder-order transitions are of the first order, and the bicritical point disappears. Below we show that this model qualitatively confirms results of the RG theory.

Let us consider the case $\tau_1^* = \tau_2^*$ when the vertex w is greater than both vertices g_1 and g_2, which means that the inequality $\Delta < 0$ is satisfied. One can show that the mixed phase, $\phi_{o1} \neq 0$, $\phi_{o2} \neq 0$, cannot be realized if $\Delta < 0$. There are two different physical solutions of the set Eq. (8.64) corresponding to the transition into the phase where one order parameter, say $\phi_{o1} \neq 0$, is finite and the other one is zero. They are

$$
\begin{aligned}
y_1 &= 0, \\
y_{2\pm}^{1/2} &= \tfrac{-k_2}{g_1} \pm \sqrt{\left(\tfrac{k_2\Delta}{g_1}\right)^2 + \tfrac{\tau(g_1-w)}{g_1}}, \\
\phi_{01\pm}^2 &= \tfrac{2wk_2}{g_1} y_{2\pm}^{1/2} - \tfrac{\tau}{g_1}.
\end{aligned}
\tag{8.66}
$$

The free energy corresponding to these solutions can be written as

$$
\mathcal{F}_{\pm} = -\frac{\tau^2}{2g_1} + \frac{2\tau(w-g_1)k_2}{g_1} y_{2\pm}^{1/2} + \frac{2\Delta k_2^2}{g_1} y_{2\pm} + \frac{2}{3} k_2 y_{2\pm}^{3/2}. \tag{8.67}
$$

As the energy $\mathcal{F}_+$ is always lower than $\mathcal{F}_-$ only the solution ϕ_{01+} corresponds to the phase $\phi_{o1} \neq 0$, $\phi_{o2} = 0$. This solution differs from the corresponding solution of the mean field theory. In fact, if the inequalities

$$
k_2^2 g_2(w-g_2) > k_1^2 g_1(w-g_1), \quad w/(w-g_2) > \frac{g_1}{(w-g_1)}
$$

are satisfied, the lowest free energy corresponds to the phase transition with the temperature dependence of the order parameter defined by the relationship

$$
\phi_{01}^2 = \left| \frac{2k_2^2\Delta(w^2 - 2g_1 w + g_1 g_2)}{g_1(w-g_1)} \right| + \frac{2k_2 w}{g_1} \sqrt{\frac{g_1 - w}{g_1} \tilde{\tau}} \tag{8.68}
$$

where $\tilde{\tau} = \tau - \tau_c^{(1)}$ and $\tau_c^{(1)} = k_2^2\Delta^2/g_1(w-g_1)$. At $\tilde{\tau} = 0$ the order parameter ϕ_{01} has a jump and, hence, the phase transition is the first-order transition. The above inequalities bound the domain in the space of vertices g_1, g_2, and w where this transition occurs. There are no bicritical points on the phase diagram in this case. In the other case when the following inequalities are satisfied

$$\frac{w}{w-g_2} < \frac{g_1}{w-g_1}, \quad k_1^2\Delta^2 < 4k_2^2 w g_2 (w-g)^2,$$

the second-order transition to the phase $\phi_{01} \neq 0$, $\phi_{02} = 0$ occurs at the temperature $\tau_c^{(2)} = 4k_2^2 w(w - g_2)$ with the order parameter

$$\phi_{01}^2 = -\frac{\bar{\tau}}{g_1}\left[1 + 2k_2^2 w^2 \frac{w - g_1}{g_1^2}\right], \tag{8.69}$$

where $\bar{\tau} = \tau - \tau_c^{(2)}$. The ordering to the phase $\phi_{01} = 0$, $\phi_{02} \neq 0$ can be considered in the same manner by exchanging the indices 1 and 2.

The above considered first-order phase transition is induced by fluctuations. This transition disappears when $g_2 = g_1 = w$. In this case we have an isotropic ϕ^4-model with a two-component order parameter. From Eq. (8.66) it follows that if $g_2 = g_1 = w$, $y_{2\pm} = 0$ and there is no jump in the value $\phi_{01\pm}$. In this limit the solution, Eq. (8.66) as well as the free energy, Eq. (8.67) are reduced to the mean-field theory values: $\phi_{01}^2 = -\tau/g_1$ and $\mathcal{F} = -\tau^2/2g_1$.

Since the above phase transitions are due to the interaction of fluctuations, in the case when fluctuations are suppressed, we must obtain the regular mean-field transitions. As discussed above, the influence of fluctuations can be suppressed in systems with long-range interactions. One can suppress fluctuations by setting $k_i \to 0$, as the parameters $k_i \propto c_i^{-3/2} \propto G_i^{1/2}$. Hence, when the interaction radii increase, the order parameter jumps decrease and in the limit of the infinite range of interaction the first-order phase transition disappears. From Eqs. (8.66) and (8.67) it follows that in this limit the solutions corresponding to the first-order transitions are reduced to the results of the mean-field theory. This shows that the order parameter jumps are really induced by fluctuations. Thus the results obtained in the framework of the model with reduced interactions are similar to those obtained in the fluctuation theory. It seems natural that one can make the following statement: if the model leads to the fluctuation induced first-order transition, the analogous result should be expected in a more general situation. Hence, the anticipation of discontinuous condensation in the RG analysis, made on the basis of the stability study of the quatric form, corresponds to real physics.

8.5 RG and Reduced Interactions

In this section we are going to apply the RG analysis to the model with reduced interactions following the work by Ivanchenko *et al* (1990 d). One may think that the RG analysis is a valuable tool only when the problem cannot be solved by other methods. Of course, such an attitude is justified owing to the complications and sometimes uncertainties inherent in the RG approach. Nonetheless, application of RG methods to exactly solvable models is not only instructive from the purely mathematical point of view, but may also reveal the physical sense of a number of phenomena encountered in the RG approach, as well as serve as starting points for new approximations. In all those cases when we have an exactly solvable model, one may expect that there should be an RG formulation with an exact solution of the main RG equation. Unfortunately, the latter has only been demonstrated in the periodical literature for the model with reduced interactions. The relationship between this and the spherical models tells us that such a consideration can be easily modified to incorporate the latter into the exact RG analysis. In Appendix B we show how it can be done starting from the general RG approach developed in Chapter 3. Nonetheless, so far, the exact RG solution for the two-dimensional Ising model has not been obtained. If it were obtained, the expansion in the vicinity of $d = 2$ might lead to new results and approximation schemes applicable at $d = 3$. Such a solution cannot be obtained in the framework of the traditional formulation of RG equations, but requires a particular consideration taking into account the specific nature of the two-dimensional Ising model with short range interactions. Below, we consider a specific RG formulation for the model with reduced interactions. However, in this case the final equation can be also obtained from the generalized RG scheme developed in Chapter 5.

8.5.1 The RG Equation for the Model

Here, we start from the partition function in the form obtained in Sect. 8.3.1, i.e.

$$Z \propto \int D\phi_q dx dy \exp\left\{-\frac{V}{2}\left[\frac{1}{V}\sum_{q<\Lambda}(\boldsymbol{q}^2+y)|\phi_q|^2 - xy + F(x)\right]\right\}. \quad (8.70)$$

As usual, we restrict momenta from above by the cutoff momentum Λ. In accordance with the general idea of the RG method, at the first stage, one should exclude modes ϕ_q such that $\Lambda/s < q < \Lambda$, where $s > 1$ is an arbitrary constant. After carrying out the integration over this momentum band, one arrives at

$$Z \propto \int D\phi_q dx dy \exp\left\{ -\frac{V}{2}\left[\frac{1}{V}\sum_{q<\Lambda/s} (\boldsymbol{q}^2 + y)|\phi_q|^2 \right.\right.$$

$$\left.\left. - xy + F(x) + \Delta\Phi(y) \right]\right\}, \tag{8.71}$$

where

$$\Delta\Phi(y) = \frac{1}{V} \sum_{\Lambda/s<q<\Lambda} \ln(y + \boldsymbol{q}^2).$$

At the second stage, one should change scale q so as to make the cut-off momentum for new vectors $q' = qs$ equal to Λ again. For this purpose we change the field variables ϕ_q to provide the coincidence of the F-independent part of the functional (namely $\sum_q \boldsymbol{q}^2|\phi_q|^2$) with its trial expression. Thus one has $\phi'_{q'} = \phi_q s^{-(d+2)/2}$ and consequently

$$Z \propto \int D\phi'_{q'} dx dy \exp\left\{ -\frac{V'}{2}\left[\frac{1}{V'}\sum_{q'<\Lambda} \boldsymbol{q}'^2|\phi'_{q'}|^2 - y\left(x - as^2\right) \right.\right.$$

$$\left.\left. + s^d F(xs^{-d}) + \Delta\Phi(y) \right]\right\}. \tag{8.72}$$

The renormalized function $F'(a)$ can be obtained from Eq. (8.72) in the form

$$F'(a) = -\frac{2}{V'} \ln \int dx dy \exp\left[-\frac{V'}{2}\Omega(x,y) \right], \tag{8.73}$$

where

$$\Omega(x,y) = s^d F(xs^{-d}) - y(x - as^2) + \Delta\Phi(y).$$

It is obvious that at $V \to \infty$ the function F' is defined by the saddle point condition, i.e. $F'(a) = \min_{x,y} \Omega(x,y)$. Thus

$$F' = s^d F(xs^{-d}) - (x - as^2)F_x + \Delta\Phi(F_x). \tag{8.74}$$

In order to derive the differential equation for the function F one can use the infinitesimal transformation $s = 1 + \delta$ with $\delta \to 0$. Using the condition $\partial\Omega/\partial y = as^2 + \partial\Phi/\partial F_x - x = 0$ and expanding to the small values $o(\delta)$, one can rewrite Eq. (8.74) as

$$F'(x) = F(x) + \delta\left(dF(x) - (d-2)xF_x + \frac{\Delta\Phi}{\delta} \right). \tag{8.75}$$

Let us now define the renormalisation group "time" l in such a way

$$\frac{\partial F}{\partial l} = \lim_{\delta \to 0} \left(\frac{F(l(1+\delta);x) - F(l;x)}{\delta l} \right).$$

Then according to Eq. (8.75) the RG equation for the function F acquires the form

$$\frac{\partial F}{\partial l} = dF - (d-2)xF_x + \bar{\Phi}. \tag{8.76}$$

In this equation the function $\bar{\Phi}$ is determined by

$$\bar{\Phi} = \lim_{\delta \to 0} \frac{\Delta \Phi}{\delta} = K_d \Lambda^d \ln \left(\Lambda^2 + F_x, \right)$$

where $K_d = 1/(2^{d-1}\pi^{d/2}\Gamma(d/2))$ is the area of the d-dimensional sphere of unit radius divided by $(2\pi)^d$. Let us expand $\bar{\Phi}$ as a power series in F_x, i.e.

$$\bar{\Phi} = K_d \left[\Lambda^d \ln \Lambda^2 + \Lambda^{d-2} F_x - \Lambda^{d-4} F_x^2 + o(\Lambda^{d-6}) \right]. \tag{8.77}$$

The first two terms in Eq. (8.77) can be eliminated from Eq. (8.76) by means of rescaling $F(0)$ and x. The factor Λ^{d-4} at F_x^2 is eliminated by the replacement $\bar{F}_x = \Lambda^{d-4} F_x$. The remaining powers $\bar{F}_x^k$ acquire factors of the form $\Lambda^{2(d-2)(1-k/2)}$. Now the cutoff momentum can be turned to infinity. The terms proportional to $\Lambda^{2(d-2)(1-k/2)}$ will vanish at $k > 2$ and this equation is simplified to

$$\frac{\partial F}{\partial l} = dF - (d-2)xF_x - F_x^2. \tag{8.78}$$

The simplicity of this equation accounts for the exact solvability of the model. Therefore, one should expect that the RG equation can be solved exactly. However, certain difficulties arise from the fact that Eq. (8.78) is nonlinear and contains partial derivatives with respect to l and x. This difficulty can be overcome with the help of Euler's method. Following this method we have to introduce new variables, X, Y, and Z defined by the relations

$$X = F_x, \qquad Y = l, \qquad Z = xF_x - F. \tag{8.79}$$

Using this variables one can transform Eq. (8.78) to the form

$$\frac{\partial Z}{\partial Y} = dZ - 2X\frac{\partial Z}{\partial X} + X^2. \tag{8.80}$$

Now we have a linear, first order, partial differential equation. This equation can be integrated. The full integral of Eq. (8.80) can be written with the help of an arbitrary function $Q(t)$ in the from

$$Z = \frac{X^2}{4-d} + |X|^{d/2} Q\left(|X|^{1/2} e^{-Y}\right).$$

Now turning back to the variables (x, l, F) we arrive at

$$F = xF_x - \frac{F_x^2}{4-d} + F_x^{d/2} Q(F_x e^{-2l}). \tag{8.81}$$

This integral may not seem essentially better than the initial form Eq. (8.78). However, its advantages can be revealed if one changes the variables and considers the function F_x as an independent variable, i.e. introduces $x = x(F_x, l)$. Differentiating Eq. (8.81) with respect to x we arrive at

$$F_{xx} \left\{ x - \frac{2}{4-d} F_x + F_x^{(d-2)/2} \left[\frac{d}{2} Q + F_x e^{-2l} \frac{\partial Q}{\partial (F_x e^{-2l})} \right] \right\} = 0, \tag{8.82}$$

where F_{xx} is the second derivative with respect to x. As is seen from Eq. (8.82) there are two branches of solutions. One branch $F_{xx} = 0$ corresponds to the Gaussian functional, giving either high temperature (noncritical) solution $F \neq 0$ or critical solution $F = 0$ stable at $d > 4$. The other branch gives a solution applicable in the range $4 > d > 2$ this solution can be written as

$$x = \frac{2}{4-d} F_x + F_x^{(d-2)/2} \bar{Q}(F_x e^{-2l}) = 0, \tag{8.83}$$

where $\bar{Q}$ is obviously an arbitrary function too. It is interesting to note that Eq. (8.83) defines x as a function of F_x and l and it does not depend on the function F. The latter may have direct relation to the fact that analytical calculation of the partition function (see Sect. 8.3.1) has led to the use of F_x as an independent variable (see Eq. (8.41)) and does not contain the function F itself.

8.5.2 Critical Exponents

As usual the correlation length critical exponent ν should be obtained as an inverse value of a maximum (positive) eigenvalue λ_1 of the linear RG. According to physical sense, the function F_x should be finite at any point x in the limit $l \to \infty$. Therefore the fixed point equation can be defined from Eq. (8.83) as

$$x = \frac{2}{4-d} F_x^* + (F_x^*)^{(d-2)/2} \bar{Q}(0). \tag{8.84}$$

At an arbitrary d the presence of the second term in Eq. (8.84) leads to non-analytical expansion of F in powers x. As is usually done in the theory of critical phenomena we must have only analytical fixed points. This leads to the constraint $\bar{Q}(0) = 0$, so that we arrive at $F_x^* = (4-d)x/2$. Now we have to linearize Eq. (8.78) in the vicinity of the found fixed points (trivial $F^* = 0$ and nontrivial $F_x^* \neq 0$). In the vicinity of the trivial point we have

$$\lambda\Psi = d\Psi - (d-2)x\Psi_x, \tag{8.85}$$

where $\Psi = F - F^*$. The solution of Eq. (8.85) can be obtained in the form $\Psi \propto x^{(d-\lambda_k)/(d-2)}$. Taking into account that the deviation of F from F^* should be obtained as a power series of x, i.e. $\Psi = \sum_k c_c x^k$ we arrive at the eigenvalues $\lambda_k = d - k(d-2)$, i.e. the spectrum coincides with the well-known spectrum for the Gaussian fixed point: $\lambda_1 = 2$, $\lambda_2 = \epsilon$, and so on. When $d < 4$ this point is unstable and we have to return to the solution of Eq. (8.84), which leads to the following linear equation

$$\lambda\Psi = d\Psi - [(d-2)x + 2F_x^*]\Psi. \tag{8.86}$$

In order to obtain every solution of this equation it is convenient to eliminate x. This can be done with the help of Eq. (8.78), which at the fixed point has the form

$$dF^* = (d-2)xF_x^* + (F_x^*)^2. \tag{8.87}$$

Differentiation of Eq. (8.87) with respect to F_x^* leads to

$$(d-2)x + 2F_x^* = 2F_x^* \frac{dx}{dF_x^*},$$

which helps to reduce Eq. (8.85) to the form

$$(d-\lambda)\Psi = 2F_x^* \frac{d\Psi}{dF_x^*} \tag{8.88}$$

From this equation one gets

$$\Psi_k \propto (F_x^*)^{(d-\lambda_k)/2} \propto x^{(d-\lambda_k)/2}.$$

Bearing in mind that $\Psi = \sum_k c_k x^k$, one finally finds $\lambda_k = d - 2k$. The beginning of the sequence λ_k has values $\lambda_1 = d-2$, $\lambda_2 = -\epsilon$, and so on. Therefore, this fixed point is stable when $d < 4$ and the exponent $\nu = 1/\lambda_1 = 1/(d-2)$.

Here a natural question arises concerning the domain of attraction to the found fixed point. This question is closely related to another question about the type of the phase transition. In order to answer these questions, we have to take into account that when an initial function $F(l = 0; x)$ (which can also be defined as $x_0(F_x) \equiv x(l = 0; F_x)$) is stated, the function $\bar{Q}(F_x)$ should be defined from Eq. (8.83) as

$$\bar{Q}(F_x) = \left[x_0(F_x) - \left(\frac{2}{4-d}F_x\right)\right] F_x^{(2-d)/2}.$$

Consequently, Eq. (8.83) can be rewritten in the form

$$x = 2F_x\left(1 - e^{(4-d)l}\right)/(4-d) + x_0\left(F_x e^{-2l}\right)e^{(d-2)l}.$$

When $l \to \infty$ the fixed point $F_x^* = (4-d)x/2$ can be reached $(2 < d < 4)$ only if x_0 vanishes as $(F_x e^{-2l})^a$, where the exponent $a > (d-2)/2$. In other words, the expansion of F in powers of x must begin from the power x^k with $k < d/(d-2)$. Since $d < 4$, the latter is satisfied by any trial Ginzburg-Landau functional whose expansion begins from a power not higher than ϕ^4. This can be considered as natural for an isotropic system where the sole event changing the critical behavior to the tricritical is nullification of a factor at the ϕ^4 term. It should also be mentioned that the above restriction on the type of $x_0(F_x)$ automatically leads to $\bar{Q}(0) = 0$.

Since the critical exponents satisfy the scaling relations, it is sufficient to only find one exponent in addition to ν. This can be easily done for the exponent η which determines the critical asymptotics of the two-point correlation function $G_q = \langle\phi_q\phi_{-q}\rangle \propto q^{-2+\eta}$. As is explicitly seen from Eq. (8.72) the RG procedure performed for this model does not give rise to new (as opposed to $q^2|\phi_q|^2$) types of non-localities of the functional vertices. Therefore, in this case $G_q \propto q^{-2}$, which means that the exponent η is zero. This result can be, naturally, verified by means of direct calculation of G_q for the model.

Using $\eta = 0$ one gets $\gamma = 2\nu = 2/(2-d)$; $\delta = (d+2)/(d-2)$, i.e. the known exponents for the systems belonging to the spherical model universality class. Thus we see that the critical asymptotics obtained both as a result of the exact solution and with the help of the RG analysis coincide.

In conclusion we would like to mention the following. As was shown by Stanley (1968), the spherical model corresponds to the limit of the infinite number of components $(n \to \infty)$ of the vector $\boldsymbol{\phi}$. It is interesting to compare the RG equation specified here with the limit for the local version of the generalized approach considered in Chapter 5. As is shown in Appendix B this equation has the form $(n \to \infty)$

$$\frac{\partial F}{\partial l} = dF - (d-2)xF_x - xF_x^2, \tag{8.89}$$

which differs from Eq. (8.78) by the factor x at the last term of the right-hand side part. However, one can verify that Eq. (8.89) also leads to the exponent $\gamma = 2/(d-2)$ (see Appendix B). Thus, though the model with reduced interaction is different from the spherical one, both these models belong to the same universality class.

Chapter 9

Application to Copper Oxides

9.1 Introduction

In this chapter we demonstrate how the methods described in this book can be applied to materials used in experiment using the example of copper oxide systems. The interest in the study of oxides was ignited by the discovery of high-T_c superconductivity by Bednortz and Muller in 1986. The first works on high-T_c were performed on La_2CuO_4 based samples ($T_c \approx 30 \div 40$ K). Lately, large families of other oxide systems were found like $YBa_2Cu_3O_{7-x}$ (Y-123) with $T_c \approx 90$ K, different modifications of BiSrCaCuO systems (T_c ranging up to 120 K). In 1993 Chu *et al.* reported results for Hg-based compounds $HgBa_2Ca_2Cu_3O_{8+x}$ (Hg-1223) (see Chu 1994 and Chen *et al.* 1994). These compounds under pressure of 150 kbar show essential increase in T_c up to 153 K from the ambient $T_c \approx 130$ K.

The physics of oxide cuprate systems is interesting not only due to the presence of superconducting transitions but also due to their unusual behavior in nonsuperconducting states. At present there is a widespread opinion that clues to understanding high-T_c superconductivity are to be found from comprehensive knowledge of the normal state. This encourages us to consider different aspects of such systems in this chapter and thus to demonstrate the rich physics related to the topic without interference with details and specificity of compound compositions, leaving also untouched questions related to the unknown (so far!) nature of superconductivity.

Practically all high-T_c materials can be in a variety of phase states including: superconductivity, normal state, different magnetic states, and coexistent phases. In Sect. 9.2 an example is given of magnetic ordering in LaCuO systems which has been actively studied. Usually, study of any particular topic in the high-T_c area creates controversial opinions and claims. Sometimes such controversy is a result of poor quality of samples used in experiments, but most frequently it is caused by the inherent variability of properties of copper oxides. All these materials may accept different amounts of nonstoichiometric oxygen, which drastically affects their properties. Thus, as a rule, experimentalists deal with a nonequilibrium material with a frozen content of oxygen. In Sect. 9.3 the problem of oxygen ordering is addressed on the base of the approach developed in Chapter 8, i.e. the method of reduced, exactly-solvable models. Even though this approach cannot give complete quantitative description, it gives a very good qualitative characterization, revealing specific features which may be lost in microscopic first-principle theories based on perturbation approaches. Thus, the data obtained in Sect. 9.3 are relevant to the technology of sample preparation, revealing physical processes which affect low-temperature behavior of such compounds.

Section 9.4 is dedicated to phase transitions of d-paired superconductors. The question of the superconducting pairing mechanism, and, in particular, the symmetry of the pairing state, was addressed practically from the start of the high-T_c saga. In principle many symmetries are allowed for the paired state. However, at present there is strong experimental evidence that the spin pairing is singlet, which allows for the orbital states with even angular momentum. The simplest situation corresponding to the traditional s-wave pairing is not completely ruled out, notwithstanding the fact that a number of recent (1993 – 1994) experimental results make d-pairing very likely. Measurements of the specific heat, thermal conductivity, Raman background spectra, magnetic field penetration depth, and microwave surface resistance in many cases are consistent with the assumption of d-wave. However, the situation is aggravated by the presence of anisotropy which makes theoretical interpretation arguable.

The material in this chapter does not touch upon mechanisms of superconductivity in oxides but is restricted to the subjects that can be treated on the basis of general symmetry consideration and known experimental facts.

9.2 La_2CuO_4 Systems

The choice of La_2CuO_4 systems is not casual. This material has been extensively studied experimentally. After the discovery of the high-T_c superconductivity in La_2CuO_4 compositions, extensive information has been gained related to the superconducting as well as to the normal states (see for instance Cheong *et al.* (1989), and B. Batlog (1990)). The nonsuperconducting state of La_2CuO_4 seems to be even more enigmatic that the superconducting. In particular, the magnetic behavior of this compound is in many respects a unique phenomenon and its study is in progress even now. At present it is proven that the stoichiometric material represents a four-sublatticed Heisenberg antiferromagnet (AFM) with $S = 1/2$. The CuO_2 planes lead to a quasi two-dimensional magnetic ordering, which is related to an extremely small intra-plane exchange. The ratio of the intra- to inter-exchange integrals is $\sim 3,7 \cdot 10^{-5}$ (Izumov *et al.* (1989), Borodin *et al.* (1991)). A very weak anisotropic exchange as well as antisymmetric Dzyaloshinsky-Moria term cause an easy-plane "chess board," noncollinear structure. In this structure the magnetic moments are ordered in the base plane along the *c*-axis ($a < b < c$, the space group C_{mca}) with small angular deviations ($\sim 0.17°$) in the *bc*-plane. The group theoretical analysis of this structure has been done by Izumov *et al.* (1989) and lately by Doroshev *et al.* (1992). Theoretical calculations of magnetic characteristics by Doroshev *et al.* (1992) are in a good agreement with experiment. These calculations are based on the four-sublatticed AFM model in the noninteracting spin-wave approximation and are valid only in the low-temperature region at $T \ll T_N$ (where T_N is the Neel temperature). In the close vicinity of the critical point there are a number of inconsistencies and essential deviations from experiment. These, as is shown by Borodin *et al.* (1991), can be attributed to the influence of strong critical fluctuations. Neutron measurements of the structure showed strong two-dimensional spin correlations in CuO_2 planes above T_N. However, the long-range magnetic order is three-dimensional (see Izumov *et al.* (1989)). Experimental studies of T_N in $La_2CuO_{4+\delta}$ are not simple and interpretation of experimental results is also aggravated by the exceptional sensitivity to the content of the over-stoichiometric oxygen. The value of T_N changes from 300 to 0 K when δ varies in the range 0 - 0.04. This latter finding led to a number of controversial claims with regard to the nature of the transition. In earlier works by Kitaoka (1987), Watanabe (1987), Nishihara (1987), Lutgemeier and Pieper (1987), and Ziolo (1988) one can find disagreement even in the definition of the type of the transition. A detailed study of the temperature dependent magnetization of nearly-stoichiometric La_2CuO_4 with the help of ^{139}La in nuclear quadruple resonance (NQR) by Borodin *et al.* (1991) showed that in the vicinity of T_N two phases, antiferromagnetic and paramagnetic, coexist in a wide temperature range (see Fig. 9.1). The latter does not allow one to find the exact value for T_N. The value of the critical exponent $\beta \approx 0.3$ corresponds to 3D second-order transition. However, there also has been found a substantial jump of magnetization ($\sim 30\%$) and

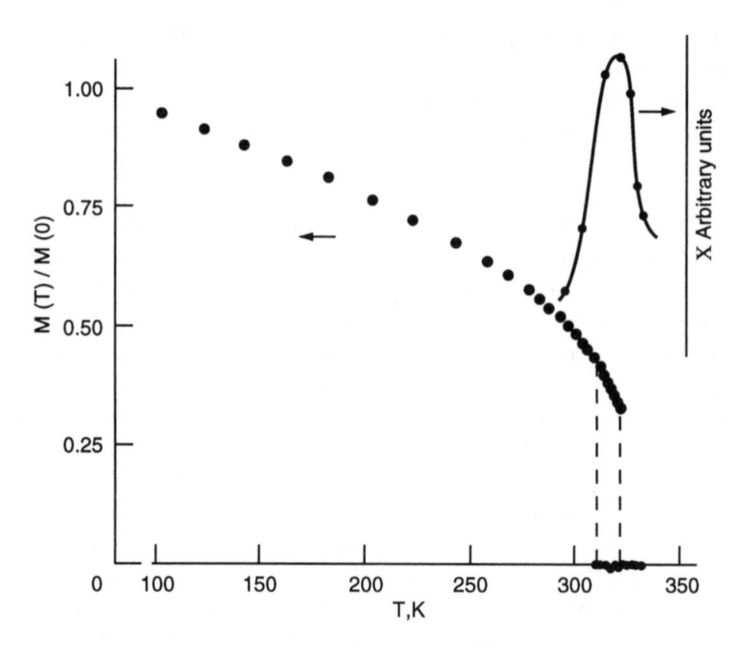

Figure 9.1: The experimental temperature dependence of the sublattice magnetization in La_2CuO_4 obtained from the NQR data. The dashed lines restrict the coexistence region for the AFM and PM phases.

a temperature hysteresis. The results of this work have been confirmed recently by MacLaughlin *et al.* (1994).[1] In the article by Borodin *et al.* (1991) a preliminary theoretical analysis has also been done based on the assumption that the phase transformation occurs due to a fluctuationally induced (weak) first order transition. A detailed theoretical study of the critical behavior in La_2CuO_4 was made in 1993 by Ivanchenko *et al.* The consideration given below is based on the latter work.

9.2.1 The Ginzburg-Landau Functional

The Ginzburg-Landau functional for the considered material can be derived following methods described in Chapter 2. So far, as the derivation is restricted to a finite temperature region, there is no need to explicitly consider commutation relations for spin operators. One can start immediately from the Hamiltonian of a classical system of two kinds of interacting magnetic moments $\boldsymbol{n}_l$ and $\boldsymbol{m}_l$

$$H = -\frac{1}{2}\sum_{l,l'}\left\{J_{ll'}\boldsymbol{n}_l\cdot\boldsymbol{n}_{l'} + I_{ll'}\boldsymbol{m}_l\cdot\boldsymbol{m}_{l'} + \boldsymbol{n}_l\cdot\boldsymbol{m}_{l'}\right\}, \tag{9.1}$$

where $\boldsymbol{n}_l$ and $\boldsymbol{m}_l$ are in general case n-component vectors with $|\boldsymbol{n}_l|^2 = |\boldsymbol{m}_l|^2 = 1$. The partition function for this system can be written

$$Z \propto \int D\boldsymbol{n}_l D\boldsymbol{m}_l \exp(-H). \tag{9.2}$$

According to Chapter 2, in order to describe the critical behavior, one has to replace the variables $\boldsymbol{n}_l$ and $\boldsymbol{m}_l$, defined on the unit hypospheres, by new continuous variables $\boldsymbol{\phi}$ and $\boldsymbol{\eta}$ defined in the whole space. As usual, the procedure of this replacement should be performed with the help of the Hubbard-Stratonovich transformation. This is more convenient to apply after the following simplifications. Let us diagonalize the bilinear form in Eq. (1.1). This can be easily done with the help of the Fourier transformation

$$\boldsymbol{n}_l = \frac{1}{\sqrt{N}}\sum_q \boldsymbol{n}_q \exp(\boldsymbol{q}\cdot\boldsymbol{l}); \qquad \boldsymbol{m}_l = \frac{1}{\sqrt{N}}\sum_q \boldsymbol{m}_q \exp(\boldsymbol{q}\cdot\boldsymbol{l}), \tag{9.3}$$

which reduces the Hamiltonian Eq. (1.1) to

$$H = -\frac{1}{2}\sum_q J_q\boldsymbol{n}_q\cdot\boldsymbol{n}_{-q} + I_q\boldsymbol{m}_q\cdot\boldsymbol{m}_{-q} + K_q\boldsymbol{m}_q\cdot\boldsymbol{m}_{-q}, \tag{9.4}$$

[1]In this work an attempt is made to interpret the transition as second order with very small $\beta \approx 0.1$

where the Fourier transformants J_q, I_q and K_q are defined as

$$J_{ll'} = \frac{1}{N} \sum_q J_q \exp[\boldsymbol{q} \cdot (\boldsymbol{l} - \boldsymbol{l}')]. \tag{9.5}$$

The matrix diagonalizing Eq. (1.4) can be written in the form

$$\hat{S}_q = \frac{1}{\sqrt{2}(1+|P_q|^2)^{1/4}(1+\sqrt{1+|P_q|^2})^{1/2}} \times \begin{pmatrix} 1+\sqrt{1+|P_q|^2} & -P_q \\ P_q^* & 1+\sqrt{1+|P_q|^2} \end{pmatrix}, \tag{9.6}$$

where $P_q = 2K_q/(J_q - I_q)$. The matrix $\hat{S}_q$ is defined so as to satisfy the condition $\det \hat{S} = 1$. The latter warrants that the transformation Jacobian is equal to unity in the partition function expressed in new variables. The eigenvalues $\lambda_q^{\pm}$ of the new Hamiltonian are equal to

$$\lambda_q^{\pm} = \frac{J_q + I_q}{2} \pm \left[\left(\frac{J_q - I_q}{2} \right)^2 + |K_q|^2 \right]^{1/2}. \tag{9.7}$$

Now the use of the Hubbard-Stratonovich identity is straight forward. Omitting a number of traditional transformations, described in Chapter 2, we arrive at the following expression for the partition function

$$Z \propto \int D\boldsymbol{\phi}_q D\boldsymbol{\eta}_q \exp\left\{ -\frac{1}{2T} \sum_q \left[\left\{ (\lambda_q^+)^{-1} - \frac{1}{nT} \right\} |\boldsymbol{\phi}_q|^2 + \left\{ (\lambda_q^-)^{-1} - \frac{1}{nT} \right\} |\boldsymbol{\eta}_q|^2 \right] + \frac{1}{2n^2(n+2)T^3} \sum_{q_i} \left[\delta \left(\sum_i \boldsymbol{q}_i \right) \times \prod_{i=1}^{4} \left\{ [(S_{11q_i} + S_{21q_i})\,\boldsymbol{\phi}_{q_i} + (S_{12q_i} + S_{22q_i})\,\eta_{q_i}]^2 \right\} \right] \right\}. \tag{9.8}$$

This equation contains all modes $\boldsymbol{\phi}_q$ and $\boldsymbol{\eta}_q$.

However, with the decrease of temperature the positive definiteness of quadratic terms $[(\lambda_q^+)^{-1} - 1/nT]|\boldsymbol{\phi}_q|^2$ and $[(\lambda_q^-)^{-1} - 1/Nt]|\boldsymbol{\eta}_q|^2$ will be first violated for the modes adjacent to the maxima of the functions $\lambda_q^{\pm}$. As is discussed in Chapter 2 (see Sect. 2.2.2) the structure of the ordered state is completely defined by the positioning of these maxima in q-space. Those modes that are not in the immediate vicinity of the "dangerous points" do not give any essential contribution to fluctuating behavior and they can be eliminated with the help of the Gaussian integration procedure. In applications to a given physical system a particular behavior of the functions $\lambda_q^{\pm}$ is of crucial importance. Having in mind La_2CuO_4, let us consider two interacting, equivalent square sub-lattices shifted with respect to each other. Restricting consideration only to nearest and next to the nearest neighbor interactions, one has

$$\begin{aligned} J_q &= \tfrac{J}{2}(\cos q_1 a + \cos q_2 a), \\ |K_q|^2 &= \tfrac{K^2}{4}(1 + \cos q_1 a + \cos q_2 a + \cos q_1 a \cos q_2 a), \end{aligned} \tag{9.9}$$

where a is the lattice constant, and $|K|$ in accordance with experimental data should satisfy inequality $|K| \ll J, I$. As each of the sublattices is ordering into an antiferromagnetic structure, one should consider only the case when $J, I < 0$. The physical equivalence of layers makes reasonable assumption $J = I$. As a result, the functions $\lambda_q^{\pm}$ are reduced to $\lambda_q^{\pm} = J_q \pm |K_q|$. In Fig. 9.2 one can see that in both functions $\lambda_q^{\pm}$ have maxima at the points where $q_{1,2} = \pm\pi/a$, which naturally corresponds to antiferromagnetic ordering (see Chapter 2). On the other hand the function $|K_q|$ tends to zero in these points (see Fig. 9.3). The latter reflects the fact that due to the reciprocal compensation of the vectors $\boldsymbol{n}_l$ in the 2D-antiferromagnetic structure of the sub-lattices, the exchange interaction between them tends to zero.

Representing the functions $\lambda_q^{\pm}$ as a power series of $\boldsymbol{q}$ in the vicinity of the maxima $\lambda_{q0}^{+} = \lambda_{q0}^{-} = \lambda_0$ and eliminating nonessential modes $\boldsymbol{\phi}_q$ and $\boldsymbol{\eta}_q$ by integration, one can obtain the GFL in the from of two noninteracting subsystems

$$\mathcal{F} = \frac{1}{2}\int d\boldsymbol{r}\left\{(\boldsymbol{\nabla}\boldsymbol{\phi})^2 + (\boldsymbol{\nabla}\boldsymbol{\eta})^2 + \tau(\boldsymbol{\phi}^2 + \boldsymbol{\eta}^2) + \frac{g}{4}\left[(\boldsymbol{\phi}^2)^2 + (\boldsymbol{\eta}^2)^2\right]\right\}, \tag{9.10}$$

where $\tau = (T - \lambda_0)/T$ and $g = \lambda^2/T^2$. In this equation the following substitutions have also been performed

$$\boldsymbol{\phi}_q\sqrt{a/T\lambda_0} \to \boldsymbol{\phi}_q; \qquad \boldsymbol{\eta}_q\sqrt{a/T\lambda_0} \to \boldsymbol{\eta}_q;$$

$$n_1 = n_2; \quad \boldsymbol{\phi}^2 = \phi_1^2 + \phi_2^2; \quad \boldsymbol{\eta}^2 = \eta_1^2 + \eta_2^2.$$

The initial Hamiltonian Eq. (1.1) does not take into account uniaxial crystallographic anisotropy which defines orientation of the magnetic moments $\boldsymbol{n}_l$ and $\boldsymbol{m}_l$ with respect to the lattice. In particular for tetragonal systems the expression of the GFL given in Eq. (1.10) should be completed by the anisotropic invariant $v(\phi_1^4 + \phi_2^4 + \eta_1^4 + \eta_2^4)$. A little more complex situation

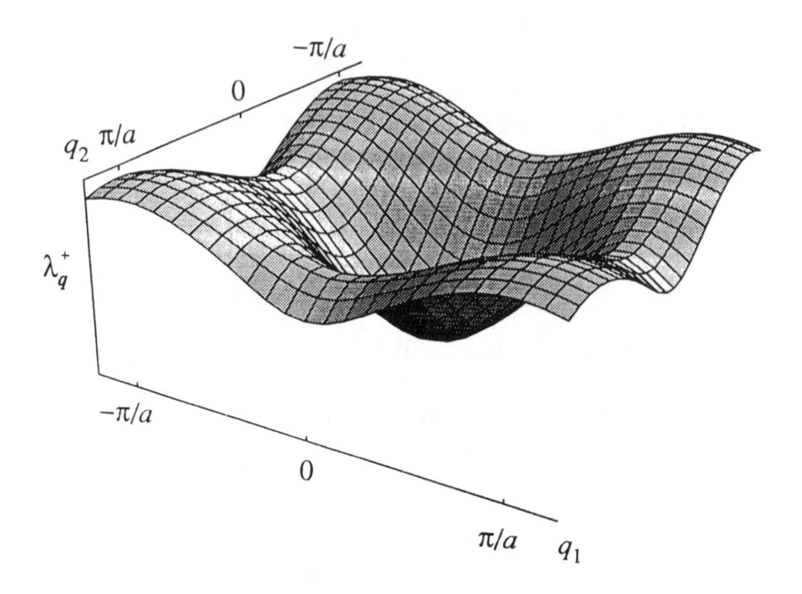

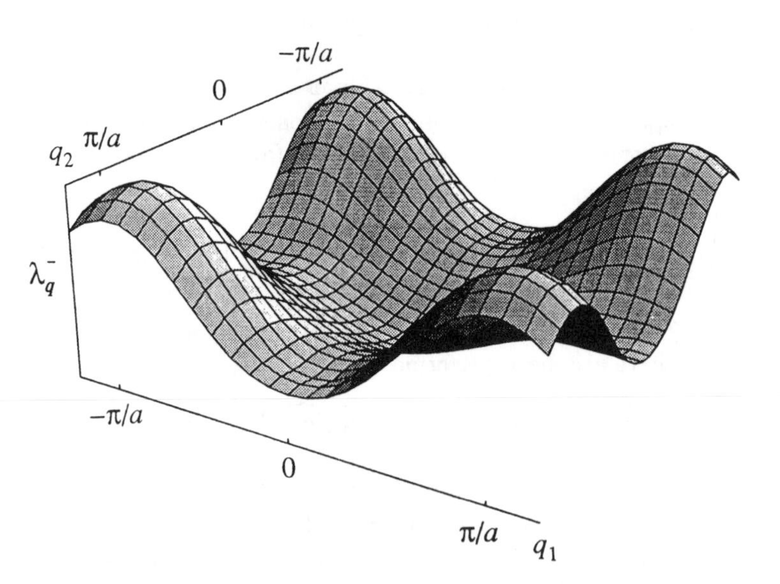

Figure 9.2: The functions λ_q^+ (a) and λ_q^- (b).

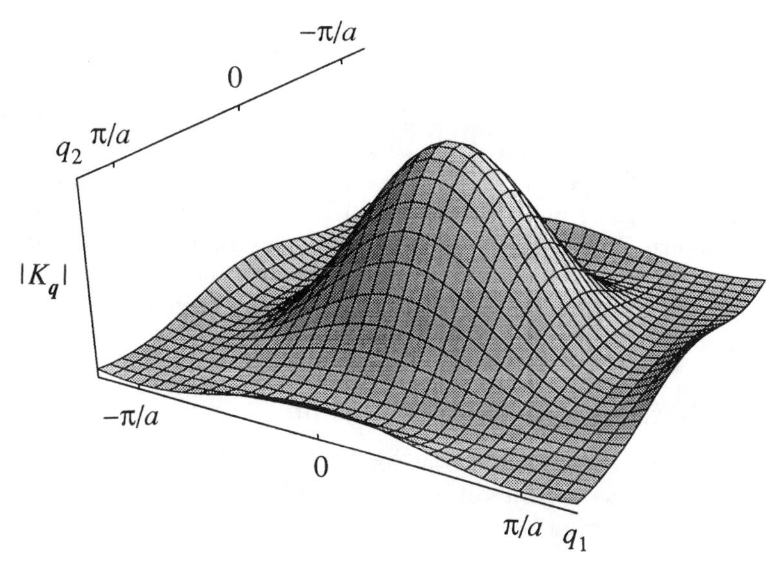

Figure 9.3: The function $|K_q|$.

is encountered with interactions of different magnetic subsystems ϕ and η. Though there are no explicit interactions in the Hamiltonian Eq. (1.1), nonetheless, such interactions in a real situation are introduced implicitly with the help of lattice deformations, influence of electronic degrees of freedom, and others. Most of these effects renormalize expansion coefficients in $\mathcal{F}$ leading to dependencies on $\boldsymbol{\eta}$ in interaction vertices related to $\boldsymbol{\phi}$ and *vice versa*, i. e. $\tau(\boldsymbol{\eta})\boldsymbol{\phi}^2 = \tau(1+\alpha\boldsymbol{\eta}^2+\cdots)\boldsymbol{\phi}^2$ and so on. In the framework of the ϕ^4-model such interactions can be taken into account with the help of the $u_3\boldsymbol{\phi}^2\boldsymbol{\eta}^2$ invariant. As a result the GLF for La_2CuO_4 system should be represented in the form

$$\mathcal{F} = \frac{1}{2}\int d\boldsymbol{r}\left\{(\boldsymbol{\nabla}\boldsymbol{\phi})^2 + (\boldsymbol{\nabla}\boldsymbol{\eta})^2 + \tau(\boldsymbol{\phi}^2+\boldsymbol{\eta}^2) + \frac{u_1}{4}\left(\phi_1^4 + \phi_2^4 + \eta_1^4 + \eta_2^4\right)\right.$$

$$\left. + \frac{u_2}{2}\left(\phi_1^2\phi_2^2 + \eta_1^2\eta_2^2\right) + \frac{u_3}{2}\left(\phi_1^2\eta_1^2 + \phi_2^2\eta_1^2 + \phi_1^2\eta_2^2 + \phi_2^2\eta_2^2\right)\right\}, \qquad (9.11)$$

where $u_1 = g + v$, $u_2 = g$, and $u_3 \ll u_1, u_2$. Thus, neglecting small orthorhombic lattice distortions, but taking into account AFM subsystems in CuO_2 layers one can reduce the GLF to the ditetragonal form. The field-theoretical analysis of critical behavior of ditetragonal systems was done by Mukamel *et al.* (1976) and Toledano and Michel (1985). They obtained RG equations in the second ϵ-approximation and described fixed points of these equations. It was found that one of the fixed points is a stable fixed point. The latter encouraged these authors to derive the conclusion that the phase transition is of the second order and to calculate the consequent critical exponents. Such an approach is really justified provided the phase flow lines (see Chapter 7) enter to the stable fixed point. However, that does not take place in the whole region of trial vertices of the GLF for these systems. With regard to this situation an especially interesting case has been found experimentally by Borodin *et al.* (1991), where a weak first-order magnetic transition occurs in the La_2CuO_4 compound used as a basic material for high-T_c superconductors.

9.2.2 Mean Field Approximation

In this section we analyze the situation in the Landau approximation. According to Chapter 1 (see Sections 1.2.1 and 1.2.2) the free energy in this approximation is given by

$$\mathcal{F}_L = \mathcal{F}(\boldsymbol{\phi}_o, \boldsymbol{\eta}_o), \tag{9.12}$$

where $\boldsymbol{\phi}_o$ and $\boldsymbol{\eta}_o$ are equilibrium values of the order parameters defined from the conditions

$$\frac{\delta \mathcal{F}}{\delta \boldsymbol{\phi}} = 0; \qquad \frac{\delta \mathcal{F}}{\delta \boldsymbol{\eta}} = 0. \tag{9.13}$$

The dependence of the free energy on values of vertices, corresponding to differently ordered phases, can be easily obtained with the help of Eq. (9.13). It can be written as follows:

1. The phase with $\phi_i \neq 0$, $\eta_i = 0$ (or $\boldsymbol{\phi} \to \boldsymbol{\eta}$)

$$\mathcal{F}_I = -\frac{\tau^2}{u_1 + u_2}; \tag{9.14}$$

2. The phase with $\phi_i \neq 0$, $\phi_j = \eta_{1,2} = 0$

$$\mathcal{F}_{II} = -\frac{\tau^2}{2u_1}; \tag{9.15}$$

3. The phase with $\phi_i \neq 0$, $\eta_i \neq 0$, $\phi_{j \neq i} = 0$, $\eta_{j \neq i} = 0$,

$$\mathcal{F}_{III} = -\frac{\tau^2}{u_1 + u_3}; \tag{9.16}$$

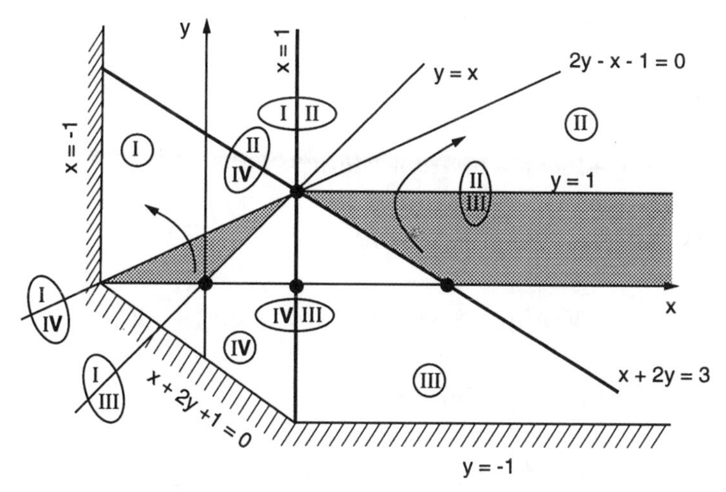

Figure 9.4: Separation of different phases in the framework of the mean field approximation.

4. The phase with $\phi_i \neq 0$, $\eta_i \neq 0$,

$$\mathcal{F}_{IV} = -\frac{\tau^2}{u_1 + u_2 + 2u_3}; \tag{9.17}$$

Using these equations one can obtain boundaries of equilibrium between these phases in the plane $x = u_2/u_1$ and $y = u_3/u_1$. One can see the results in Fig. 9.4. In this figure the straight line $x = 1$ divides the phases $\phi_i \neq 0$, $\eta = 0$ (or $\phi \rightarrow \eta$) and $\phi_i \neq 0$, $\phi_j = \eta_{1,2} = 0$. Similarly, the line $y = 1$ demarcates between the phases II and III and $2y = x + 1$ between I and IV.

9.2.3 RG Analysis

According to Chapter 6, in the RG approach the critical behavior of the system is defined by the evolution of the trial vertices of the functional Eq. (1.11) when one approaches the critical region. For the system of ditetragonal symmetry the RG equations were derived by Mukamel *et al.* (1976) (see also Toledano and Michel (1985)) in the second ϵ-approximation.

These equations can be written in the form

$$\begin{aligned}
\frac{du_1}{dl} &= u_1 - \tfrac{9}{2}u_1^2 - \tfrac{1}{2}u_2^2 - u_3^2 \\
&\quad +108u_1^3 + 16u_1u_2^2 + 24u_1u_3^2 + 8u_2^3 + 16u_3^3, \\
\frac{du_2}{dl} &= u_2 - 2u_2^2 - 3u_1u_2 - u_3^2 \\
&\quad +36(u_1^2u_2 + 2u_1u_2^2) + 24u_2u_3^2 + 20u_2^3 + 16u_3^3, \\
\frac{du_3}{dl} &= u_3 - 2u_3^2 - 3u_1u_2 - u_3u_2 \\
&\quad +36(u_1^2u_3 + 2u_1u_3^2) + 24u_3^2u_2 + 12u_3u_2^2 + 24u_3^3.
\end{aligned} \tag{9.18}$$

As usual, one has to start analysis of these equations by considering fixed points. The set of fixed points contains in the first ϵ-approximation a marginal point when $n = n_1 + n_2 = 4$. All other fixed points are unstable. The presence of the marginal point compels one to perform analysis of the critical behavior in the second order of ϵ-expansion. In this order the isotropic fixed point is stable. In the case where flow lines reach this point one should have the second-order phase transition with the critical exponent $\beta = 0.39$. However, as is shown below, the attraction region to this point is narrow, covering only a small fraction of the whole phase space. Thus, the analysis of other choices becomes very important in this case. This analysis can be done just as well with the help of the first ϵ-approximation which essentially simplifies the RG set, namely

$$\begin{aligned}
\frac{du_1}{dl} &= u_1 - \tfrac{9}{2}u_1^2 - \tfrac{1}{2}u_2^2 - u_3^2, \\
\frac{du_2}{dl} &= u_2 - 2u_2^2 - 3u_1u_2 - u_3^2, \\
\frac{du_3}{dl} &= u_3 - 2u_3^2 - 3u_1u_2 - u_3u_2.
\end{aligned} \tag{9.19}$$

Even this simplified system of nonlinear equations is too complex for analytical considerations. Here we are again compelled to use the method of analysis introduced in Chapter 6. As only types of phase transitions and structures of ordered phases are of interest, one can reduce the number of equations by rewriting the previous RG set in terms of ratios $x = u_2/u_1$ and $y = u_3/u_1$. In these variables the RG set can be written as

$$\begin{aligned} \frac{1}{u_1}\frac{dx}{dl} &= \tfrac{3}{2}x - 2x^2 + \tfrac{1}{2}x^3 + xy^2 - y^2, \\ \frac{1}{u_1}\frac{dy}{dl} &= y\left(\tfrac{3}{2} - x - 2y + y^2 + \tfrac{1}{2}x^2\right). \end{aligned} \quad (9.20)$$

This set of equations defines the phase portrait of the system on the (x, y)-plane. The choice of the vertex u_1 as a normalizing variable is only a matter of convenience for analysis and comparison with the results of the mean field approximation. All other normalization choices would lead to a more complex analysis. The full set of fixed points on the (x, y)-plane contains the following four points

$$\mu_0^* : (0,0); \quad \mu_1^* : (3,0); \quad \mu_2^* : (1,0); \quad \mu_3^* : (1,1). \quad (9.21)$$

Taking into account the results of the analysis (see for instance Mukamel *et al.* 1976) in the second ϵ-approximation, we should treat the point μ_3^* as a stable fixed point. In order to find separatrices demarcating scenarios of different critical behavior one can represent the phase portrait analytically in the form

$$2y = (2-C)x + C \pm \left[[(2-C)x + C]^2 - 2x(3-x)\right]^{1/2}. \quad (9.22)$$

This equation defines two $(\pm)$ families of flow lines. Each line is defined by the choice of arbitrary constant C. The isocline of verticals $(dx/dl = 0)$ help to define directions in the parametric space. The equation for the isocline when $x \neq 1$ is the ellipse

$$\left(x - \frac{3}{2}\right)^2 + 2y^2 = \frac{9}{4},$$

at $x = 1$, the isocline is $y = 1$. From the structure of the functional Eq. (1.11) one can see that at $u_1 \geq 0$, as a result, from the equation $dy/dl = u_1 y(y-1)^2$ $(x = 1)$ follows that $dy/dl \geq 0$ when $y \geq 0$ and $dy/dl \leq 0$ when $y \leq 0$. All these comments help to elicit the topology of the phase portrait of Eq. (1.22), which is shown in Fig. 9.5. As is seen from this figure, the flow lines originating in all regions from I to IV cannot reach the stable fixed point and leave the stability regions of the GLF. This means that in all cases, except when the trial parameters of the functional Eq. (1.11) are bound to the region V, the second-order transition is impossible in the magnetic subsystem of La_2CuO_4. According to analyses of

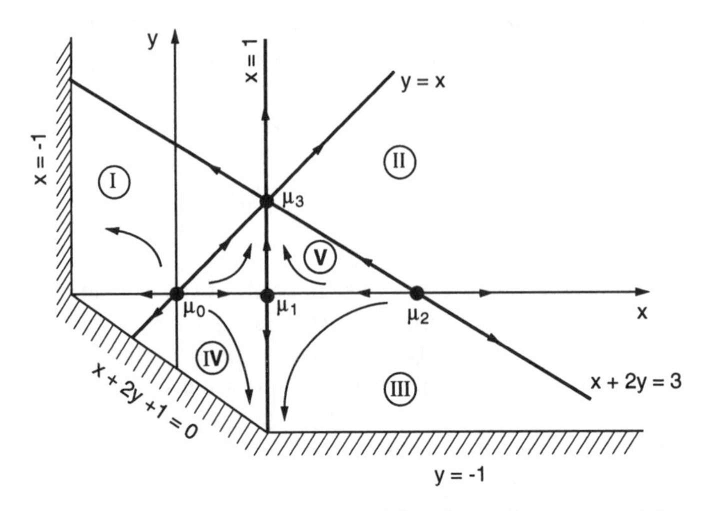

Figure 9.5: The phase portrait of flow lines in x, y variables.

similar situations considered in Chapter 7, such a statement derived from the RG analysis should be treated as a most probable event and in the general case it requires independent additional confirmation.[2]

Now let us return to Fig. 9.4 derived from the mean field consideration. The straight line $x = 1$ demarcating the phases I and II from phases III and IV coincides with the separatrix in the RG approach. On the other hand, one can elicit an important statement from the comparison of these two approaches. Namely, on the flow lines portrait (see Fig. 9.5) there are regions (regions A and B in Fig. 9.4) from which flow lines cross the phase equilibrium boundaries of the mean field treatment.[3] One can expect that in these cases there will be induced phases which are unfavorable in the mean field approach, as a result of fluctuation interactions. However, the phase anticipated from the mean field treatment is stable out of the fluctuation region. Hence, with the decrease of temperature it should be reestablished with the suppression of fluctuations. Thus, in the low temperature region the traditional mean field picture will be restored.

[2]Such a confirmation can be done, for example, with the methods considered in Chapter 8. In fact, this statement has been proven for a reduced exactly solvable model (see Sect. 8.42).

[3]The straight line $2y - x - 1 = 0$ demarcating the phases I and IV and the straight line $y = 1$ separating the phases II and III in Fig. 9.4 can be crossed.

9.2.4 The Influence of Impurities

In real systems one always encounters some amount of inhomogeneities and impurities which can essentially influence the picture of critical behavior. Randomly distributed impurities can be taken into account by the introduction of random variations of parameters of the GLF, in the same manner, as is considered in Chapter 7. As usual, the main contributions come from the variations of $\tau \to \tau + \delta\tau(\boldsymbol{r})$ which is considered as a Gaussian distribution with the correlators

$$\overline{\tau(\boldsymbol{r})} = 0, \qquad \overline{\tau(\boldsymbol{r})\tau(\boldsymbol{r}')} = \Delta\delta(\boldsymbol{r} - \boldsymbol{r}'),$$

where Δ is a value proportional to the concentration of impurities. With the help of the technique developed in Sect. 7.4 one can derive the following RG set of equations

$$\begin{aligned} \frac{du_1}{dl} &= u_1 - \tfrac{9}{2}u_1^2 - \tfrac{1}{2}u_2^2 - u_3^2 + 6\Delta u_1, \\ \frac{du_2}{dl} &= u_2 - 2u_2^2 - 3u_1u_2 - u_3^2 + 6\Delta u_2, \\ \frac{du_3}{dl} &= u_3 - 2u_3^2 - 3u_1u_2 - u_3u_2 + 6\Delta u_3, \\ \frac{d\Delta}{dl} &= \Delta\left(1 - 3u_1 - u_2 - 2u_3 + 4\Delta\right). \end{aligned} \tag{9.23}$$

In the derivation of this set we have taken into account the fact that the components of the order parameters ϕ_i and η_i have the same physical origin and, hence, they should be renormalized under the influence of impurities in the same manner. As one may expect, this set can be reduced to the set Eq. (9.19) in the limit $\Delta \to 0$ corresponding to the ideal system.

The analysis of this set can be performed similarly to the previous case (see Sect. 9.2.3). Let us introduce three ratios, the variables $x = u_2/u_1$, $y = u_3/u_1$, and a new variable $z = \Delta/u_1$. In terms of these variables the RG set can be rewritten as

$$\begin{aligned} \frac{1}{u_1}\frac{dx}{dl} &= x\left(\tfrac{3}{2} - 2x + \tfrac{1}{2}x^2 + y^2\right) - y^2, \\ \frac{1}{u_1}\frac{dy}{dl} &= y\left(\tfrac{3}{2} - x - 2y + y^2 + \tfrac{1}{2}x^2\right), \\ \frac{1}{u_1}\frac{dz}{dl} &= z\left(\tfrac{3}{2} - x - 2y + y^2 + \tfrac{1}{2}x^2 - 2z\right). \end{aligned} \tag{9.24}$$

From this set one can see that the impurity vertex z does not influence the equations for the x and y variables and the first two equations coincide with Eq. (9.20). Thus, the equation for dz/dl is separated from the other two and it has the structure of the Bernoulli equation. This leads to the doubling of the number of fixed points in an impure system. The full set of fixed points can be represented by

$$\mu_0^*(0,0,0); \qquad \overline{\mu_0^*}(0,0,3/4);$$

$$\mu_1^*(1,0,0); \qquad \overline{\mu_1^*}(1,0,1/2);$$

$$\mu_2^*(3,0,0); \qquad \overline{\mu_2^*}(3,0,3/2);$$

$$\mu_3^*(1,1,0); \qquad \overline{\mu_3^*}(1,1,0).$$

The impure fixed points $\overline{\mu_i^*}$ have the same projections on the (x, y)-plane as the points μ_i^* of the ideal system with $\Delta = 0$. More than that, the pure point μ_3^* does not split in the perturbation theory under the influence of impurities, i.e. $\mu_3^* = \overline{\mu_3^*}$. The equation for dz/dl can be easily integrated. However, in this case the explicit solution does not bear much information as the separatrices of attraction to the point μ_3^* constitute a straight prism, orthogonal to the (x, y)-plane. Thus, the regions of physical parameters of the GLF, from which the fluctuationally induced first-order transition occurs, do not change under the influence of impurities. On the other hand the consideration of this section is evidently not valid at large impurity concentrations as one cannot derive reliable RG equations in this case. Therefore, one can expect that with the increase of concentrations, the impurity influence should have a threshold-like behavior.

In conclusion, the discussed results agree well with the experimental observations by Borodin *et al.* (1991) and by MacLaughlin *et al.* (1993) and explain the contradictory data of earlier works.

9.3 Oxygen Ordering in $ABa_2Cu_3O_{6+x}$

This class of oxide high-T_c superconductors includes a variety of compounds with the general chemical formula $ABa_2Cu_3O_{6+x}$. In this formula A can be any of the following elements:Y, La, Nd, Sm, Eu, Gd, Dy, Ho, Er, Tm, Yb, Ce, Pr, Tb, Lu. An important physical property of this class is the ability of a compound to accept practically any value of oxygen in the range $0 \leq x \leq 1$. It has been found that the superconducting properties are essentially influenced by the structural phase transition at a temperature of the order of 700 °C. The lattice cell of the compound can be imagined with the help of three perovskite cubes placed on each other along the c-axis. These cubes are centered with the elements Ba, A, and Ba, respectively. The copper atoms are at the corners of the cubes and the oxygen atoms are positioned between Cu. If such a perovskite structure could be synthesized it would have the chemical formula $ABa_2Cu_3O_9$. However, the real compounds have a number of oxygen vacancies positioned on the basal plane. If one starts with three vacancies (O_6) then in order to increase x additional oxygen should be incorporated. The latter is easily provided with the decrease of temperature below the sintering point of the powder ground earlier to form a compound. Structural measurements (see for instance, Sukharevsky *et al.* 1987 on the use of X-rays; and Jorgensen *et al.* on neutrons 1987) show that the transition occurs from a tetragonal phase (the symmetry group P4/mmm) to an orthorhombic (Pmmm) phase. The symmetry group Pmmm of the low-temperature phase is a subgroup of P4/mmm and according to the phenomenologic Landau approach one should expect a second-order transition. The orthorhombic phase is superconducting with quite high critical temperature (up to $T_c \approx 90$ K). The specimens which were quenched from $T \approx 900$°C had a tetragonal lattice cell and either had low T_c or were completely nonsuperconducting at all (see Ng *et al.* 1988). Thus to understand properties of high-T_c superconductivity one should have a sufficiently correct description of the structural phase transition. This transition is characterized by a number of unusual properties. Thus, in the tetragonal phase, oxygen atoms and vacancies are distributed uniformly on the basal plane. On the other hand, below the structural phase transition point, the appearance of orthorhombic distortions is accompanied by an alignment of oxygen chains along the b-axis and by an increase in the number of oxygen vacancies along the a-axis of the basal plane. At sufficiently low temperature there are almost no oxygen atoms along the a-axis (see Schuller *et al.* 1987). More than that, with a temperature decrease from the tetragonal phase the total oxygen concentration x increases, but on the $x(T)$ curve one cannot distinguish the critical point. The latter property is inherently present in a number of compounds. However, the critical temperature, cell parameters, concentrations of oxygen atoms along the a- and b-axes, and, finally, the curves $x(T)$ not only change from a compound to compound, but also depend on the method of sample preparation and, in particular, on the oxygen partial pressure.

The study of electrical resistivity of YBaCuO by Freitas and Plaskett in 1987 revealed a considerable width of the fluctuation region in the vicinity of the structural phase transition. The resistance substantially deviates from the mean field value and shows a typical fluctuation peak over a temperature range of the order of 50 K. This fact shows that any consistent theory of structural transition in these compounds should take the fluctuation effects into account. However, microscopic first-principle theories appear to be too complicated to justify their use for this purpose. In this section we follow the theory based on the exactly solvable models approach developed in Chapter 8. The consideration given below is due to Ivanchenko *et al.* (1990 e). Despite its rather conventional character, the model used in this section describes quite well the structural transition in the high-T_c compounds of YBaCuO class reflecting, in particular the above peculiarities seen in experiment.[4]

9.3.1 Description of the Model

Let us consider the basal CuO plane of the YBaCuO structure. According to the structural data, the oxygen atoms can occupy only two sites along a- and b-axes. In the following, the interactions between nearest and next-to-nearest oxygen atoms are taken into account. One also has to consider interactions of oxygen vacancies which are assumed to interact due to screened noncompensated charge between themselves and with the present oxygen atoms. As a result, the Hamiltonian of this system can be written as

$$H = -\frac{1}{2}\sum_{i,j=1}^{2}\sum_{ll'}\left[A_{ll'}^{ij}n_l^i n_{l'}^j + B_{ll'}^{ij}(1-n_l^i)(1-n_{l'}^j) + C_{ll'}^{ij}(1-n_l^i)n_{l'}^j\right], \tag{9.25}$$

where n_l^i are oxygen occupation numbers equal to either 0 or 1 for free and occupied sites respectively, l and l' are position vectors of periodic cells, and the indices i and j indicate the oxygen positions inside the cell ($i = 1$ for the site on the b-axis and $i = 2$ for the site on the a-axis). The Hamiltonian can be rewritten in the following equivalent form

$$H = -\frac{1}{2}\sum_{i,j=1}^{2}\sum_{ll'} J_{ll'}^{ij} n_l^i n_{l'}^j - h\sum_{i=1}^{2}\sum_{l} n_l^i, \tag{9.26}$$

where

$$J_{ll'}^{ij} = A_{ll'}^{ij} + B_{ll'}^{ij} - C_{ll'}^{ij}, \qquad h = \frac{1}{2}\sum_{j=1}^{2}\sum_{l\prime}\left[2B_{ll'}^{ij} + C_{ll'}^{ij}\right].$$

In this form H is equivalent to the Hamiltonian of the two-dimensional Ising model for the two-sublattice system with a field. This equivalency immediately disappears if one recalls the fact that the Hamiltonian of Eq. (9.25)

[4]Below we write the symbol YBaCuO for the whole class of ABaCuO compounds.

should be supplemented by the condition fixing the total number of the oxygen atoms in the specimen basal planes

$$\sum_{i=1}^{2}\sum_{l} n_l^i = (2N)\bar{x} = Nx, \tag{9.27}$$

where $\bar{x}$ is the oxygen concentration in the basal plane and $2N$ is the total number of oxygen sites.

In order to calculate the partition function of the system the following substantial simplification of the problem is used. The set of conditions $n_l^i = 0;\ 1$ is replaced by one total restriction

$$\sum_{l}\sum_{i=1}^{2}(n_l^i)^2 = Nx. \tag{9.28}$$

Changing the variables to $S_l^i = 2n_l^i - 1$ one can see that the restriction is equivalent to the condition $2N = \sum_{i=1}^{2}\sum_l (S_l^i)^2$ used in the spherical model (see Chapter 8) of phase transitions. Of course, the use of the single condition Eq. (9.28) introduces an error into the analysis, since the number of atoms per site becomes arbitrary. Nonetheless this model (just as well as the spherical model in magnetism) still makes it possible to obtain rather good agreement with experiments in oxides.[5] It should also be mentioned that in the two-dimensional spherical model the phase transition is known to be absent at finite temperature. Therefore, in particular calculations a third-dimension interaction should be introduced in the Hamiltonian of Eq. (9.25) which, however, does not result in considerable complication.

As a result the partition function of the system described by the Hamiltonian Eq. (9.25) can be written as

$$Z = \int_{-\infty}^{\infty}\prod_{i,l} dn_l^i \int_{-i\infty}^{i\infty} dv \exp\left\{\frac{v}{2}\left[Nx - \sum_{i=1}^{2}\sum_{l}(n_l^i)^2\right] + \frac{1}{2T}\sum_{i,j=1}^{2}\sum_{ll'} J_{ll'}^{ij} n_l^i n_{l'}^j + \frac{h}{T}\sum_{i=1}^{2}\sum_{l} n_l^i\right\}. \tag{9.29}$$

9.3.2 Free Energy

The partition function, Eq. (9.29) can be calculated with the help of diagonalization of the exponential functional with respect to l, l' and indices

[5] Of course, the agreement is only qualitative. One should not expect to get good quantitative description of an experiment confined only to the framework of this model.

i, j. The diagonalization over l and l' as usual can be easily provided with the help of the Fourier transformation of the function $J_{ll'}$ and variables n_l. One should also take into account the condition, Eq. (9.28) assuming that the value x is to be determined self-consistently. For the following it is convenient to introduce the oxygen atom concentration in sublattices with $i = 1$ and $i = 2$ (i.e denoting the concentrations along the a- and b-axis as $x_{i=1,2}$). Then $n^i_{q=0} = \sqrt{N} x_i$ and the condition, Eq. (9.28) is simply written as $x_1 + x_2 = x$. The additional degree of freedom is denoted below as $\Delta = x_1 - x_2$.

In the nearest and next-to-nearest neighbor approximation, the Fourier transforms of the matrix $J^i_{ll'}$ for the two-dimensional square lattice can be written as

$$\begin{aligned} J^{ii}_q &= J \cos q_i a, \\ |J^{12}_q|^2 &= K^2(1+\cos q_1 a)(1+\cos q_2 a), \end{aligned} \tag{9.30}$$

where a is the lattice constant and the interactions constants J and K for the YBaCuO system should also satisfy the inequalities $J > 0$ and $K < 0$. The diagonalization of the matrix J^{ij}_q leads to the following expressions for the eigenvalues $\lambda^{\pm}_q$

$$\lambda^{\pm}_q = -Tv + J\frac{\cos q_1 a + \cos q_2 a}{2}$$

$$\pm \sqrt{J^2\left(\frac{\cos q_1 a - \cos q_2 a}{2}\right)^2 + K^2(1+\cos q_1 a)(1+\cos q_2 a)}. \tag{9.31}$$

With the help of these expressions the partition function can be reduced to the following integral

$$Z \propto \int_{-\infty}^{\infty} d\Delta \prod_{q\neq 0} d\phi^+_q d\phi^-_q \int_{-i\infty}^{i\infty} dv \exp\left\{\frac{N}{4T}\left[2Tvx + (J+2K-Tv)x^2\right.\right.$$

$$\left.\left. + (J-2K-Tv)\Delta^2 + 4hx + \frac{2}{N}\sum_{q\neq 0}(\lambda^+_q|\phi^+_q|^2 + \lambda^-_q|\phi^-_q|^2)\right]\right\}. \tag{9.32}$$

The type of structure formed at the phase transition can be determined by the position of the absolute maximum of the functions $\lambda^{\pm}_q$. The mode corresponding to this maximum is nonergodic (it is condensed at the phase transition) and it should not be integrated in Eq. (9.32). One can show that the absolute maximum is related to the function λ^+_q and it occurs at $q = 0$

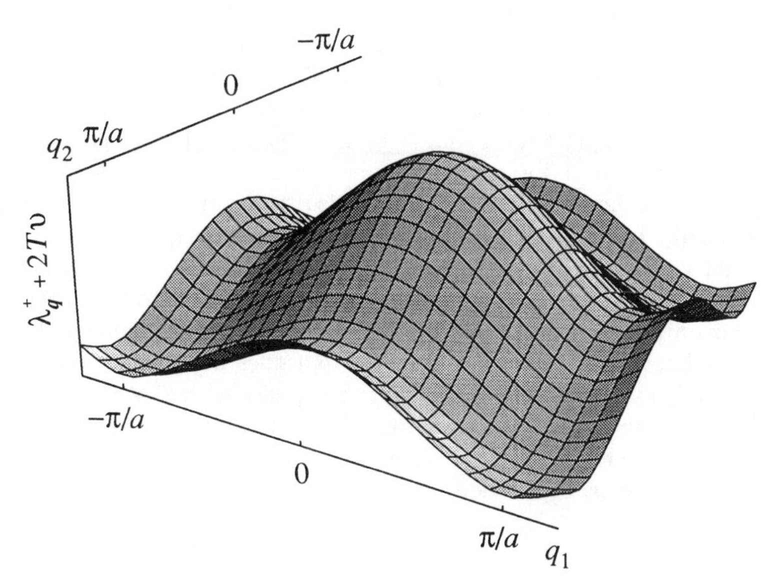

Figure 9.6: The surface $\lambda^+(q_1, q_2)$.

for any relationships between parameters J and K (provided $J > 0$). The surface $\lambda^+(q_1, q_2)$ is schematically depicted in Fig. 9.6. As the nonergodic mode in Eq. (9.32) corresponding to $q = 0$ is already isolated, integration over other modes can be easily carried out leading to the following structure of the partition function

$$Z \propto \int_{-\infty}^{\infty} d\Delta \int_{-i\infty}^{i\infty} dv \exp[-F(\Delta, v, x, t)], \tag{9.33}$$

where all parameters are made dimensionless with the help of the energy scale J ($t = T/J$, $b = -2K/J$, and $c = 2h/J$), v is replaced by v/t, and the function F is given by

$$F(\Delta, v, x, t) = -\frac{N}{4t}\Bigg[2vx + (1-b-v)x^2$$

$$+ (1+b-v)\Delta^2 + 2cx - \frac{2t}{N}\sum_{q\neq 0}\ln\left(\frac{\lambda_q^+\lambda_q^-}{J^2}\right)\Bigg]. \tag{9.34}$$

The presence of the factor $N \gg 1$ in the integral exponent in Eq. (9.33) allows one to use the saddle-point technique in carrying out the integration over v. In the thermodynamic limit ($N \to \infty$) the function $F(\Delta, v, x, t)$ with the value v defined from the saddle-point equation will be the exact nonequilibrium free energy depending on the value of the order parameter Δ. The equilibrium value of the order parameter as usual should be determined from the equation $\partial F/\partial\Delta = 0$. The disordered (tetragonal) phase corresponds to $\Delta = 0$ (i.e. oxygen is uniformly distributed along the a- and b-axes). Orthorhombic distortions give rise to asymmetry in the oxygen distribution. In this case one has $x_1 \neq x_2$ and correspondingly $\Delta \neq 0$. It should be noted that at $\Delta \neq 0$, oxygen forms a nonuniformly ordered (periodic) structure in the sample as a whole. At the same time each sublattice is uniformly ordered so that the mode with $q = 0$ is nonergodic. On the other hand, the presence of the "field" h does not prevent the abrupt transition from the phase with $\Delta = 0$ to the phase with $\Delta \neq 0$. In fact, the order parameter $\Delta = x_1 - x_2$, expressed in terms of $S_i = 2n_i - 1$, corresponds to the antiferromagnetic ordering. As is well known, the presence of the homogeneous field h does not affect the antiferromagnetic transition in any way.

9.3.3 Phase Transition

The saddle-point equation and the equation defining the equilibrium value of the order parameter form a set that can be written as

$$\begin{aligned} x(1 - x/2) - \Delta^2/2 - t\mathcal{A}(v) &= 0, \\ (1 + b - v)\Delta &= 0. \end{aligned} \tag{9.35}$$

The function $\mathcal{A}(v)$ in this set is determined by the expression

$$\mathcal{A}(v) = \int_0^\pi \frac{dx_1 dx_2}{\pi^2} \frac{4(\cos x_1 + \cos x_2 - 2v)}{b^2(1 + \cos x_1)(1 + \cos x_2) - 4(v - \cos x_1)(v - \cos x_2)}$$

The integration in the expression $\mathcal{A}(v)$ can be carried out explicitly. However, the explicit result is quite cumbersome. Here we reproduce only a singular (in the vicinity of critical point) part of the integral.[6]

$$\mathcal{A}(v) = \frac{4(v-1)\sqrt{v(v-1) - b^2/2}}{\pi(1-\gamma)(v+1)(b^2/2 + v - 1)} \Pi\left(\frac{\pi}{2}, \frac{-2v}{(1+v)(\gamma - 1)}, \sqrt{\frac{2}{v-1}}\right), \tag{9.36}$$

[6] A regular contribution can be incorporated as usual into the renormalization of parameters.

where $\Pi(x, y, z)$ is a third-order elliptic integral, the quantity γ is defined by

$$\gamma = \frac{v(v-1) - b^2/2}{(v-1) + b^2/2}.$$

The set Eq. (9.35) leads to two branches of solutions: $\Delta = 0$, $v = v(x,t)$ and $\Delta = \Delta(x,t) \neq 0$, $v = v_c = 1 + b$. The latter branch leads to

$$\Delta^2 = (2-x)x\left(1 - \frac{t}{t_c}\right),$$

where $t_c = (2-x)x\mathcal{A}^{-1}(v_c)$. However, when $v \to v_c$ the integral Eq. (9.36) formally diverges so that the critical temperature $t_c = 0$. This is a natural consequence of the fact that the interactions $\lambda_q^{\pm}$ have been taken above as two-dimensional and it is in agreement with the general statement for the spherical model that at $d = 2$ the phase transition is absent at finite temperatures. The finite value of $\mathcal{A}(v_c)$ can be arranged if one takes into account interaction of oxygen atoms and vacancies along the c-axis which is always present in real YBaCuO systems. In this case the value J_q should be completed by the third dimension contribution in the form $\kappa(x_3) = \zeta(1 - \cos x_3)$, where $\zeta \ll 1$ and the integration should also be carried out over the variable x_3, i.e.

$$\mathcal{A} \to \mathcal{A}\int_0^{\pi} \frac{dx_3}{\pi}\mathcal{A}(\kappa(x_3)) \equiv \int_0^{\pi}\int_0^{\pi}\int_0^{\pi} \frac{dx_1 dx_2 dx_3}{\pi^3}\phi(x_1, x_2, \kappa(x_3)), \tag{9.37}$$

where $v_c \to v_c + \kappa(x_3)$. The singular part of the integral Eq. (9.37) can be evaluated explicitly with the help of the following identical representation

$$\phi = \frac{1}{2}\left(\cos\frac{x_1}{2} + \cos\frac{x_2}{2}\right)\phi + \left[1 - \left(\cos\frac{x_1}{2} + \cos\frac{x_2}{2}\right)\right]\phi.$$

The use of this representation reduces calculations to a procedure similar to the field-theoretical subtraction technique. The integral from the second term converges when $\kappa \to 0$ thus leading to a regular contribution to $\mathcal{A}(v)$. The structure of the first term is chosen in the form which allows us to carry out calculation of $\mathcal{A}$. At $v = v_c + \kappa$ the value of γ is approximately equal to

$$\gamma \approx 1 + \frac{2\kappa}{1 + b/2}$$

so that the main contribution to $\mathcal{A}(v)$ can be written as

$$\mathcal{A} \approx \int_0^{\pi}\int_0^{\pi} \frac{dx_2 dx_3}{\pi^2} \frac{2\cos x_2/2}{\sqrt{(b^2/2+b)\,(2+b)}} \sqrt{\frac{1 + b - \cos x_2}{1 - \cos x_2 + 2k/(1 + b/2)}}.$$

Now one can see the reason of the choice of $\cos x_i/2$ in the subtraction transformation. The combination $1 - \cos x_2$ in the square root of the integrand can be represented as $2\sin^2 x_2/2$. The latter enables one to switch to the new variable $u = \sin x_2/2$ which helps to carry out integration over x_2. As a result one arrives at

$$\mathcal{A} = \frac{4}{\pi^2(2+b)} \int_0^\pi dx_3 \left[F\left(\frac{\pi}{2} - 2\sqrt{\frac{\kappa}{b(2+b)}}, 1\right) - E\left(\frac{\pi}{2} - 2\sqrt{\frac{\kappa}{b(2+b)}}, 1\right)\right],$$

where F and E are elliptic integrals of the first- and the second-order, respectively. In the limit of interest, these integrals can be represented as

$$\begin{aligned} F\left(\tfrac{\pi}{2} - 2\sqrt{\tfrac{\kappa}{b(2+b)}}, 1\right) &\approx \ln\cot\sqrt{\tfrac{\kappa}{b(2+b)}} &= -\tfrac{1}{2}\ln\kappa + o(1), \\ E\left(\tfrac{\pi}{2} - 2\sqrt{\tfrac{\kappa}{b(2+b)}}, 1\right) &\approx \cos 2\sqrt{\tfrac{\kappa}{b(2+b)}} &= 1 - o(\kappa^2). \end{aligned}$$

As a result the function $\mathcal{A}$ can be reduced to the form

$$\mathcal{A} = -\frac{2}{\pi^2(2+b)} \int_0^\pi dx_3 \ln\kappa(x_3). \tag{9.38}$$

Here we have obtained the value of $\mathcal{A}$ at the critical point $v = v_c$. We will also need to use an expression for the function $\mathcal{A}(v)$ in the vicinity of v_c, i.e. $v = v_c + \rho$. Then the value κ should be replaced by $\kappa = \zeta(1 - \cos x_3) + \rho$. The expansion in powers of ρ/ζ in Eq. (9.38) can be performed with the help of the following subtraction

$$\ln\kappa(x_3) = \cos x_3 \ln\kappa(x_3) + (1 - \cos x_3)\ln\kappa(x_3).$$

Now the integral expression defining $\mathcal{A}$ is convenient to rewrite in the form

$$\mathcal{A} = -\frac{2}{\pi(2+b)}\left[\ln 2\zeta + \frac{1}{\pi}\int_0^\pi dx_3 \ln\left|\frac{\rho}{2\zeta} + \sin^2\frac{x_3}{2}\right|\right] = -\frac{2}{\pi(2+b)}\left[\ln 2\zeta + I\right].$$

The integral I can be carried out as follows

$$\begin{aligned} I &= \frac{2}{\pi}\int_0^{\pi/2} dx \ln\left|\frac{\rho}{2\zeta} + \sin^2\frac{x}{2}\right| \left[\cos x + (1 - \cos x)\right] \\ &\approx \frac{2}{\pi}\left[\left(\ln\left|1 + \frac{\rho}{2\zeta}\right| - 2 + 2\sqrt{\frac{\rho}{2\zeta}}\arctan\sqrt{\frac{\rho}{2\zeta}}\right)\right. \\ &\left. + \int_0^{\pi/2} dx(1 - \cos x)\left[\ln\sin^2 x + \frac{\rho}{2\zeta\sin^2 x}\right]\right] \end{aligned}$$

$$\approx 2\sqrt{\frac{\rho}{2\zeta}} - 2\ln 2 + o\left(\left[\frac{\rho}{2\zeta}\right]^2\right).$$

As a result one can find the representation for the function $\mathcal{A}$ in the vicinity of the critical point in the form

$$\mathcal{A} = \mathcal{A}_c + \Delta\mathcal{A} = \frac{2}{\pi(2+b)}\left[\ln\frac{2}{\zeta} - 2\sqrt{\frac{\rho}{2\zeta}} + o\left(\left[\frac{\rho}{2\zeta}\right]^2\right)\right]. \tag{9.39}$$

It is necessary to stress that the introduction of the third dimension has resulted in the finite value of the critical temperature, i.e. $t_c = A_c^{-1}(2-x)x$.

After the substitution of $v(x,t)$ and $\Delta(x,t)$ in Eq. (9.35) one can get the expression for the free energy $F = F(x,t)$.

9.3.4 Change of Oxygen Concentration

Up to now the contribution to the free energy from the oxygen reservoir surrounding the YBaCuO specimen has not been taken into account. To incorporate this contribution one must equate the oxygen chemical potential in the specimen to that in the reservoir. For a gas of two-atomic O_2 molecules with partial pressure P_{O_2} in the reservoir one has

$$\mu_{O_2} = \ln P_{O_2} - C_pT - \xi T + \epsilon_O,$$

where ξ and ϵ_O are standard chemical constants (see for instance Landau and Lifshits, 1964), C_p is the gas specific heat at the constant pressure — $C_p = 9/2$. It is necessary to mention that in the calculation of the chemical potential of the oxygen absorbed by the CuO planes one has to take into account not only the contribution $\propto \partial F/\partial x$ but also the contribution caused by the oscillatory degrees of freedom. As a result one gets

$$\mu_O = J\frac{\partial F}{\partial x} + 2T\ln\frac{\hbar\omega}{T},$$

where ω is a characteristic oscillation frequency. Finally, taking into account the chemical equilibrium condition $\mu_{O_2} = 2\mu_O$, one arrives at the equation relating the variables v and x to the temperature

$$\begin{aligned}(1-x)v &= \phi(t) + (b-1)x - \bar{c} \\ &= \tfrac{t}{2}\left[\tfrac{1}{2}\ln At - \ln P_{O_2}\right] - x(1-b) - \bar{c},\end{aligned} \tag{9.40}$$

where as before $t = T/J$ and constants $\bar{c}$ and A are determined by the relations $\bar{c} = c + \epsilon_O/J$ and $A^{1/2} = \xi - 4\ln\hbar\omega + (\ln J)/2$. Now, Eq. (9.35) and Eq. (9.40) form a closed set for the variables v, Δ, and x. This set, as before, has two branches of solutions. When $\Delta = 0$ one can write

$$v = \frac{\phi(t) + (b-1)x - \bar{c}}{1-x}, \qquad x(2-x) = 2t\mathcal{A}(v). \tag{9.41}$$

In the limit $t \to \infty$ one has $v \approx \phi(t)/(1-x) \to \infty$ and the equation for $x(t)$ accepts the form

$$x(2-x) \propto \frac{1-x}{\ln At - 2\ln P_{O_2}}.$$

Since when $x \to 2$ the difference $1-x$ is negative, one has to consider only the branch corresponding to $x \to 0$. It should also be mentioned that the value x is increasing with the increase of oxygen partial pressure which is in an agreement with experimental data obtained by various authors (e.g. Jorgensen *et al.*, 1987). When $t \to 0$ the solution corresponding to this branch ($\Delta = 0$) leads to $v \to v(0) = [(b-1)x - \bar{c}]/(1-x)$ and $t\mathcal{A}(v(0)) \to 0$ (let us recall that at $x = 2$ one has $x_1 = x_2 = 1$). However, this limit is physically inaccessible as, at the critical point $t = t_c$, the solution with $\Delta \neq 0$ is thermodynamically stable. Therefore, the value $x(t)$ will switch to the new branch $\bar{x}(t)$ defined by the equation

$$b + 1 + v = 2b\bar{x}(t) + \phi(t). \tag{9.42}$$

The change of the function $\bar{x}(t)$ along this branch leads to $\bar{x}(0) = (1 + b + \bar{c})/2b$. Thus

$$\Delta^2 = \bar{x}(0)(2 - \bar{x}(0)) = \frac{(1 + b + \bar{c})(3b - 1 - \bar{c})}{4b^2}.$$

The experimental values of $\bar{x}(0) = \Delta(0) = 1$ can be well approximated by taking $\bar{c} + 1 = b$. At the critical point $t = t_c(v_c)$ Eqs. (9.40) and (9.42) coincide and $\bar{x}(t)$ begins at the same x_c which has completed the evolution of $x(t)$ above t_c. However, as is mentioned in Sect. 9.3, the interesting feature of the experimental curves $x(t)$ is the fact that, whereas $\Delta(t) \neq 0$ below t_c and it appears with an infinite derivative, the curve $x(t)$ is not only continuous at this point but does not even exhibit a bend. As is shown below, the considered model exactly describes such a behavior of functions $x(t)$ and $\Delta(t)$ with any choice of function $\phi(t)$.

Let the values x and v at above t_c ($t = t_c + \delta t$, $\delta t > 0$) be equal to

$x = x_c + \epsilon$, $v = v_c + \rho$. Then, using Eq. (9.41) one obtains

$$\begin{aligned} \rho &= \tfrac{\phi_t \delta t + 2\epsilon b}{1 - x_c}, \\ 2(1 - x_c)\epsilon &= 2\delta t \mathcal{A}_c - \tfrac{4}{\pi(2+b)} \sqrt{\tfrac{2\rho}{\zeta}}. \end{aligned}$$

From these relations, it follows that $\delta t \sim \epsilon \sim \sqrt{\rho} \gg \rho$ and hence

$$\left. \frac{dx}{dt} \right|_{t=t_c+0} = -\frac{\phi t}{2b}.$$

On the other hand, at $t = t_c - 0$ from Eq. (9.42) one can get

$$\frac{d\bar{x}}{dt} = -\frac{\phi t}{2b},$$

thus

$$\left. \frac{dx}{dt} \right|_{t=t_c+0} = \left. \frac{dx}{dt} \right|_{t=t_c-0},$$

i.e. the function $x(t)$ does not have a bend at the point $t = t_c$.

Now let us use the formula $\Delta^2(t) = x(2 - x) - 2t\mathcal{A}$. At first glance, it may seem quite obvious that at small $\delta t = t_c - t$ (here $t < t_c$) the value $\Delta^2(t)$ is proportional to δt, i.e. $\Delta(t)$ in the vicinity of t_c has an infinite derivative. However, in the relation

$$\Delta^2(t) \approx \left(\mathcal{A}_c - (1 - x_c) \left. \frac{dx}{dt} \right|_{t_c} \right) \delta t$$

the fact that the coefficient at δt is not equal to zero at $t = t_c$ should be specially proven. The simplest proof can be presented with the use of the above-proven continuity of the derivative dx/dt. Namely,

$$\mathcal{A}_c - (1 - x_c) \left. \frac{dx}{dt} \right|_{t_c} = \frac{2\sqrt{2}}{\pi(2+b)\sqrt{\zeta}} \lim_{\delta t \to 0} \left. \frac{\sqrt{\rho}}{\delta t} \right|_{t_c} = \text{const} \neq 0.$$

Thus when $\delta t \to 0$ the value Δ is proportional to $\sqrt{\delta t}$. The functions $x(t)$ and $x_{1,2}(t)$ are presented in Fig. 9.7.

These curves well reflect qualitatively the experimental ones obtained by Jorgensen *et al.* (1987) in YBaCuO systems. This shows that the model considered in this section quite adequately describes processes taking place at the structural phase transition in the high-T_c superconducting compounds ABaCuO.

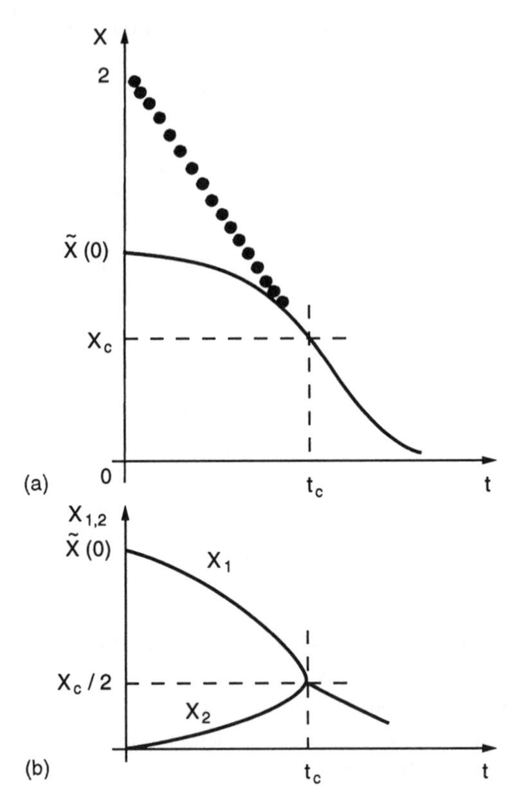

Figure 9.7: a) The function $x(t)$. The dotted line shows the solution corresponding to the branch with $\Delta = 0$. b) The functions $x_1(t)$ and $x_2(t)$.

9.4 *d*-Pairing in the Superconducting State

Weak orthorhombic distortions of crystal lattices, characteristic of ABaCuO systems stimulated interest in the nature of the superconducting phase transition in such systems. The latter was especially supported by the discovery of a fine structure (splitting) of the transition. Inderhees *et al.* (1988) and Butera (1988) measured the specific heat in the YBaCuO system and found that the main transition, with strong evidence of fluctuational contributions, is preceded by a reproducible discontinuous peak. Estimates of the Ginzburg parameter G_i show that for high-T_c materials the fluctuation region is sufficiently wide, i.e. of the order of several K. The experimental measurements of fluctuation corrections to the specific heat by Salamon *et al.* (1988) agree well with the last statement.

The first explanation of the splitting of the phase transition in superconducting ABaCuO systems was advanced by Volovik (1988). He assumed that pairing occurred in the orbital *d*-state. According to Volovik, in the case of *d*-pairing in tetragonal systems, a weak orthorhombic distortion may lead to a narrow temperature interval in which the phase transition splits into two transitions and an intermediate phase is formed. Additional confirmation of the *d*-pairing hypothesis was supported by measurements of the Knight shift of Cu ions (see Takigava *et al.*, 1988). The explanation of the splitting of the superconducting transition given by Volovik assumed that the transition is continuous within the mean field theory approach. On the other hand as was reported by Butera (1988) the first phase transition is discontinuous. The latter may happen either due to the interaction of superconducting fluctuations with the fluctuations of the gauge electromagnetic field (see Halperin *et al.*, 1974) or due to the fluctuation-induced discontinuous transition considered in this book. Below we follow the work by Ivanchenko *et al.* (1989 b,c) where such an explanation was first advanced. It is necessary to note that recently (1993) the problem of the pairing state has again attracted much attention owing to experimental activity in this area. In 1992 Monthoux *et al.* advanced a phenomenological theory based on nuclear magnetic resonance data by Barrett *et al.* (1991) (see also Martindale *et al.* (1993)) that the exchange of antiferromagnetic fluctuations could, with suitable assumptions, result in a *d*-pairing state with transition temperatures of the order observed in YBaCuO systems. Hardy *et al.* (1993) performed precision measurements of the temperature dependence of the magnetic field penetration depth in high quality YBaCuO single crystals. The data obtained in this work show a strong linear contribution in temperature from approximately 3 to 25 K. This seems to confirm the presence of nodes in the gap function (order parameter) inherent in *d*-pairing. A little later, Wollman *et al.* (1993) presented measurements of phase coherence with the help of Superconducting Quantum Interference Devices (SQUID) and tunnel junctions made from single crystals of YBaCuO and thin films of conventional s-wave superconductor Pb. Such an experiment, in principle, is capable of unambiguously determining the orbital symmetry of the pairing state in YBaCuO. This experiment is sensitive to the relative phase of the superconducting order parameter in orthogonal k-space directions and thus can distinguish the d-wave state from an anisotropic s-wave state. However, the experimental results are complicated by the influence of the trapped magnetic vortices and asymmetries in SQUID. Wollman *et al.*, nonetheless concluded that the experiment gives rather strong evidence for a phase shift of π in orthogonal directions, which is predicted for the d-pairing state. On the other hand, Klein *et al.* (1993) have addressed this problem with the microwave surface impedance measurements and found an exponential temperature dependence of the penetration depth. The latter should rule out d-wave pairing. Thus, at present this problem cannot be considered as solved. New discoveries may give a final answer which now cannot be even remotely anticipated. Nonetheless, the consideration given in this chapter is based on the symmetry properties of the Ginzburg-Landau functional and thus, has a more general meaning. The same functional can be obtained, for instance, in the study of a structural transition to the incommensurate phase of $BaMnF_4$ (see Dvorak and Fousek, 1980). A discontinuous phase transition was observed in this material by Pisarev *et al.* in 1983. This transition can be interpreted as a fluctuation-induced first-order transition since the mean field theory predicts a second-order phase transition in this case.

9.4.1 RG Approach

The Ginzburg-Landau functional for such a system was obtained by Volovik and Gor'kov in 1985. It can be written in terms of a two-component, complex, vector-order parameter $\boldsymbol{\eta}$ as

$$\mathcal{F} = \frac{1}{2} \int d\boldsymbol{r} \left\{ |\nabla \boldsymbol{\eta}|^2 + \tau |\boldsymbol{\eta}|^2 + \beta_1 |\boldsymbol{\eta}|^2 + \beta_2 |\boldsymbol{\eta}^2|^2 + \beta_3 (|\eta_x|^4 + |\eta_y|^4) \right\}. \tag{9.43}$$

In order to obtain RG equations for this functional it is convenient to rewrite complex vector components $\eta_{x,y}$ in terms of four real functions ϕ_i $(i = 1, 2, 3, 4)$ as follows

$$\eta_x = \phi_1 + i\phi_2, \qquad \eta_y = \phi_3 + i\phi_4.$$

Now the functional Eq. (9.43) can be represented in the form

$$\mathcal{F} = \frac{1}{2} \int d\boldsymbol{r} \Big\{ (\boldsymbol{\nabla}\phi)^2 + \tau\phi^2 + \frac{\beta_1 + \beta_2 + \beta_3}{2} \left[(\phi_1^2 + \phi_2^2)^2 + (\phi_3^2 + \phi_4^2)^2 \right]$$

$$+ \beta_1 \,(\phi_1^2 + \phi_2^2)(\phi_3^2 + \phi_4^2) + \beta_2 \Big[(\phi_1\phi_3 + \phi_2\phi_4)^2 \; - (\phi_1\phi_4 - \phi_2\phi_3)^2 \Big] \Big\} \tag{9.44}$$

From this representation one can see that it is convenient to consider the critical evolution of the following vertexes

$$u = \frac{\beta_1 + \beta_2 + \beta_3}{6}, \qquad \nu = \frac{\beta_1}{6}, \qquad \mu = \frac{\beta_2}{6}.$$

For these vertices, according to Chapter 6, one can write the following RG set

$$\begin{aligned} \tfrac{du}{dl} &= u - 5u^2 - \mu^2 - \nu^2, \\ \tfrac{d\nu}{dl} &= \nu - 4u\nu - 2\nu^2 - 4\mu^2, \\ \tfrac{d\mu}{dl} &= \mu(1 - 2u - 4\nu). \end{aligned} \tag{9.45}$$

This set has four stable points

$$\mu_1^* : \left\{ \frac{1}{5}, 0, 0 \right\}; \qquad \mu_2^* : \left\{ \frac{1}{6}, \frac{1}{6}, 0 \right\};$$

$$\mu_{3a}^* : \left\{ \frac{1}{10}, \frac{2}{10}, \frac{1}{10} \right\}; \quad \mu_{3b}^* : \left\{ \frac{1}{10}, \frac{2}{10}, -\frac{1}{10} \right\}.$$

The points μ_1^*, μ_{3a}^*, and μ_{3b}^* are unstable. The point μ_2^*, corresponding to $O(4)$-symmetry of the functional $\mathcal{F}$, has marginal stability in the first ϵ-approximation. This phenomenon is typical for systems with a four-component order parameter (see for instance Toledano *et al.* 1985). Using the symmetry analysis of the RG equations in the second ϵ-approximation performed by Kerszberg and Mukamel (1979), one can see that the Heisenberg point for the functional Eq. (9.44), which has the $D_2 \times D_\infty$ orthocylindrical symmetry, splits in two unstable points. As the partition function with the GLF given by the quatric-form Eq. (9.44) can only exist for the positively defined form, one should find the stability limits of the functional. They can be written as follows

$$2(\beta_1 + \beta_2) + \beta_3 = 0, \qquad \beta_1 + \beta_2 + \beta_3 = 0, \qquad 2\beta_1 + \beta_2 = 0.$$

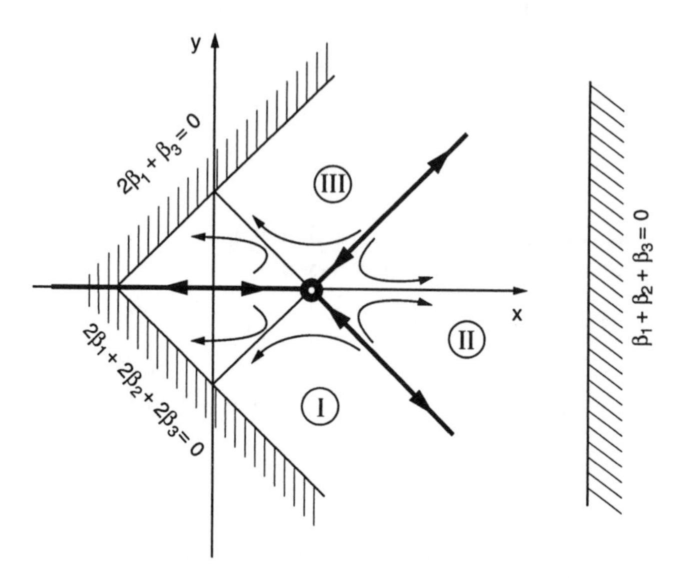

Figure 9.8: The phase portrait of the RG set. The boundaries of the stability region are hatched (the boundary $\beta_1 + \beta_2 + \beta_3 = 0$ is shown schematically). Separatrices of sectors I, II, and III are represented by solid lines.

As usual, since the structure of the ordered phase is defined by the ratios of vertices rather than by their values, one can reduce the number of equations by introducing the new variables $x = \nu/u$ and $y = \mu/u$. In terms of these variables the RG set can be rewritten as

$$\begin{aligned} \frac{1}{u}\frac{dx}{dl} &= x(x-1)^2 + (x-4)y^2, \\ \frac{1}{u}\frac{dy}{dl} &= y[(x-1)(x-3) + y^2]. \end{aligned} \tag{9.46}$$

The stability limits expressed in these variables have the following form

$$y = -(1+x), \qquad x, y \to \infty, \qquad y = 1 + x.$$

The portrait of flow lines for this set is shown in Fig. 9.8. The important feature of the phase portrait is the presence of separatrices. These lines separate regions from which flow lines escape to the stability limits $y = \pm(1+x)$ and $x, y \to \infty$. Direct substitution shows that the lines $y =$

$\pm(1+x)$ are particular integrals of the set, Eq. (9.46). In order to make clear the physical significance of the separatrices we need to briefly discuss the boundaries separating different phases within the mean field theory approach. Such a discussion was given by Volovik in 1988. We have the following three phases with different symmetries and the conditions of their stability

$$\begin{array}{llllll} \text{I.} & \phi_1 = \phi_3, & \phi_2 = \phi_4 = 0; & \beta_3 \geq 0, & \beta_2 \leq 0, & \\ \text{II.} & \phi_1 \neq 0, & \phi_{k \neq 1} = 0; & \beta_3 \leq 0, & 2\beta_2 + \beta_3 \leq 0, & \\ \text{III.} & \phi_1 = \phi_4, & \phi_2 = \phi_3 = 0; & \beta_2 \geq 0, & 2\beta_2 + \beta_3 \geq 0. & \end{array} \tag{9.47}$$

All other equivalent structures can be easily obtained by rotations $(\phi_1 \pm \phi_2)/\sqrt{2} = \kappa_{1,2}$, etc. Corresponding regions in Fig. 9.8 are denoted as sectors I, II, and III. Now one can see that the boundaries separating different phases in the mean field theory are identical with the particular integrals of RG set: $2\beta_2 + \beta_3 = 0$ and $\beta_3 = 0$. As usual, the loss of stability, i.e. the flow over the regions of stability, should be interpreted as a fluctuation-induced discontinuous phase transition. Therefore, one may conclude that the structure obtained should correspond to the structure anticipated from the Landau theory. However, there is a difference, namely, the second-order transition is replaced by the corresponding first-order transition. This difference is important when one considers weak orthorhombic lattice distortions by adding the functional

$$\delta\mathcal{F} = \frac{\Delta}{2}\int d\boldsymbol{r}[|\eta_x|^2 - |\eta_y|^2] = \frac{\Delta}{2}\int d\boldsymbol{r}[(\phi_1^2 + \phi_2^2) - (\phi_3^2 + \phi_4^2)]. \tag{9.48}$$

Such an addition, as was shown by Kertszberg and Mukamel (1979) and Blankschtein and Mukamel (1982), has a threshold effect on the nature of the transition. The phase transition is discontinuous up to a finite value Δ_{min}. At $\Delta = \Delta_{min}$ one has a tricritical point and when $\Delta > \Delta_{min}$ the transition becomes continuous and is qualitatively well described by the mean field theory.

In conclusion, it is necessary to point out that the first-order transition switches the system directly to the low-temperature phases I or III (for suitable values of β_i). For small Δ, there is no need to pass through the intermediate state II to reach these phases. This contrasts with the results of the mean field theory. Unfortunately, the RG methods do not allow us to make a reliable estimate of the magnitude of Δ_{min}. As usual, the assumption that the transition corresponding to the flow overcoming the stability limits is a discontinuous one, is a hypothesis. This hypothesis should be confirmed or refuted in each specific case by a direct calculation of the free energy. Such calculations should be based on models which allow one to take into account fluctuation effects (at least partially). In the following section we address this question by methods developed in Chapter 8.

9.4.2 An Exactly Solvable Model

To formulate an exactly solvable model, we need to modify somewhat the structure of invariants made up of the components ϕ_i in the functional, Eq. (9.44). In fact, we will use the following identity

$$|\boldsymbol{\eta}^2|^2 = (\boldsymbol{\eta}\boldsymbol{\eta}^*)^2 - [2(\phi_1\phi_4 - \phi_2\phi_3)]^2.$$

Combining the first term from this identity with the analogous term in Eq. (9.44) one can introduce the new vertex $\beta = \beta_2 + \beta_3$. The term $(\phi_1\phi_4 - \phi_2\phi_3)^2$ can be regarded as a new basic invariant of $\mathcal{F}$. It is clear that $\mathcal{F}$ now forms a sum of squares $\boldsymbol{\eta}\boldsymbol{\eta}^*$, $|\eta_x|^2$, $|\eta_y|^2$, and $2(\phi_1\phi_4 - \phi_2\phi_3)^2$ with the corresponding coefficients $(\beta/2;\ \beta_{2,3}/2)$. The essence of the model is the substitution of integrals $\int d\boldsymbol{r}A^2$ by $\left(\int d\boldsymbol{r}A\right)^2/V$. As is shown in Chapter 8 this replacement is equivalent to an approximation retaining interactions of fluctuations with the same wave vectors. The functionals $a = \int d\boldsymbol{r}A$ are quadratic in terms of the fields ϕ_i. Now using of the Hubbard-Stratonovich identity

$$\exp\left(-\frac{\beta}{2}a^2\right)\int_{-i\infty}^{i\infty} d\alpha \exp\left[\frac{\alpha^2}{2\beta}\right] = \int_{-i\infty}^{i\infty} d\alpha \exp\left[\frac{\alpha^2}{2\beta} - \alpha a\right] \qquad (9.49)$$

helps to linearize the argument of the exponential in the partition function

$$Z = \int D\phi_i \exp[-\mathcal{F}\{\phi_i\}]$$

with respect to the quadratic functional. The latter allows one to explicitly carry out the integration in Z over the fields ϕ_i.

Let us denote the variables of integration in Eq. (9.49) corresponding to $a = (\boldsymbol{\eta}\boldsymbol{\eta}^*)$, $a = |\eta_{x,y}|^2$, and $a = 2(\phi_1\phi_4 - \phi_2\phi_3)$ by α, $\rho_{x,y}$, and ξ. Then, the quadratic in ϕ_i form in the partition function can be written as

$$R = (\tau_x + \alpha + \rho_x)\phi_1^2 + 2\xi\phi_1\phi_4 + (\tau_y + \alpha + \rho_y)\phi_4^2$$

$$+(\tau_x + \alpha + \rho_x)\phi_2^2 - 2\xi\phi_2\phi_3 + (\tau_y + \alpha + \rho_y)\phi_3^2 = \phi^+\hat{A}\phi,$$

where $\phi^+ = (\phi_1, \phi_4, \phi_2, \phi_3)$. Now, one has to diagonalize this quadratic form. As the matrix $\hat{A}$ is a quasi-diagonal matrix with two identical blocks (except for the signs of the off-diagonal elements), it has a pair of doubly degenerate eigenvalues $\lambda^\pm$. As a result it can be represented in a diagonal form as follows

$$R = \sum_\sigma \lambda^\sigma \left[(\chi_1^\sigma)^2 + (\chi_2^\sigma)^2\right],$$

where $\sigma = \pm$ and

$$\lambda^\pm = \alpha + \tau + \frac{\rho_x + \rho_y}{2} \pm \sqrt{\left(\Delta + \frac{\rho_x - \rho_y}{2}\right)^2 + \xi^2},$$

$$\tau = \frac{\tau_x + \tau_y}{2}, \qquad \Delta = \frac{\tau_x - \tau_y}{2}. \tag{9.50}$$

It is convenient to combine the fields $\chi_{1,2}^\pm$ into two two-component vectors $\boldsymbol{\chi}^\pm = (\chi_1^\pm, \chi_2^\pm)$ and represent the form as $R = \sum_\sigma \lambda^\sigma (\boldsymbol{\chi}^\sigma)^2$. The integration with respect to the fields $\boldsymbol{\chi}^\pm$ can be easily carried out and one arrives at the following representation for Z

$$Z \propto \int d\boldsymbol{\chi}_o^\pm \int d\alpha \int d\rho_x d\rho_y d\xi$$

$$\times \exp\left\{-V\left[-\frac{\alpha^2}{2\beta} - \frac{\rho_x^2 + \rho_y^2}{2\beta_3} + \frac{\xi^2}{2\beta_2} + \sum_\sigma \left(\Phi^\sigma + \lambda^\sigma (\boldsymbol{\chi}_o^\sigma)^2\right)\right]\right\}, \tag{9.51}$$

where

$$\Phi^\pm = \frac{1}{V}\left(\sum_q \ln(\lambda^\pm + q^2)\right)_R = -\frac{(\lambda^\pm)^{3/2}\Theta(\lambda^\pm)}{6\pi}$$

with the definition of the regularized sum $(\sum)_R$ is given in Chapter 8 (see Eq. (8.37)), $\Theta(\lambda^\pm)$ is Heaviside's step function. As is seen, Eq. (9.51) contains the integration over the variables $\boldsymbol{\chi}_o^\pm \equiv \boldsymbol{\chi}_{q=o}^\pm/\sqrt{V}$ which have been left unintegrated as they may become nonergodic in the vicinity of phase transition points and may play the role of order parameters in the free energy $F = -\ln Z$. Thus they should be defined as solutions of the equations of state $\partial F/\partial \boldsymbol{\chi}_o^\pm = 0$. The integrals in Eq. (9.51) can be evaluated exactly in the thermodynamic limit ($V \to \infty$) by the steepest descent method. The set of equations for the corresponding saddle points (naturally, including the equations of state) has the following form

$$\begin{aligned}
\tfrac{1}{V}\tfrac{\partial F}{\partial \chi_o^{\pm}} = 2\chi_o^{\pm}\lambda^{\pm} &= 0; \\
-\tfrac{\alpha}{\beta} + \textstyle\sum_{\sigma}[(\chi_o^{\sigma})^2 - 2f^{\sigma}] &= 0; \\
-\tfrac{\rho_{x,y}}{\beta_3} + \textstyle\sum_{\sigma}[(\chi_o^{\sigma})^2 - 2f^{\sigma}]\tfrac{\partial \lambda^{\sigma}}{\partial \rho_{x,y}} &= 0; \\
\tfrac{\xi}{\beta_2} + \textstyle\sum_{\sigma}[(\chi_o^{\sigma})^2 - 2f^{\sigma}]\tfrac{\partial \lambda^{\sigma}}{\partial \xi} &= 0,
\end{aligned} \tag{9.52}$$

where

$$f^{\pm} = \frac{\partial \Phi^{\pm}}{\partial \lambda^{\pm}} = -\frac{\sqrt{\lambda^{\pm}}\Theta(\lambda^{\pm})}{4\pi}.$$

The derivatives from the functions $\lambda^{\pm}$ can be directly calculated. They are equal to

$$\frac{\partial \lambda^{\pm}}{\partial \rho_{x,y}} = \frac{1}{2}\left[1 \pm \frac{2\Delta + \rho_x - \rho_y}{\lambda^+ - \lambda^-}\right], \qquad \frac{\partial \lambda^{\pm}}{\partial \xi} = \frac{\pm 2\xi}{\lambda^+ - \lambda^-}.$$

The set of Eq. (9.52) can be rewritten in a more compact form as follows

$$\begin{aligned}
\chi_o^{\pm}\lambda^{\pm} &= 0; \\
\tfrac{\rho_x + \rho_y}{\beta_3} &= \tfrac{\alpha}{\beta} = \Sigma; \\
\tfrac{\rho_x - \rho_y}{\beta_3} &= D\tfrac{2\Delta + \rho_x - \rho_y}{\lambda^+ - \lambda^-}; \\
\xi\left[\tfrac{1}{\beta_2} + \tfrac{2D}{\lambda^+ - \lambda^-}\right] &= 0,
\end{aligned} \tag{9.53}$$

where

$$\Sigma = \sum_{\sigma}[(\chi_o^{\sigma})^2 - 2f^{\sigma}], \qquad D = [(\chi_o^{+})^2 - 2f^{+}] - [(\chi_o^{-})^2 - 2f^{-}]$$

From this set one can see that there are two types of solutions: 1) $\chi_o^{\pm} \neq 0$, $\lambda^{\pm} = 0$; 2) $\chi_o^{+} = 0$ $\chi_o^{-} \neq 0$ for $\lambda^{+} \neq 0$, $\lambda^{-} = 0$ (and vice versa: $+ \leftrightarrow -$). Moreover, each solution has two branches determined by the choice

$\xi = 0$ or $\xi \neq 0$. The required quantities can be obtained for each branch with the use of straightforward but lengthy algebra. The final results are:

1*a*) $\chi_o^+ \neq 0, \quad \chi_o^- \neq 0, \quad \lambda^+ = 0, \quad \lambda^- = 0, \quad \xi = 0.$

$$\frac{1}{2}(\chi_o^\pm)^2 = -\frac{\tau}{\beta+\beta_3} \pm \frac{\Delta}{\beta_3}, \tag{9.54}$$

$$F_{I_a} = -\frac{1}{4}\left[\frac{\tau^2(2\beta+\beta_3)}{(\beta+\beta_3)^2} + \frac{\Delta^2}{\beta_3}\right]. \tag{9.55}$$

1*b*) $\chi_o^+ \neq 0, \quad \chi_o^- \neq 0, \quad \lambda^+ = 0, \quad \lambda^- = 0, \quad \xi \neq 0.$

$$(\chi_o^\pm)^2 = -\tau\frac{\beta+\beta_3}{\beta(2\beta+\beta_3)}, \tag{9.56}$$

$$F_{I_b} = -\frac{1}{4}\left[\frac{\tau^2}{2\beta+\beta_3} + \frac{\Delta^2}{2\beta_2+\beta_3}\right]. \tag{9.57}$$

2*a*) $\chi_o^+ = 0, \quad \chi_o^- \neq 0, \quad \lambda^+ \neq 0, \quad \lambda^- = 0, \quad \xi = 0.$

$$\sqrt{\lambda_{1,2}^+} = \sqrt{\lambda_a^+}\left[1 \pm \sqrt{\left(1 + \frac{(2\beta+\beta_3)\Delta}{(\beta+\beta_3)\lambda_a^+}\right)\left(1 - \frac{\tau}{\tau_a}\right)}\right], \tag{9.58}$$

$$\sqrt{\lambda_a^+} = \frac{\beta_3(2\beta+\beta_3)}{4\pi(\beta+\beta_3)}, \quad \tau_a = -\frac{\beta+\beta_3}{\beta_3}\left[\frac{2\beta+\beta_3}{\beta+\beta_3}\Delta + \lambda_a^+\right], \tag{9.59}$$

$$(\chi_{1,2}^-)^2 = -\frac{1}{\beta+\beta_3}\left[\tau - \Delta + \frac{\beta\sqrt{\lambda_{1,2}^+}}{2\pi}\right], \tag{9.60}$$

$$F_{II} = -\frac{1}{4}\left[\frac{(\lambda^+ - 2\tau)^2}{2\beta+\beta_3} + \frac{(\lambda^+ - 2\Delta)^2}{\beta_3}\right] + 2\Phi^+. \tag{9.61}$$

2*b*) $\chi_o^+ = 0, \quad \chi_o^- \neq 0, \quad \lambda^+ \neq 0, \quad \lambda^- = 0, \quad \xi \neq 0.$

$$\sqrt{\lambda_{1,2}^+} = \sqrt{\lambda_b^+}\left[1 \pm \sqrt{\left(1 - \frac{\tau}{\tau_b}\right)}\right], \tag{9.62}$$

$$\sqrt{\lambda_b^+} = -\frac{\beta_2(2\beta+\beta_3)}{\pi(2\beta_1+\beta_3)}, \quad \tau_b = -\frac{2\beta_1+\beta_2}{4\beta_2}\lambda_b^+, \tag{9.63}$$

$$(\chi_{1,2}^-)^2 = \frac{\sqrt{\lambda_{1,2}^+}}{2\pi\beta_2}\left[\pi\sqrt{\lambda_{1,2}^+}+\beta_2\right], \tag{9.64}$$

$$F_{III} = -\frac{1}{4}\left[\frac{(\lambda^+-2\tau)^2}{2\beta+\beta_3} - \frac{(\lambda^+)^2}{2\beta_2} + \frac{(2\Delta)^2}{2\beta_2+\beta_3}\right] + 2\Phi^+. \tag{9.65}$$

The first pair of solutions, Eqs. (9.54) — (9.57) describes a second-order phase transition at $\tau = 0$ (or $\tau \sim o(\Delta)$). The second pair of solutions is obtained for τ_c satisfying the condition that the free energy F_{II} (or F_{III}) displays an inflection point corresponding to $\sqrt{\lambda_{1,2}^+} = \sqrt{\lambda_c^+}$, which further develops into a pair of extrema (determined by $\sqrt{\lambda_1^+}$ and $\sqrt{\lambda_2^+}$) at the points $\chi_{1,2}^-$. In other words, the structure of F_{II} and F_{III} is typical of first-order transitions.

9.4.3 Comparison with RG

Now we are going to compare the results obtained in this exact approach with those deduced from the renormalization group analysis.

Tetragonal Approximation

Here we restrict the discussion to the limit $\Delta = 0$. Let us first consider only the second pair of solutions (Eqs. (9.58) — (9.65)). It should be noted that the ordering $\chi = 0$, $\chi_o^- \neq 0$ for $\xi = 0$ corresponds to the structure II (see Eq. (9.47)) with $\phi_1 \neq 0$, $\phi_{k\neq 1} = 0$ whereas, for $\xi \neq 0$ (and $\xi = \sqrt{\lambda^+/2}$) we have ordering corresponding to the structure III with $\phi_1 = \phi_4$ and $\phi_2 = \phi_3 = 0$. It follows from inequalities, Eq. (9.47) that these structures have the lowest energy within the mean field theory in the sectors II and III (Fig. 9.8). We will show that this result also takes place in the exactly solvable model. The Ginzburg-Landau functional has an additional symmetry with respect to simultaneous substitution of the vertices $-\beta_3 \leftrightarrow 2\beta_2$ and rotations

$$\frac{\phi_1 \pm \phi_4}{\sqrt{2}} = \psi_{1,4}; \qquad \frac{\phi_2 \pm \phi_3}{\sqrt{2}} = \psi_{2,3} \tag{9.66}$$

when $\Delta = 0$. The rotations described by Eq. (9.66) transform the structure II into III and vice versa, whereas the substitution $-\beta_3 \leftrightarrow 2\beta_2$ leads to inversion of the sectors II and III about the line $2\beta_2 + \beta_3 = 0$, which separates these two sectors. One can show that the solutions Eqs. (9.58) — (9.61) and Eqs. (9.62) — (9.65) satisfy the aforementioned symmetry (at $\Delta = 0$) and transform into one another under the corresponding transformations. Therefore, it is sufficient to discuss only one of these solutions and investigate its stability, choosing the sign of the inequality $2\beta_2 + \beta_3 > 0$ or $2\beta_2 + \beta_3 < 0$. The free energy has a minimum at the point χ^- determined by the solution $\sqrt{\lambda_1^+} = \sqrt{\lambda_c^+}[1 + \sqrt{1 - \tau/\tau_c}]$. This point is first reached (as an inflection point) on the overheating spinodal for $\tau = \tau_c$

$$(\chi_c^-)_{II} = \frac{\beta_3\sqrt{2\beta + \beta_3}}{4\pi(\beta + \beta_3)} \qquad (II \leftrightarrow III \quad \text{with} \quad -\beta_3 \leftrightarrow 3\beta_2). \tag{9.67}$$

The minimum is separated from the origin ($\chi_0^- = 0$) by a barrier whose position χ_2^- is determined by the second root $\sqrt{\lambda_2^+}$. The barrier disappears on the supercooling spinodal when χ_2^- vanishes at a temperature given by

$$(\tau_{cr})_{II} = -\frac{\beta\beta_3}{(2\pi)^2} \qquad (II \leftrightarrow III). \tag{9.68}$$

The corresponding discontinuity of the order parameter χ_{cr}^- is given by

$$(\chi_{cr}^-)_{II} = -\frac{\sqrt{\beta}\beta_3}{2\pi(\beta + \beta_3)} \qquad (II \leftrightarrow III). \tag{9.69}$$

To be specific, let us assume that $2\beta_2 + \beta_3 > 0$. Then using the aforementioned symmetry property, one can show that the nontrivial solution III is obtained earlier (i.e. $(\tau_c)_{III} > (\tau_c)_{II}$ for both $\tau_c > 0$) and it reaches the supercooling spinodal for a higher value of τ_{cr} ($(\tau_{cr})_{III} > (\tau_{cr})_{II}$ when $\tau_{cr} > 0$). In the situation where the system is between the overheating and supercooling spinodals, the minimum F_{III} lies below the minimum F_{II}. In other words, phases II and III have the lowest energy in the same sectors as in the mean field theory or in the renormalization group theory. Since $\tau_{cr} > 0$ is met when $\beta_2 > 0$ and $\beta_3 < 0$ for both these sectors, the transition to either of these two structures takes place for $\tau > 0$ and, hence, the solutions 1a and 1b cannot compete with the solutions 2a and 2b in the framework of this model. When the two above-mentioned inequalities are reversed (i.e. $\beta_2 < 0$ and $\beta_3 > 0$) τ_{cr} becomes negative in both cases,[7] and the solutions for $\sqrt{\lambda_{1,2}^+}$ and $\chi_{1,2}^-$ become unphysical. The valid solutions are those predicting a second-order phase transition. In order to determine the

[7]This means that the parameters β_i lie in sector I in Fig. 9.8.

range of values of the parameters β_i in which the branches 1a and 1b can be realized, one needs to compare the free energies F_{Ia} of Eq. (9.55) and F_{Ib} of Eq. (9.57). It is interesting that in the limit $\Delta \to 0$, one has $\xi \neq 0$, but on the other branch, this quantity tends to zero ($\xi \sim \Delta$). Formally, this is similar to the approach to the solution 1a. However, the energies F_{Ia} and F_{Ib} differ by a finite quantity when $\Delta = +0$ and one has

$$\left|\frac{F_{Ia}}{F_{Ib}}\right| = \left(\frac{2\beta + \beta_3}{\beta + \beta_3}\right)^2 .$$

It occurs that for $\beta > 0$ (i.e. $\beta_1 > -\beta_2$ or $x > -y$ in Fig. 9.8) the solution is 1a. In the opposite ($\beta < 0$) case one can find the solution 1b.

Orthorhombic Distortions

In a more general situation with $\Delta \neq 0$ we first discuss different branches of solutions. As the inequalities $\beta_2 < 0$ and $\beta_3 > 0$ are met in the sector I, from Eqs. (9.55) and (9.57) it follows that branch 1a has lower energy than branch 1b. On the other hand, the quantities $(\chi_o^{\pm})^2$ are simultaneously not equal to zero only beginning from $\tau = -(\beta + \beta_3)|\Delta|/\beta_3 < 0$, whereas the nontrivial solution 1b occurs even for $\tau = 0$. One can, therefore, choose the parameters Δ and β_i to obtain at $\tau = 0$ a structure with equal χ_o^- and χ_o^+ which at lower temperatures will lead to a phase with $\chi_0^+ \neq \chi_0^-$. However, since these results disagree with the results of the renormalization group approach for this sector, it is most likely that such an ordering is a specific feature of the model. The results for sectors II and III are more convincing, since they confirm the RG analysis. The above-mentioned symmetry between solutions 2a and 2b disappears for $\Delta \neq 0$. For case 2a, if $\Delta \gg \lambda_c^+$ one can derive from Eq. (9.59) that τ_c tends to $\tau_c = (2\beta + \beta_3)\Delta/(-\beta_3)$. Consequently, Eq. (9.60) yields

$$(\chi_c^-)^2 \to -\frac{1}{\beta + \beta_3}[\tau_c - \Delta] = \frac{2\Delta}{\beta_3} < 0.$$

This means that for arbitrary values of the parameters β_i one can find in sector II a value Δ_{min} at which the inflection point of the free energy occurs for $(\chi^-)^2 = 0$. When $\Delta > \Delta_{min}$ a phase transition to the state with $\phi_{k\neq 1} = 0$ and $\phi_1 \neq 0$ is now a second-order phase transition.

Now let us discuss case 2b. In this case Eqs. (9.62) — (9.64) do not contain the parameter Δ and it might seem that the difference between the quantities τ_x and τ_y does not affect this branch. However, this conclusion is incorrect. When $\Delta \neq 0$, one can find that the value $\xi \neq 0$ is determined by

$$\xi^2 = \left(\frac{\lambda^+}{2}\right)^2 - \Delta^2 \left(\frac{2\beta_2}{2\beta_2 + \beta_3}\right)^2$$

For large Δ, ξ becomes imaginary. Recall that the fields $\chi^\pm$ were derived from the diagonalization of a quadratic (in the fields ϕ_i) form with the eigenvalues defined by Eq. (9.50). The matrix of the diagonalization transformation is quasidiagonal and contains two identical blocks (they differ only by the substitution $\xi \to -\xi$)

$$\hat{S} = \frac{1}{\sqrt{2B(1+B)}} \begin{pmatrix} 1+B & -P \\ P & 1+B \end{pmatrix}, \qquad \det \hat{S} = 1,$$

where

$$P = 2\xi/[\Delta + (\rho_x - \rho_y)/2], \qquad B = \sqrt{1+P^2}.$$

Returning to the specific situation where only the field χ_0^- is not equal to zero, one can find that the fields ϕ_1 and ϕ_4 have the form $\phi_1 \sim \xi\chi_0^-$ and $\phi_4 \sim \chi_0^-$. When ξ is imaginary, one of the variables ϕ_i is also imaginary. However, according to the initial definition, all ϕ_i in the functional given by Eq. (9.44) are real. Therefore, at large Δ, the solution with $\xi \neq 0$ and $\chi_0^+ = 0$ becomes incorrect. At the same time, at $\Delta = (2\beta_2 + \beta_3)\lambda^+/4\beta_2$, the quantity ξ first vanishes before it becomes imaginary and, hence, a transition to the branch with $\xi = 0$, $\phi_i \neq 0$, and $\phi_{j\neq i} = 0$ is possible. As already noted, such a solution leads to a smooth transformation of a discontinuous transition to a continuous transition.

9.4.4 Conclusion

The predictions of the exactly solvable model agree well with the results of the RG analysis and also with the mean field considerations. As is shown the fluctuation-induced first-order transition takes place up to a threshold value Δ_{min}. Next, if one has the low-temperature phase described by the order parameter $\boldsymbol{\eta} = (1, 0)$, then the considered transition is replaced at $\Delta = \Delta_{min}$ by a continuous transition to the same phase. Assuming that the low-temperature structure is described by the order parameter $\boldsymbol{\eta} = (1, i)$, one can find that a direct transition to this phase can only be discontinuous. For large Δ when fluctuations of one of the fields $\eta_{x,y}$ are strongly suppressed, such a transition should be split into two transitions, as anticipated in the mean field theory. Thus, summarizing the results of all three approaches (the RG, the model calculations, and the mean field results), one can describe the phase transition to a superconducting state in systems

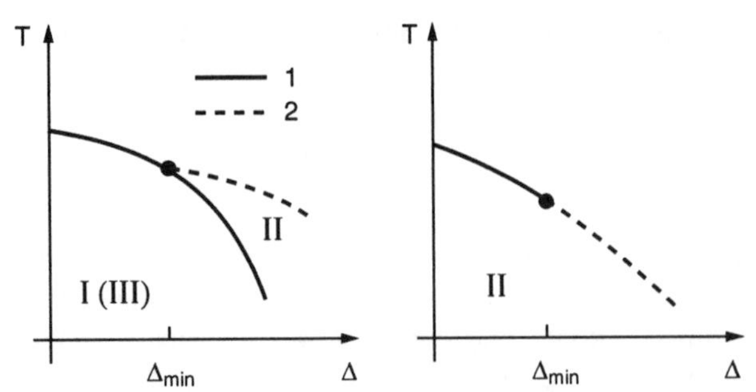

Figure 9.9: Phase diagrams on $T-\Delta$-plane: *a*) low-temperature phases of types I or III; *b*) the same for phase II. Solid lines correspond to a first-order transition, broken lines denote second-order transitions.

with d-pairing and with weak orthorhombic distortions by the phase diagrams shown in Fig. 9.9. The renormalization group equations were also derived by Golub and Mashtakov (1988) for the functional, Eq. (9.43) with the addition of an anisotropy quadratic in gradient terms

$$\delta\mathcal{F} = \gamma \int d\boldsymbol{q} q_x q_y (\eta_x \eta_y^* + \eta_y \eta_x^*).$$

Such an addition leads to the following corrections to Eq. (9.45)

$$\begin{aligned}
\frac{du}{dl} &= u - 5u^2 - \mu^2 - \nu^2 - 2u(\mu + 2\nu)\delta, \\
\frac{d\nu}{dl} &= \nu - 4u\nu - 2\nu^2 - 4\mu^2 - 2(\nu^2 + 2\mu\nu + 2u^2)\delta, \\
\frac{d\mu}{dl} &= \mu(1 - 2u - 4\nu) - (u^2 + \nu^2 + 5\mu^2)\delta,
\end{aligned}$$

where the parameter δ is proportional to the anisotropy γ^2. However, these corrections do not change qualitatively the phase diagram of the system under consideration. In particular, it can be shown that the separatrix separating sectors II and III $(2\beta_2 + \beta_3 = 0)$ retains its form for this generalized set of RG equations. Therefore, the consideration remains qualitatively valid in the presence of this additional anisotropy.

Appendix A

Evaluation of Integrals

A.1 Gaussian Integral

The Schwinger's parametrical representation (see Eq. (1.22)) helps to reduce calculation of any diagram in the field theoretical RG to a multifold Gaussian integral. In the general case such an integral can be written as

$$J = \int \prod_{k=1}^{L} d^d \boldsymbol{q}_k \exp\left\{ -\frac{1}{2} \sum_{kj} \boldsymbol{q}_k A_{kj} \boldsymbol{q}_j + \sum_k \boldsymbol{q}_k \cdot \boldsymbol{B}_k \right\}, \tag{A.1}$$

where A_{kj} is a symmetric matrix. Its eigenvalues must have positive real parts, otherwise the integral, Eq. (A.1) would diverge. Let us make the substitution

$$\boldsymbol{q}_j = \sum_k \left(A^{-1}\right)_{jk} \boldsymbol{B}_k + \boldsymbol{p}_j.$$

Now the integral, Eq. (A.1) takes the form

$$J = \exp\left(\sum_{kj} \boldsymbol{B}_k \left(A^{-1}\right)_{kj} \boldsymbol{B}_j \right) \int \prod_{k=1}^{L} d^d \boldsymbol{p}_k \exp\left\{ -\frac{1}{2} \sum_{kj} \boldsymbol{p}_k A_{kj} \boldsymbol{p}_j \right\}. \tag{A.2}$$

After diagonalizing the bilinear form, the integral part in Eq. (A.2) factorizes into one-dimensional Gaussian integrals which can be easily carried out, and the result of integration can be represented as

$$J = \frac{(2\pi)^{Ld/2}}{(det A)^{1/2}} \exp\left(\sum_{kj} \boldsymbol{B}_k \left(A^{-1}\right)_{kj} \boldsymbol{B}_j \right). \tag{A.3}$$

A.2 A Typical Diagram

Let us now evaluate the integral that appears in the diagram U_1 (see Sect. 4.5.3) According to the Rules of Evaluation (see Sect. 1.2.3) this diagram contains the integral

$$\bar{I} = \int \frac{d^d \boldsymbol{q}}{(2\pi)^d} \frac{1}{q^2(\boldsymbol{q}+\boldsymbol{p})^2}. \tag{A.4}$$

Since the factor $S_d/(2\pi)^d = 1/[2^{d-1}\pi^{d/2}\Gamma(d/2)]$ is usually adsorbed into the redefinition of the interaction constant, we can rewrite Eq. (A.4), introducing notation $I = (2\pi)^d \bar{I}/S_d$ and using Eq. (1.22), in the form

$$I = 2^{d-1}\pi^{d/2}\Gamma(d/2) \int_0^\infty d\alpha_1 d\alpha_2 \int \left(\frac{d\boldsymbol{q}}{2\pi}\right)^d \exp[-\alpha_1 q^2 - \alpha_2(\boldsymbol{q}+\boldsymbol{p})^2]. \tag{A.5}$$

Applying Eq. (A.3) to this representation, one can obtain

$$I = 2^{d-1}\pi^{d/2}\Gamma(d/2) \int_0^\infty \frac{d\alpha_1 d\alpha_2}{(2\pi)^d} \left(\frac{\pi}{\alpha_1+\alpha_2}\right)^{d/2} \exp\left(-\frac{\alpha_1\alpha_2}{\alpha_1+\alpha_2}p^2\right). \tag{A.6}$$

Carrying out the integrations over α_1 and α_2 we arrive at

$$I = \frac{p^{d-4}}{2\Gamma(d-2)}\Gamma(2-d/2)\Gamma^2(d/2-1)\Gamma(d/2). \tag{A.7}$$

This expression allows one to make the ϵ-expansion. The following properties of the Γ-function have to be used in the derivation of the first two leading terms in the ϵ-series

$$\Gamma(1+x) = x\Gamma(x)$$

and

$$\frac{\prod_{i=1}^m \Gamma(1+\alpha_i)}{\prod_{i=1}^n \Gamma(1+\beta_i)} \quad \text{if} \quad \sum_{i=1}^m \alpha_i - \sum_{i=1}^n \beta_i = 0.$$

These properties enable one to write the answer in the form

$$I \approx \frac{1}{\epsilon}\left(1+\frac{\epsilon}{2}\right).$$

Appendix B

Local RG

The general RG equation, describing the evolution of the Ginzburg-Landau functional in the critical region, is very complicated. The main complication of its analytical study arises from the variational nature of this equation. General methods of analysis of such equations have not been developed as yet. The theory of ordinary differential equations is on a better footing. Thus, all endeavors in RG equation analysis are related to a reduction of this functional equation to a set of differential equations. In its turn, such a set is usually solved in the framework of perturbation theories over vertexes or $\epsilon = 4 - d$. Sometimes, it makes sense to retain vertexes in the Ginzburg-Landau functional at all powers of field variables $\boldsymbol{\phi}(\boldsymbol{r})$. This compels one to search for other ways of approximate descriptions of the RG process not related to perturbation theories. One such approximation is the use of a local form for the Ginzburg-Landau functional.

B.1 General Consideration

Let us consider the RG equation in the form given by Eq. (5.32):

$$\begin{aligned}\frac{\partial H_I}{\partial l} &= dV\frac{\partial H_I}{\partial V} - \frac{V}{2}\int_q \eta(q) + \frac{1}{2}\int_q \eta(q) G_0^{-1}(q)|\boldsymbol{\phi}_q|^2 \\ &+ \int_q \left[\frac{d+2-2\eta(q)}{2}\boldsymbol{\phi}_q + \boldsymbol{q}\cdot\frac{\partial \boldsymbol{\phi}_q}{\partial \boldsymbol{q}}\right]\cdot\frac{\delta H_I}{\delta \boldsymbol{\phi}_q} \\ &+ \int_q h(\boldsymbol{q})\left[\frac{\delta^2 H_I}{\delta \boldsymbol{\phi}_q \cdot \delta \boldsymbol{\phi}_{-q}} - \frac{\delta H_I}{\delta \boldsymbol{\phi}_q}\cdot\frac{\delta H_I}{\boldsymbol{\phi}_{-q}}\right] \equiv \hat{R}_g\{H_I\}. \qquad \text{(B.1)}\end{aligned}$$

The functional H_I in the general case is a nonlocal form over the powers of the vector $\boldsymbol{\phi}$. Even in the case where the initial functional H_I is a local form, the nonlocalities are generated according to Eq. (B.1). One can try to find H_I as the sum $H_I = H_I^{(0)} + H_I^{(1)}$, choosing $H_I^{(0)}$ in the local form $H_I^{(0)} = \int d\boldsymbol{r} F\{\boldsymbol{\phi}(\boldsymbol{r})\}$. Eq. (8.1) now can be rewritten as

$$\begin{aligned} \frac{\partial H_I}{\partial l} &= \frac{\partial H_I^{(0)}}{\partial l} + \frac{\partial H_I^{(1)}}{\partial l} \\ &= \hat{R}_g\{H_I^{(0)}\} + \hat{R}_g\{H_I^{(1)}\} - 2\int_q \frac{\delta H_I^{(0)}}{\partial \boldsymbol{\phi}_q} \cdot \frac{\delta H_I^{(1)}}{\partial \boldsymbol{\phi}_{-q}}. \end{aligned} \tag{B.2}$$

One can easily verify that the action of the operator $\hat{R}_g$ on $H_I^{(0)}$ generates nonlocal contributions. On the other hand the function $\eta(\boldsymbol{q})$ contains information related to the nonlocal vertexes of the functional (see Chapter 5). This influences the derivative $\partial H_I^{(0)}/\partial l$. The maximum value of $\eta(\boldsymbol{q})$ is reached at $\boldsymbol{q} = 0$, but the value $\eta(0)$ is small (exponent η). It seems, that by neglecting terms of the order of $\eta(0)$ in Eq. (B.1), we do not essentially change the evolution of the local functional. A more complicated situation is encountered when looking for direct contributions to the local functional from the term $\hat{R}_g H_I^{(1)}$. Really, $H_I^{(1)}$ contains a function of the kind $\overline{H}_I^{(1)} = H_I^{(10)}\{\int A(\boldsymbol{q})|\boldsymbol{\phi}_q|^2\}$ in the general case, where $A(\boldsymbol{q})$ is a function of $\boldsymbol{q}$ and $H_I^{(10)}$ is a local functional. The action of the operator containing the second variational derivative on the functional $H_I^{(10)}$ leads to the generation of contributions to $\partial H_I^{(0)}/\partial l$. The conjecture that all nonlocal vertexes are small of the order of η, seems plausible but has never been proven. Nonetheless, if one assumes that this conjecture is correct and neglects the contributions discussed above, then the equation for H_I^0 separates and we arrive at

$$\frac{\partial H_I^{(0)}}{\partial l} = \hat{R}_g^0\{H_I^{(0)}\},$$

where $\hat{R}_g^0 = \hat{R}_g(\eta = 0)$. Taking into account that $H_I^{(0)} = \int d\boldsymbol{r} F\{\boldsymbol{\phi}(\boldsymbol{r})\}$, one can represent this equation in terms of the function F and its derivatives as

$$\frac{\partial F}{\partial l} = dF - \frac{d-2}{2}\boldsymbol{\phi} \cdot \boldsymbol{\nabla}_\phi F + \boldsymbol{\nabla}_\phi^2 F - (\boldsymbol{\nabla}_\phi F)^2 \equiv \hat{D}F. \tag{B.3}$$

In this equation the action of the operator $\boldsymbol{\nabla}$ is defined as

$$\boldsymbol{\nabla} F = \sum_{s,\alpha} e_s^\alpha \frac{\partial F}{\partial \phi_s^\alpha},$$

where e_s^α is a unit vector in the direction of the vector ϕ_s^α (s denotes different fluctuating fields, α denotes components of the s-field). The equation obtained has a universal form.[1] The latter is not influenced by the symmetry and applicable for any number of fluctuating fields ϕ_s. In contrast with the functional equation (see Eq. (B.1)), this equation is an ordinary equation in partial derivatives. The traditional analysis of this equation helps to reveal a number of quite general results. A detailed study of the physical branch of solutions for this equation was made by Filippov and Breus in 1991. The results discussed in the next two sections were obtained by Lisyansky *et al* in 1991.

B.2 Spectral Theorems

Fixed points of Eq. (B.3) are, as usual, defined by the equation $\partial F/\partial l = 0$, i.e.

$$\hat{D}F = 0. \tag{B.4}$$

We are not going to restrict ourselves to consideration of a particular symmetry of the solution $F^*(\phi)$. The only restriction which it is reasonable to assume is a finite value of F^* at any ϕ except in the limit ($|\phi| \to \infty$). Let us now consider the equation defining eigenvalues λ of the linear RG. This equation can be obtained from Eq. (B.3) in the form

$$\lambda\Psi = d\Psi - \frac{d-2}{2}\phi\cdot\nabla_\phi\Psi + \nabla_\phi^2\Psi - 2\nabla_\phi F^*\cdot\nabla_\phi\Psi, \tag{B.5}$$

where $\Psi(\phi) = F(\phi) - F^*(\phi)$ is a linear deviation from the fixed-point solution. There are a number of natural questions not resolved in the theory of critical behavior based on the RG approach. Some of them are: Is the spectrum of the linearized RG operator always discrete? Is it bounded from above? Are all the eigenvalues real? The positive answers to these questions are welcome in the theory of critical behavior, while in the opposite case, we encounter serious contradictions beween the initial postulates and the results of concrete calculations. In the framework of the local RG approach, one can obtain exhaustive answers to these questions using Eq. (B.5).

Let us represent the function Ψ as

$$\Psi(\phi) = \chi(\phi)\exp\left\{\frac{d-2}{8}\phi^2 + [F^*(\phi)]^2\right\}.$$

The equation for the function χ can be obtained from Eq. (B.5). It has the form of Schrödinger's equation

$$\nabla^2\chi + [(d-\lambda) - q(\phi)]\chi = 0 \tag{B.6}$$

[1] The equation for the local functional in the form, Eq. (B.3) was obtained by Tokar in 1984 with the help of the direct derivation implying restriction on the appearance of nonlocalities.

with the "potential"

$$q(\phi) = \left(\frac{d-2}{4}\phi\right)^2 - \frac{d-2}{4}m + dF^*(\phi),$$

defined in the m-dimensional space ($m = \sum_s n_s$, where n_s is the number of components of the s-th field). The spectral structure of this equation is defined by the behavior of the "potential" $q(\phi)$ when $\phi^2 \to \infty$. It is necessary to notice that, as follows from Eq. (B.4), $F^*(\phi) \sim \phi^2$ when $\phi^2 \to \infty$. Thus, $q(\phi) \to \infty$ when $\phi^2 \to \infty$. This condition results in the following spectral theorems (see, for instance, Titchmarsh (1958)):

1. All eigenvalues λ_k are real.
2. The spectrum $\{\lambda_k\}$ is bounded from above satisfying inequality $\lambda_k \leq d - \min q(\phi)$.
3. The spectrum is discrete (the condition $q(\phi) \to \infty$, when $\phi^2 \to \infty$, is of importance starting from this item only).
4. There is a limited number of eigenvalues $\lambda_k > 0$.
5. The eigenfunctions Ψ_k have exactly k-zeros.

For the trivial (Gaussian) fixed point the functions Ψ_k can be explicitly determined. One can verify that they are Laguerre's polynomials $\Psi_k = L_k^{(m/2-1)}(\phi)$ with the eigenvalues $\lambda_k = \varepsilon_k = d + (2-d)k$, and the statements 1 — 5 are selfevident.

The integration of Eq. (B.3) is not a trivial problem and possibly can be carried out only for particular symmetries of the function $F(\phi)$. Of course, the simplest case should correspond to $\mathcal{O}(m)$ symmetry. Let us show that Eq. (B.3), at least, has a solution corresponding to this symmetry. For this purpose we rewrite Eq. (B.3) in "spherical" coordinates as

$$\frac{\partial F}{\partial l} = \left(\hat{D}_\phi + \frac{1}{\phi^2}\hat{D}_\theta\right) F, \tag{B.7}$$

where $\hat{D}_\phi$ is the "radial" (depending only on the absolute value of ϕ) part of the operator $\hat{D}$:

$$\hat{D}_\phi F = dF - \frac{d-2}{2}\phi F_\phi + F_{\phi\phi} + \frac{m-1}{\phi}F - F_\phi^2. \tag{B.8}$$

The action of the angular operator in Eq. (B.7) is defined as

$$\hat{D}_\theta F = \nabla_\theta^2 F - (\nabla_\theta F)^2. \tag{B.9}$$

Here the value θ denotes a set of angles in the m-dimensional space $\{\phi_s^\alpha\}$. The operator ∇_θ is the angular part of the operator $\boldsymbol{\nabla}$ and the operator ∇_θ^2 is the angular part of the Laplacian operator ($\nabla_\theta^2 \equiv \Delta_\theta$). The explicit equations defining actions of these operator can be written as

$$\begin{aligned} (\nabla_{\theta(m)})^2 &= \sum_{j=1}^{m-1} \frac{1}{q_j} \left(\frac{\partial F}{\partial \theta_j} \right)^2, \\ \Delta_{\theta(m)} F &= \sum_{j=1}^{m-1} \frac{1}{q_j \sin^{m-j-1} \theta_j} \frac{\partial}{\partial \theta_j} \left(\sin^{m-j-1} \theta_j \frac{\partial F}{\partial \theta_j} \right), \end{aligned} \tag{B.10}$$

where

$$q_1 = 1, \quad q_j = \left(\prod_{i=1}^{j-1} \sin \theta_i \right)^2, \quad j \geq 2,$$

$$\phi_1 = \phi \cos \theta_1, \quad \phi_j = \phi \cos \theta_j \left(\prod_{i=1}^{j-1} \sin \theta_i \right), \quad \phi_m = \phi \left(\prod_{i=1}^{j-1} \sin \theta_i \right).$$

When the starting functional $H_I^{(0)}$ has the symmetry $\mathcal{O}(m)$, the RG equation does not contain the operator ∇_θ, i.e.

$$\frac{\partial F_0}{\partial l} = \hat{D}_\phi F_0. \tag{B.11}$$

If there is a solution for Eq. (B.11) and we have an isotropic fixed point (this assumption is one of the main hypotheses of the RG approach), i.e. $\hat{D}_\phi F_0^* = 0$, this solution should inevitably be the solution of Eq. (B.7) and the fixed point F_0^* is one of the fixed points of this equation. In other words, the local equation in any case has, in addition to the trivial point, one more isotropic ($\mathcal{O}(m)$ symmetry) fixed point. If this fixed point is stable, the symmetry of the initial functional is replaced in the critical region by the $\mathcal{O}(m)$ symmetry. This means that the symmetry of the fluctuating system is essentially higher than the symmetry of the nonordered phase. This phenomenon, also considered in Chapter 6, is called "asymptotic symmetry."

B.3 Local Equation in the Limit $m \to \infty$

The equation of the local RG can be exactly integrated in the limit $m \to \infty$. This is quite natural, since, according to Stanley (1968), this limit leads to the exactly solvable spherical model. The analysis of this case is methodically very instructive.

Let us consider the isotropic local equation

$$\frac{\partial F}{\partial l} = dF - \frac{d-2}{2}\phi F_\phi + \frac{m-1}{\phi}F_\phi - F_\phi^2.$$

As the function F depends only on ϕ^2 it is convenient to use the new independent variable $\rho = \phi^2$ transforming the above equation to

$$\frac{\partial F}{\partial l} = dF - (d-2)\rho F_\rho + 4\rho F_{\rho\rho} + 2(m-1)F_\rho - 4\rho F_\rho^2.$$

In order to explicitly separate terms of the order of $o(1/m)$ in the right-hand side of this equation, let us substitute $\rho = 2mx$ and $F = my$. We then find

$$\frac{\partial y}{\partial l} = dy - (d-2)xy_x + y_x - 2xy_x^2 + \frac{1}{m}(2xy_{xx} - y_x).$$

Shifting the variable x on the value $2/(d-2)$ in the limit $m \to \infty$ one arrives at

$$\frac{\partial y}{\partial l} = dy - (d-2)xy_x - 2\left[x + \frac{2}{d-2}\right]y_x^2.$$

The equations defining fixed points and eigenvalues can be written as

$$dy = (d-2)xy_x + 2\left[x + \frac{2}{d-2}\right]y_x^2, \qquad \text{(B.12)}$$

$$(d-\lambda)\Psi = \left\{(d-2)x + 4\left[x + \frac{2}{d-2}\right]y_x\right\}\Psi. \qquad \text{(B.13)}$$

Differentiating Eq. (B.12) with respect to x and multiplying the result by the derivative dx/dy_x we arrive at

$$y_x(y_x - 1)\frac{dx}{dy_x} + \left(\frac{d-2}{2} + 2y_x\right)x + \frac{4}{d-2}y_x = 0. \qquad \text{(B.14)}$$

The integral of this equation can be written as

$$x = \frac{|y_x|^{(d-2)/2}}{|y_x - 1|^{(d+2)/2}}\left[c - \frac{4}{d-2}\int dy_x |y_x - 1|^{d/2}|y_x|^{1-d/2},\right] \qquad \text{(B.15)}$$

where c is a constant of integration. Taking Eq. (B.14) into account, one can transform Eq. (B.13) into the following equation with separating variables (when $y_x \neq 0$)

$$(\lambda - d)\Psi = 2(y_x - 1)y_x\frac{d\Psi}{dy_x}.$$

This equation can be easily integrated, leading to

$$\Psi = A\left|\frac{y_x - 1}{y_x}\right|^{(\lambda-d)/2}, \tag{B.16}$$

where A is an arbitrary constant.

Now, let us recall that according to the physical meaning, the function $F(\phi)$ is the free energy density and the derivative $\partial F/\partial\phi^2$ gives the renormalized difference $(T - T_c)/T$ when $\phi^2 \to 0$. Therefore, at the critical point $\partial F^*/\partial\phi^2|_{\phi^2\to 0} \to 0$ and consequently the value $y_x \to 0$. In this limit Eqs. (B.15) and (B.16) can be replaced by

$$\begin{array}{rcl} x & \approx & c|y_x|^{(d-2)/2} - \frac{8}{(4-d)(d-2)} y_x + o(y_x^{d/2}), \\ \Psi & \approx & A|y_x|^{(d-2)/2}. \end{array} \tag{B.17}$$

Since the physically reasonable must be analytical expansions for the function $y(x)$, i.e.

$$y(x)|_{\tau=0} = a + bx^2 + o(x^2),$$

one can see that only the solution of Eq. (B.17) with $c = 0$ is acceptable. This means that $x \approx -8y_x/(4-d)(d-2)$. The deviation from this solution when $\tau \neq 0$ should be searched as $\Psi = a + \tau x + Bx^2 + o(x^3)$, so that in the expression $\Psi \sim |y_x|^{(d-2)/2} \sim x^{(d-\lambda)/2}$ one should have only integer powers of the variable x. As a result we obtain $\lambda_k = d - 2k$. Taking into account that when $\eta = 0$ the exponent γ is given by $\gamma = 2/\lambda_1$, we arrive at $\gamma = 2/(d-2)$, which is a well known value for the spherical model. Therefore, the local RG equation in the limit $m \to \infty$ belongs to the spherical model universality class.

Bibliography

[1] Abe, R. (1972). *Prog. Theor. Phys.* **48**, 1414.

[2] Abrikosov, A.A. (1957). *JETP* **47**, 720. Abrikosov, A.A., L.P. Gorkov, and I.E. Dzyaloshinski (1963). "Methods of Quantum Field Theory in statistical physics." Dover Publications, New York.

[3] Aharony, A. (1976). *In* "Phase Transitions and Critical Phenomena, Vol. 6." (C. Domb and M.S. Green, eds.), pp. 357–424, Academic Press, London.

[4] Aharony, A. and M.E. Fisher (1973). *Phys. Rev. B* **8**, 3323.

[5] Amit, D. (1984). "Field Theory, the Renormalization Group and Critical Phenomena." World Scientific, Singapore.

[6] Anisimov, M.A., A.T. Berestov, V.P. Voronov. *et al* (1979). *JETP* (1976) **76**, 1661.

[7] Andrews, T. (l869). *Phil. Trans. R. Soc.* **159**, 575.

[8] Aslamazov, L.G. and A.I. Larkin (1968). *Sov. Phys. - Solid State* **10**, 875.

[9] Bagnuls, C. and C. Bervillier (1981). *Phys. Rev. B* **24**, 1226.

[10] Bagnuls, C. and C. Bervillier (1984). *J. Phys. Lett. (Paris)* **45**, L95.

[11] Bagnuls, C. and C. Bervillier (1985). *Phys. Rev. B* **32**, 7209.

[12] Bagnuls, C., C. Bervillier, and Y. Garrabas (1984). *J. Phys. Lett. (Paris)* **45**, L95.

[13] Bak P. (1976). *Phys. Rev. B* **14**, 3980.

[14] Bak P., S. Krinsky, and D. Mukamel (1976). *Phys. Rev. Lett.* **36**, 52.

[15] Bak P., D. Mukamel, and S. Krinsky (1976). *Phys. Rev. B* **13**, 5065.

[16] Baker, G.A., Jr., B.G. Nickel, and D.I. Meiron (1978). *Phys. Rev. B* **17**, 1365.

[17] Baker, Jr., G.A. (1990). "Quantative Theory of Critical Phenomena." Academic Press, New York. Barber, M.N. and M.E. Fisher (1973). *Ann. Phys.* **77**, 1.

[18] Barkev, J.A. and D. Henderson (1967). *Ann. Phys.* **47**, 4714.

[19] Barrett, S.E., J.A. Martindale, D.J. Durand, C.H. Pennington, C.P. Slichter, T.A. Friedmann, J.P. Rice, and D.M. Ginsberg (1991). *Phys. Rev. Lett.* **66**, 108.

[20] Batlog, B. (1990). *In* "High Temperature superconductivity: Proceedings of the Los Alamos Symposium," (K. S. Bedell, D. Coffey, D. E. Meltzer, D. Pines, and J. R. Schrieffer, eds.), p.37, Addison Wesley, Reading, Mass.

[21] Baxter, R.J. (1982). "Exactly Solvable Models in Statistical Physics." Academic Press, New York.

[22] Bednortz, J.G. and K.A. Muller (1986), *Z. Phys. B* **64**, 189.

[23] Bell, T.L. and K.G. Wilson (1975). *Phys. Rev. B* **11**, 3431.

[24] Berezinsky, V.L. (1970). *JETP*, **59**, 907.

[25] Berezinsky, V.L. (1971). *JETP*, **61**, 1144.

[26] Berlin, T.H. and M. Kac (1952). *Phys. Rev.* **86**, 821.

[27] Beysens, D. and A. Bourgon (1979). *Phys. Rev. A* **19**, 2407.

[28] Beysens, D. (1982). *In* "Phase Transitions." Plenum Press, New York.

[29] Blankschtein, D. and D. Mukamel (1982). *Phys. Rev. B* **25**, 6939.

[30] Bogoliubov, N.N. (1946). "Problems of Dynamical Theory in Statistical Physics." Nauka, Moscow *[In Russian]*.

[31] Bogoliubov, N.N. and D.V. Shirkov (1955). *Dokl. Acad. Sci. [In Russian]* **103**, 391.

[32] Bogoliubov, N.N. and D.V. Shirkov (1956). *JETP* **30**, 77.

[33] Bogoliubov, N.N. and D.V. Shirkov (1959). "Introduction to the Theory of Quantized Fields." Interscience Publishers, New York.

[34] Borodin, V.A., V.D. Doroshev, Yu.M. Ivanchenko, M.M. Savosta, and A.E. Filippov (1991). *Sov. Phys. -JETP Lett.* **52**, 1073.

[35] Brézin, E., J.C. Le Guillou, and J. Zinn-Justin (1976). *In* "Phase Transitions and Critical Phenomena, Vol. 6." (C. Domb and M.S. Green, eds.), pp. 125-247, Academic Press, London.

[36] Bruce, A.D. and R.A. Cowley (1981). "Structural Phase Transitions." Taylor & Francis, London.

[37] Buckingham, C. and W.H. Fairbank (1961). *In* "Progress of Low Temperature Physics, Vol. 3." North-Holland, Amsterdam.

[38] Burchardt, T.W. and J.M.J. Leeuwen (1988). "Real Space Renormalization." Springer–Verlag, Berlin.

[39] Butera, R.A., (1988). *Phys. Rev. B* **37**, 5909.

[40] Cardy, J.L. (1988). "Finite-Size Scaling." North-Holland, Amsterdam.

[41] Chen, F., L. Gao, R.L. Meng, Y.Y. Xue, and C.W. Chu (1994). *Bull. APS* **39**, 239.

[42] Chen, M.E., M.E. Fisher, and B.G. Nickel (1982). *Phys. Rev. Lett.* **48**, 630.

[43] Cheong, S.W., J.D. Thompson, and Z. Fisk (1989). *Physica C* **158**, 109.

[44] Chu, C.W., (1994). *Bull. APS* **39**, 172.

[45] Coleman, S. and E. Weinberg (1973). *Phys. Rev. D* **7**, 1888.

[46] Collot, J.-L., J.A.C. Loodts, and R. Brout (1975). *J. Phys. A* **8**, 594.

[47] Domany, E. D. Mukamel, and M.E. Fisher (1977). *Phys. Rev. B* **15**, 5432.

[48] Di Castro, C. and G. Jona-Lasinio (1969). *Phys. Lett. A* **29**, 322.

[49] Di Castro, C. (1972). *Lett. Nuovo Cimento* **5**, 69.

[50] Di Castro, C. and G. Jona-Lasinio (1976). *In* "Phase Transitions and Critical Phenomena, Vol. 6." (C. Domb and M. S. Green, eds.), pp. 508–558, Academic Press, London.

[51] Domb, C. and D.W. Wood (1965). *Proc. Phys. Soc.* **86**, 1.

[52] Doroshev, V.D., V.N. Krivoruchko, M.M.Savosta. A.A. Shestakov, and D.A. Yablonsky (1992). *Sov. Phys. -JETP* **101**, 190.

[53] Dvorak, V. and J. Fousek (1980). *Phys. Stat. Sol.* **61**, 99.

[54] Emery, V.J. (1975). *Phys. Rev. B* **11**, 239.

[55] Filippov, A.E. and S.A. Breus (1991). *Phys. Lett. A* **158**, 300.

[56] Fisher, D.S. and D.A. Huse (1985). *Phys. Rev. B* **32**, 247.

[57] Fisher, M.E. (1974). *Rev. Mod. Phys.* **46**, 597.

[58] Fisher, M.E. (1975). *Phys. Rev. Lett.* **34**, 1634.

[59] Fisher, M.E., S.-K. Ma, B.G. Nickel (1972). *Phys. Rev. Lett.* **29**, 917.

[60] Fisher, M.E. and D.R. Nelson (1974). *Phys. Rev. Lett.* **32**, 1350.

[61] Fisher, M.E. and P. Pfeuty (1972). *Phys. Rev. B* **6**, 1889.

[62] Fomichov, S.V. and S.B. Khokhlachev (1974). *JETP* **66**, 983.

[63] Freitas, P.P. and T.S. Plaskett (1987). *Phys. Rev. B* **36**, 5723.

[64] Freitas, P.P., C.C. Tsuei, and T.S. Plasket (1988). *Phys. Rev. B* **36**, 1833.

[65] Gao L., Y.Y. Xue, F. Chen, Q. Xiang, R.L. Meng, and C. W. Chu (1994). it Bull. APS **39**, 239.

[66] Gauzzi, A. and D. Pavuna (1995). Preprint.

[67] Gell-Mann, M. and F.E. Low (1954). *Phys. Rev.* **95**, 1300.

[68] Ginzburg, S.L. (1975). *JETP* **68**, 273.

[69] Ginzburg, V.L. (1960). *Sov. Phys. - Solid State* **2**, 1824.

[70] Ginzburg, V.L. and L.D. Landau (1950). *Sov. Phys - JETP* **20**, 1064.

[71] Goldstone, J., A. Salam, and S. Weinberg (1962). *Phys. Rev.* **127**, 965.

[72] Golner, G.R. (1986). *Phys. Rev. B* **33**,7863.

[73] Golner, G.R. and E.K. Riedel (1975). *Phys. Rev. Lett.* **34**, 171.

[74] Golner, G.R. and E.K. Riedel (1975). *Phys. Rev. Lett.* **34**, 856.

[75] Golner, G.R. and E.K. Riedel (1976). *Phys. Lett. A* **76**, 11.

[76] Golub, A.A. and Yu. Mashtakov (1988). *in* "Proc. First Conf. on High-Temperature Superconductivity," p.62, Kha'rkov *[In Russian]*.

[77] Gorishny, S.G., S.A. Larin and F.V. Tkachov (1984). *Phys. Lett. A* **101**, 120.

[78] Gorkov, L.P. (1958). *JETP* **34**, 735.

[79] Gorkov, L.P. (1959). *JETP* **36**, 1918; **37**, 1407.

[80] Greer, S.C. and M.P. Moldover (1981). *Annu. Rev. Phys. Chem.* **32**, 233.

[81] Grinstein, G. and A. Luther (1976). *Phys. Rev. B* **13**, 1329.

[82] Guinta G., C. Caccamo, and P.V. Giaquinta (1985). *Phys. Rev. A* **31**, 2471.

[83] Halperin, B.I., T.C. Lubensky and S.-K. Ma (1974). *Phys. Rev. Lett.* **32**, 292.

[84] Hardy, W.N., D.A. Bonn, D.C. Morgan, Ruixing Liang, and Kuan Zhang (1993). *Phys. Rev. Lett.* **70**, 3999.

[85] Harris, A.B. (1974). *J. Phys. C* **7**, 1974.

[86] Hohenberg, P.C. (1967). *Phys. Rev.* **158**, 383.

[87] Inderhees, S.E., M.B. Salamon, N. Goldenfeld *et al.* (1988). *Phys. Rev. Lett.* **60**, 1178.

[88] Ivanchenko, Yu.M. and A.E. Filippov (1984). *Sov. Phys. - Solid State* **26**, 80.

[89] Ivanchenko, Yu.M., A.E. Filippov, and A.V. Radievsky (1993). *Sov. Low Temp. Phys.* **19**, 655.

[90] Ivanchenko, Yu.M. and A.A. Lisyansky (1983). *Phys. Lett. A* **98**, 115.

[91] Ivanchenko, Yu.M. and A.A. Lisyansky (1984). *Theor. Math. Phys.* **58**, 97.

[92] Ivanchenko, Yu.M. and A.A. Lisyansky (1992). *Phys. Rev. A* **45**, 8525.

[93] Ivanchenko, Yu.M., A.A. Lisyansky, and A.E. Filippov (1984). *Sov. Phys. - JETP* **60**, 582.

[94] Ivanchenko, Yu.M., A.A. Lisyansky, and A.E. Filippov (1986a). *Theor. Math. Phys.* **66**, 183.

[95] Ivanchenko, Yu.M., A.A. Lisyansky, and A.E. Filippov (1986b). *Phys. Lett. A* **119**, 55.

[96] Ivanchenko, Yu.M., A.A. Lisyansky, and A.E. Filippov (1987a). *Theor. Math. Phys.* **71**, 649.

[97] Ivanchenko, Yu.M., A.A. Lisyansky, and A.E. Filippov (1988b). *Theor. Math. Phys.* **72**, 786.

[98] Ivanchenko, Yu.M., A.A. Lisyansky, and A.E. Filippov (1989a). "Fluctuation Effects in Systems with Competing Interactions." Naukova Dumka, Kiev *[in Russian]*.

[99] Ivanchenko, Yu.M., A.A. Lisyansky, and A.E. Filippov (1989b). *Phys. Lett. A***136**, 171.

[100] Ivanchenko, Yu.M., A.A. Lisyansky, and A.E. Filippov (1990b). *Sov. Phys. - Solid State* **31**, 1767.

[101] Ivanchenko, Yu.M., A.A. Lisyansky, and A.E. Filippov (1990a). *J. Stat. Phys.* **58**, 295.

[102] Ivanchenko, Yu.M., A.A. Lisyansky, and A.E. Filippov (1990b). *Theor. Mat. Phys.* **84**, 223.

[103] Ivanchenko, Yu.M., A.A. Lisyansky, and A.E. Filippov (1990c). *Phys. Lett. A* **150**, 100.

[104] Ivanchenko, Yu.M., A.A. Lisyansky, and A.E. Filippov (1990d). *J. Phys. A* **23**, 91.

[105] Ivanchenko, Yu.M., A.A. Lisyansky, and A.E. Filippov (1990e). *Phase Trans.* **22**, 31.

[106] Ivanchenko, Yu.M., A.A. Lisyansky, and A.E. Filippov (1992). *J. Stat. Phys.* **66**, 1159.

[107] Izumov, Yu.A., N.M. Plakida and Yu.N. Skryabin (1989). *Sov. Phys. - Uspekhi* **159**, 621.

[108] Jacobson, H.H. and D.J. Amit. (1981). *Ann. Phys.* **133**, 57.

[109] Jorgensen, J.D., M.A. Beno, D.G. Hinks, L. Soderholm, K.J. Volin, R.L. Hitterman, J.D. Grace, and I.K. Schuller (1987). *Phys. Rev. B* **36**, 3608.

[110] Joyce, C.S. (1966). *Phys. Rev.* **146**, 349.

[111] Joyce, C.S. (1972). *In* "Phase Transitions and Critical Phenomena, Vol. 2." (C. Domb and M.S. Green, eds.), pp. 152–280, Academic Press, New York.

[112] Kadanoff, L.P. (1966). *Physics* **2**, 263.

[113] Kerszberg, M. and D. Mukamel (1978). *Phys. Rev. B* **23**, 3943.

[114] Kerszberg, M. and D. Mukamel (1979). *Phys. Rev. Lett.* **43**, 293.

[115] Kerszberg, M. and D. Mukamel (1981). *J. Appl. Phys.* **52**, 1929.

[116] Khmelnitsky, D.E. (1975). *JETP* **68**, 1960.

[117] Kierstead, H.A. (1967). *Phys. Rev.* **162**, 153.

[118] Kitaoka, Y. (1987). *J. Phys. Soc. Japan* **56**, 3024.

[119] Klein, N., N. Tellmann, H. Schulz, K. Urban, S.A. Wolf, and V. Z. Kresin, (1993). *Phys. Rev. Lett.* **71**, 3355.

[120] Kosterlitz, J.M., D.R. Nelson, and M.E. Fisher (1976). *Phys. Rev. B* **13**, 412.

[121] Kosterlitz, J.M. and D.J. Thouless (1973). *J. Phys. C* **6**, 1181.

[122] Landau, L.D. (1937). *Phys. Zurn. Sowjetunion* **11**, 26, 545.

[123] Landau, L.D. and E.M. Lifshitz (1976). "Statistical Physics, 3rd Ed." Pergamon Press, New York.

[124] Larkin, A.I. and D.E. Khmelnitsky (1969). *JETP* **56**, 627.

[125] Le Guillou, J.C. and J. Zinn-Justin (1980). *Phys. Rev. B* **21**, 3976.

[126] Levaniuk, A.P. (1959). *Sov. Phys. - JETP* **36**, 571.

[127] Lipatov, L.N. (1977). *JETP* **72**, 411.

[128] Lisyansky A.A. and A.E. Filippov (1986a). *Sov. Phys. - Solid State* **28**, 886.

[129] Lisyansky A.A. and A.E. Filippov (1986b). *Theor. Math. Phys.* **68**, 923.

[130] Lisyansky A.A. and A.E. Filippov (1987a). *Phys. Lett. A* **125**, 335.

[131] Lisyansky A.A. and A.E. Filippov (1987b). *Ukr. Fiz. Zh. [Russian]* **32**, 626.

[132] Lisyansky A.A., Yu.M. Ivanchenko, and A.E. Filippov (1992). *J. Stat. Phys.* **66**, 1667.

[133] Lubensky T.C. (1975). *Phys. Rev. B* **11**, 3573.

[134] Luksutov, I.F. and V.L. Pokrovskii (1975). *JETP Lett.* **21**, 22.

[135] Lutgemeier, H. and M.W. Pieper (1987). *Sol. State Commun.* **64**, 267.

[136] Ma, S.-K. (1972). *Phys. Rev. Lett.* **29**, 131.

[137] Ma, S.-K. (1976). "Modern Theory of Critical Phenomena." Benjamin, Reading, Mass.

[138] MacLaughlin, D.E., J.P. Vithayathil, H.B. Brom, J.C.J.M. de Rooy, P.C. Hammel, P.C. Canfield, A.P. Reyes, Z. Fisk, J.D. Thompson, and S.W. Cheong (1994). *Phys. Rev. Lett.* **72**, 760.

[139] Martindale, J.A., S.E. Barrett, K.E. O'Hara, C.P. Slichter, W.C. Lee, and D.M. Ginsberg, (1993). *Phys. Rev. B* **47**, 9155.

[140] Martinetz, V.G. and E.V. Matizen (1974). *JETP* **67**, 607.

[141] Mayer, I.O. (1984). *Theor. Math. Phys.* **60**, 476.

[142] Mermin, N.D. and H. Wagner (1966). *Phys. Rev. Lett.* **17**, 1133.

[143] Monthoux, P., A. Balatsky, and D. Pines, (1992). *Phys. Rev. B* **46**, 14803.

[144] Mukamel, D., S. Krinsky, and P. Bak (1976). *Phys. Rev. B* **13**, 5078.

[145] Newman, K.E. and E.K. Riedel (1984). *Phys. Rev. B* **30**, 6615.

[146] Ng, H.K., H. Mathias, W.G. Moulton, K.K. Pan, S.J. Pan, C.M. Rey, and L.R. Testardi (1988). *Solid St. Comm.* **65**, 36.

[147] Nicoll, J.F. (1981). *Phys. Rev. A* **24**, 2203.

[148] Nishihara, H. (1987). *J. Phys. Soc. Japan* **56**, 4559.

[149] Onsager, L. (1944). *Phys. Rev.* **65**, 117.

[150] Parizi, G. (1980). *J. Stat. Phys.* **23**, 49.

[151] Patashinskii, A.Z. and V.L. Pokrovskii (1966). *JETP* **50**, 439.

[152] Patashinskii, A.Z. and V.L. Pokrovskii (1964). *JETP* **46**, 994.

[153] Patashinskii, A.Z. and V.L. Pokrovskii (1979). "Fluctuation Theory of Phase Transitions." Pergamon Press, Oxford.

[154] Pawly, G.S., R.H. Swendsen, D.J. Wallace, and K.G. Wilson (1984). *Phys. Rev. B* **29**, 4030.

[155] Pestak, M.W. and M.H.W. Chan (1984). *Phys. Rev. B* **30**, 1274.

[156] Pfeuty, P., D. Jasnow, and M.E. Fisher (1974). *Phys. Rev. B* **10**, 2088.

[157] Pisarev, R.V., B.B. Krichevtsov, P.A. Markovin, *et al.* (1983). *Phys. Rev. B* **28**, 2677.

[158] Privman, V. (1990). "Finite Size Scaling and Numerical Simulation of Statistical Systems." World Scientific, Singapore.

[159] Privman, V., P.C. Hohenberg, and A. Aharony (1991). *In* "Phase Transitions and Critical Phenomena, Vol. 6." (C. Domb and J.L. Lebovitz, eds.), pp. 1–677, Academic Press, London.

[160] Riedel, E.K., G.R. Golner, and K.E. Newman, (1985). *Ann. Phys. (N.Y.)* **161**, 178.

[161] Rudnick, J. (1975). *Phys. Rev. Lett.* **34**, 438.

[162] Rudnick, J. (1978). *Phys. Rev. B* **18**, 1406.

[163] Rushbrooke, G.S. and P.J. Wood (1955). *Proc. Phys. Soc. A* **68**, 1161.

[164] Salamon, M.B., S.E. Inderhees, J.P. Rice *et al.* (1988). *Phys. Rev. B* **38**, 885.

[165] Sarbach, S. and T. Shneider (1975). *Z. Phys. B* **20**, 399.

[166] Schuller, I.K., D.G. Hicks, M.A. Beno, G.W. Gapone, L. Soderholm, J.P. Loequet, Y. Baruynseraede, C.U. Serge, and K. Zang (1987). *Solid St. Comm.* **63**, 385.

[167] Shapira, Y. (1984). *In* "Multicritical Phenomena." Plenum Press, New York.

[168] Shukla, P. M.S. Green (1974). *Phys. Rev. Lett. 33*, 1263.

[169] Shukla, P. M.S. Green (1975). *Phys. Rev. Lett. 34*, 436.

[170] Shneider, T., E. Stoll, and H. Beck (1975). *Physica A* **79**, 201.

[171] Sokolov, A.I. (1979). *JETP* **77**, 1598.

[172] Sokolov, A.I. (1983). *Sov. Phys. - Solid State* **25**, 552.

[173] Sokolov, A.I. and A.K. Tagantsev (1979). *JETP* **76**, 181.

[174] Spenser, A.J.M. (1971). "Theory of invariants." New York.

[175] Stanley, H.E. (1968). *Phys. Rev.* **176**, 718.

[176] Stephen, M.J. and E. Abrahams (1973). *Phys. Lett. A* **44**, 85.

[177] Storer, R.G. (1969). *Austral. J. Phys.* **22**, 747.

[178] Stueckelberg, E.C.G. and A. Petermann (1953). *Helv. Phys. Acta.* **25**, 499.

[179] Sukharevsky, B.Ya., G.E. Shatalova, S.E. Khohlova, P.N. Mikheenko, V.G. Ksenofontov, I.V. Zhikharev, I.V. Vilkova, and E.N. Malyshev (1987). *Sov. J. Low Temp. Phys.* **13**, 566.

[180] Svidzinsky, A.V. (1982). "Space-Inhomogeneous Problems in Theory of Superconductivity." Nauka, Moscow *[In Russian]*.

[181] Swendsen, R.H. (1982). *In* "Monte Carlo Renormalization." (T.W. Burkhardt and J.M.J. van Leeuwen, eds.), p.57, Springer, New York.

[182] Swendsen, R.H. (1984). *Phys. Rev. Lett.* **52**, 2321.

[183] Symanzik, K. (1973). *Lett. Nuovo Cimento* **8**,771.

[184] Takigava, H., P.C. Hammel, R.H. Heffner, and Z. Fisk (1988). *Preprint*, Los Alamos National Laboratory, Los Alamos, NM.

[185] Tarazona, P. (1985). *Nuovo Cimento* **31**, 2672.

[186] Theumann, W.K. (1969). *J. Chem. Phys.* **51**, 3484.

[187] Titchmarsh, E.C. (1962). "Eigenfunction Expansions Associated with Second-Order Differential Equations." Oxford University Press, New York.

[188] Tokar, V.A. (1984). *Phys. Lett. A* **104**, 135.

[189] Toledano J.-C., L. Michel, P. Toledano, and E. Brezin (1985). *Phys. Rev. B* **31**, 7171.

[190] Tsuneto, T. and E. Abrahams (1973). *Phys. Rev. Lett.* **30**, 217.

[191] Van der Waals, J.D. (1873). "Over de continuiteit van den gas-en vloeistoftoestand, Thesis." Leiden.

[192] Vause, C. and J. Sak (1980). *Phys. Rev. A* **21**, 2099.

[193] Volovik, G.E. (1988). *JETP Lett.* **48**, 41.

[194] Volovik, G.E. and L.P. Gor'kov (1985). *Sov. Phys. JETP* **61**, 843.

[195] Wallace, D.J. (1973). *J. Phys. C* **6**, 1390.

[196] Wallace, D.J. (1976). *In* "Phase Transitions and Critical Phenomena, Vol. 6." (C. Domb and M. S. Green, eds.), pp. 293–356, Academic Press, London.

[197] Wallace, D.J. and R.K.P. Zia (1974). *Phys. Let A* **48**, 325.

[198] Watanabe, I. (1987). *J. Phys. Soc. Japan* **56**, 3028.

[199] Weeks, J.D., D. Chandler, and H.C. Anderson (1971). *J. Chem. Phys.* **54**, 5237.

[200] Wegner, F.J. (1972). *Phys. Rev. B* **5**, 4529.

[201] Wegner, F.J. (1976). *In* "Phase Transitions and Critical Phenomena, Vol. 6." (C. Domb and M. S. Green, eds.), pp. 7–124, Academic Press, London.

[202] Wegner, F.J. and A. Houghton (1973). *Phys. Rev. A* **8**, 401.

[203] Weiss, P. (1907). *J. Phys. Radium (Paris)* **6**, 667.

[204] Widom, B. (1965). *J. Chem. Phys.* **43**, 3998.

[205] Wilson, K.G. (1971). *Phys. Rev. B* **4**, 3174, 3184.

[206] Wilson, K.G. and J. Kogut (1974). *Phys. Rept. C* **12**, 75.

[207] Wilson, K.G. and M.E. Fisher (1972). *Phys. Rev. Lett.* **28**, 240.

[208] Wollman, D.A., D.J. Van Harlingen, W.C. Lee, D.M. Ginsberg, and A.J. Leggett (1993). *Phys. Rev. Lett.* **71**, 2134.

[209] Ziman, J.M. (1972). "Principles of the Theory of Solids." Univ. Press, Cambridge.

[210] Zinn-Justin, J. (1981). *J. Phys. (Paris)* **42**, 783.

[211] Zinn-Justin, J. (1989). "Quantum Field Theory and Critical Phenomena." Claredon Press, Oxford.

[212] Ziolo, J. (1988). *Physica C* **153–155**, 725.

[213] Zwanzig, R.W. (1964). *J. Chem. Phys.* **22**, 1420.

Index